MOLECULAR ANALYSIS OF PLANT ADAPTATION TO THE ENVIRONMENT

Springer Handbook Series of Plant Ecophysiology

Volume 1

Series Editors:

Luit J. De Kok and Ineke Stulen

University of Groningen,
The Netherlands

Aims & Scope:

The Kluwer Handbook Series of Plant Ecophysiology comprises a series of books that deals with the impact of biotic and abiotic factors on plant functioning and physiological adaptation to the environment. The aim of the Plant Ecophysiology series is to review and integrate the present knowledge on the impact of the environment on plant functioning and adaptation at various levels of integration: from the molecular, biochemical, physiological to a whole plant level. This Handbook series is of interest to scientists who like to be informed of new developments and insights in plant ecophysiology, and can be used as advanced textbooks for biology students.

Molecular Analysis of Plant Adaptation to the Environment

Edited by

Malcolm J. Hawkesford

and

Peter Buchner

Agriculture and Environment Division,
Institute of Arable Crop Research,
Harpenden, United Kingdom

SPRINGER-SCIENCE+BUSINESS MEDIA, B.V.

A C.I.P. Catalogue record for this book is available from the Library of Congress.

ISBN 978-90-481-5826-3 ISBN 978-94-015-9783-8 (eBook)
DOI 10.1007/978-94-015-9783-8

Printed on acid-free paper

Contents

Contributors

Jonas Borch
Department of Plant Biology, Royal Veterinary and Agricultural University
1871-Frederiksberg C, Denmark.
borch@biobase.dk

Peter Buchner
Agriculture and Environment Division, IACR-Rothamsted, Harpenden AL5 2JQ, UK.
peter.buchner@bbsrc.ac.uk

David Collinge
Department of Plant Biology, Royal Veterinary and Agricultural University
1871-Frederiksberg C, Denmark.
dbc@kvl.dk

Malcolm J. Hawkesford
Agriculture and Environment Division, IACR-Rothamsted, Harpenden, Hertfordshire, AL5 2JQ, UK.
malcolm.hawkesford@bbsrc.ac.uk

Holger Hesse
Freie Universität Berlin, Institut für Biologie, Angewandte Genetik, Albrecht-Thaer-Weg 6, 14195 Berlin, Germany.
hesse@mpimp-golm.mpg.de

Rainer Höfgen
Max-Planck-Institut für Molekulare Pflanzenphysiologie, Am Mühlenberg 1, 14476 Golm, Germany.
hoefgen@mpimp-golm.mpg.de

Kenneth Madriz-Ordeñana
Centro de Investigación en Biología Celular y Molecular (CIBCM). Universidad de Costa Rica San José, Costa Rica.
kmadriz@cibcm.ucr.ac.cr

Mari-Anne Newman
Department of Plant Biology, Royal Veterinary and Agricultural Universit. 1871-Frederiksberg C, Denmark.
mari@kvl.dk

Jean-Louis Prioul
Institut de Biotechnologie des Plantes, Bat. 630, Université de Paris-Sud, 91405 ORSAY Cedex, France.
Jean-Louis.Prioul@ibp.u-psud.fr

Andy Pereira
Plant Research International, Business unit Genomics, 6700AA Wageningen, The Netherlands.
A.Pereira@plant.wag-ur.nl

Claudine Thévenot
Institut de Biotechnologie des Plantes, Bat. 630, Université de Paris-Sud, 91405 ORSAY Cedex, France.
Claudine.Theveno@ipb.u-psud.fr

David Salt
Chemistry Department, Northern Arizona University, Flagstaff, AZ 86011, USA.
david.salt@nau.edu

John G. Scandalios
Department of Genetics, North Carolina State University, Raleigh, NC 27695-7614, USA.
jgs@unity.ncsu.edu

Frank W. Smith
CSIRO Tropical Agruculture, Long Pocket Laboratories, Indooroopilly, Qld 4068, Australia
frank.smith@pi.csiro.au

Gareth J. Warren
School of Biological Sciences, Royal Holloway, University of London, Egham, Surrey TW20 0EX, United Kingdom.
g.warren@rhbnc.ac.uk

Ilga Winicov
Department of Plant Biology, PO Box 871601, Arizona State University, Tempe, AZ 85287, USA.
winicov@asu.edu

Preface

The aim of this volume is to bridge a perceived gap between scientists working in the rapidly advancing and increasingly specialist field of molecular biology and biologists concerned with plant physiology at the ecosystem level.

The book is divided into two sections, consisting of chapters introducing some principle methods (Chapters 2-5) and a series of chapters (Chapters 6-11) which review specific biotic and abiotic stresses which impact on plant viability with an emphasis on application of molecular methods. Such a broad remit does not intend to provide comprehensive analysis of the selected topics but to give an impression of recent trends and progress. Neither is it the intention to cover all factors which influence plant adaptation to the environment. Of particular importance is the desire to highlight what experimental possibilities exist, and how diverse disciplines can interact and complement one another. It is clear that many major advantages arise form such cross-fertilisation of ideas and methods. The only result can be the furthering of knowledge of fundamental aspects of plant physiology.

Gene isolation and identification has been possible for some decades, and molecular biological techniques have been applied to every aspect of biology. The simple premise of isolating a gene and analysing its expression in relation to environmental parameters is easily grasped. However few biologists have the time to keep fully aquainted with the increasing sophistication of the technology, which improves at an ever-increasing rate. Progress in technology has been facilitated by application in the biotechnology industry, particularly in the biomedical field. Additionally, recent growth has been seen in the agro-biotechnolgy sector, with applications for at stress-resistant crops with improved yields and quality. Many stress situations limit plant growth, resulting in crop production difficulties. Understanding the molecular basis of these plant-environment interactions is a major step toward harnessing the innate mechanisms for crop production using gene technology for marker assisted breeding or gene transformation. At the very least an understanding of the principles involved in plant resistance to environmental stress enables optimisation of agronomic practices to optimise production.

A serious consequence of modern society is the negative impact on the natural environment. Whilst a major goal of plant sciences is improved agriculture, paradoxically agronomic practices have also contributed to environmental pollution and habitat destruction. Pollutants arising from industrial and/or mining activities have also left a legacy of damaged environments. Ideally plants may be used for bio-remediation, employing naturally tolerant species or transferring relevant traits across species. In this instance, as in many others, it is clear that one of the most valuable tools for plant improvement is bio-diversity, ironically one of the main casualties of human influence on the biosphere. Many of the approaches outlined in this volume depend upon and exploit variation between individuals and populations, and therefore presentation of biodiversity is imperative.

We are grateful for the efforts of the contributing authors for their time and effort. We hope that this book will be of interest to all scientists who wish to be kept informed of new developments and insights relevant to plant eco-physiology, and will be used as an advanced textbook for biology students, irrespective of their primary discipline.

Malcolm J. HAWKESFORD and Peter BUCHNER, Rothamsted,

June 2001.

Chapter 1

INTRODUCTION: THE MOLECULAR ANALYSIS OF PLANT ADAPTATION TO THE ENVIRONMENT

Malcolm J. Hawkesford

Agriculture and Environment Division, IACR-Rothamsted, Harpenden, Hertfordshire, AL5 2JQ, UK.
malcolm.hawkesford@bbsrc.ac.uk

INTRODUCTION

A major challenge for biologists is to understand the underlying mechanisms which enable a plant to adapt to its environment and perform optimally in as broad a range of conditions as possible. A complete understanding is only obtained by the integration of many disciplines of research from ecology to the molecular biology of individual genes.

An array of new technologies has arisen in recent years as a result of the rapid development of modern molecular biology, particularly the various genome-sequencing projects and ever more sensitive means to analyse gene expression. These technologies are now being applied to studies of plant ecophysiology, and specifically the study of the importance of biotic and abiotic factors on plant functioning and physiological adaptation to the environment.

The aim of this volume is to survey some of the molecular approaches that are currently being developed and illustrate their application in a number of areas of plant responses to specific environmental factors. Whilst ecology adequately describes populations within the environment and ecophysiology explores the underlying mechanisms of adaptation, the combination with molecular approaches offers a more complete understanding of fundamental principles of plant function within the environment. The link is gene expression and its control (Figure 1). Much of molecular biology has sought

M.J. Hawkesford and P. Buchner (eds.),
Molecular Analysis of Plant Adaptation to the Environment, 1–15.

to deliver tools for analysis of gene expression and ecophysiology can set these analyses in the appropriate context. Knowledge gained from such studies not only benefits our basic understanding of plant function but is clearly of benefit to agronomists and biotechnologists seeking to develop crops which can be grown optimally in a wide range of environmental conditions.

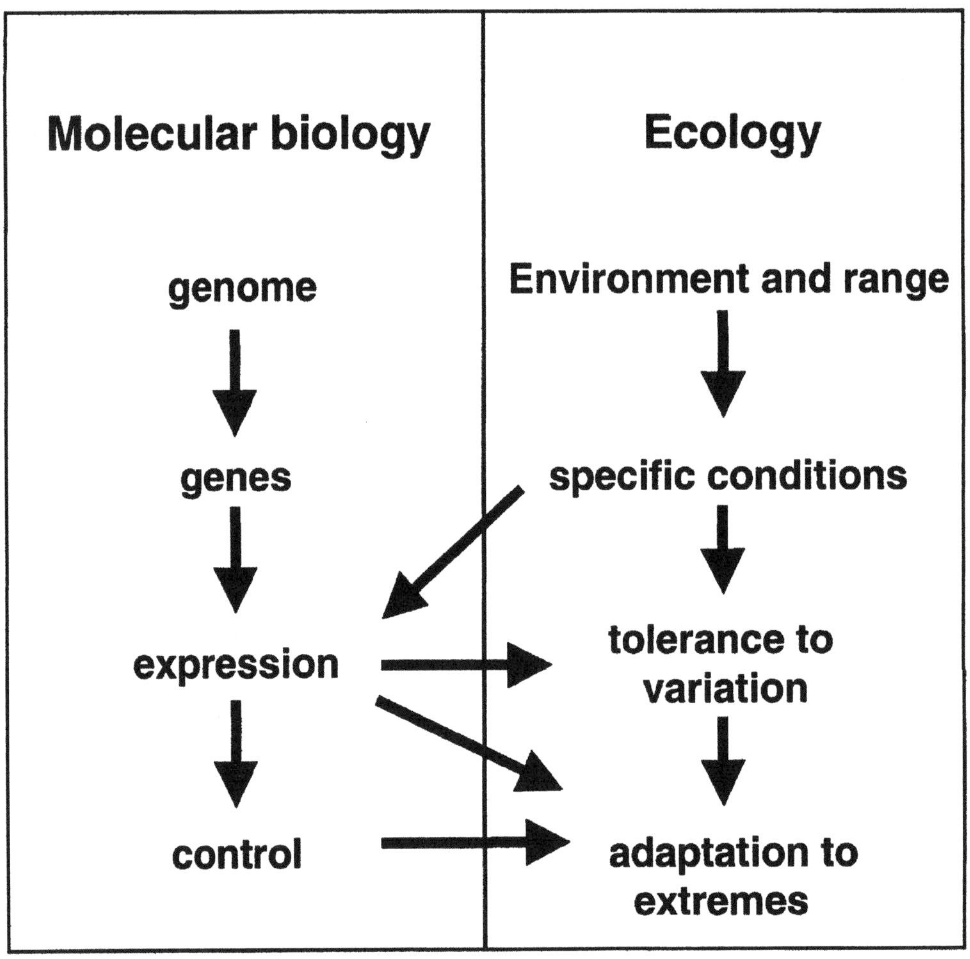

Figure 1. Gene expression and the mechanisms that control gene expression are the basis of plant adaptation to the environment as determined by ecological and physiological studies.

A range of abiotic and biotic factors (see Table 1) contribute to limiting the successful exploitation of the environment by an individual plant, and ultimately delimits the ecological range in which a species occurs. Many topical reviews and collections of papers have been published in recent years on specific stresses (see Table 1) or general plant stress responses (Vierling and Kimpel, 1992; Lerner, 1999; Smallwood *et al.*, 1999; Wilkinson, 2000).

Table 1. Environmental factors which limit plant exploitation of the environment.

Type	References
Light	Smith *et al.*, 2000
Oxidative stress	Chapter 9; Noctor and Foyer 1998
Cold	Chapter 10, Pearce, 1999
Heat	Howarth and Ougham, 1993
Nutrition	Chapter 11
Water	Chapters 6, 10; Ingram and Bartels, 1996
Salinity	Chapter 6, Yeo, 1998
Toxic concentrations of metals	Chapter 8
Pathogens	Chapter 7

Adaptation to environment or stress response

Plants are able to grow in most natural environments adapting to a huge range of conditions, generally however species are limited to particular environments and this defines their range or niche. Due to the evolutionary pressures exerted by these conditions species have evolved to tolerate wider extremes of conditions. Plant responses to environmental conditions, both abiotic and biotic factors, can be short-term physiological responses or may be longer-term adaptation brought about by evolutionary mechanisms. What are perceived as stress one day, may be optimum conditions following evolutionary adaptation and this is the process by which organisms evolve to fill new ecological niches.

Plants can tolerate a range of conditions utilising physiological responses (phenotypic plasticity, see Via *et al.*, 1995) and this is often referred to as tolerance to stress. Plants may alter metabolism or morphology to maintain growth, vigour and fecundity and this may be interpreted as successful resistance or tolerance to stress. Adaptation may result in decreased growth, but allow survival and reproduction in an otherwise inhospitable environment; these plants may be stressed, but they are nevertheless adapted. Stressed plants may be defined as not being adapted to the prevailing conditions and so fail to survive and reproduce. Mechanisms for adaptation may be elucidated at the molecular level and many examples may be found in other chapters of this volume. Adaptation may have evolved with single mutations, for example, leading to a more thermo-stabile enzyme, or may be more elaborate and complex, with the development of pathways to produce novel molecules such as osmo-protectorants.

The interacting factors, which define a species and its tolerance to stress, are summarised in Figure 2. The range of a species may be defined as that area with conditions where growth is optimal, or an area where, with appropriate responses, growth and reproduction can occur, even at sub-optimal levels. Eventually conditions become too extreme and no growth or even survival can occur.

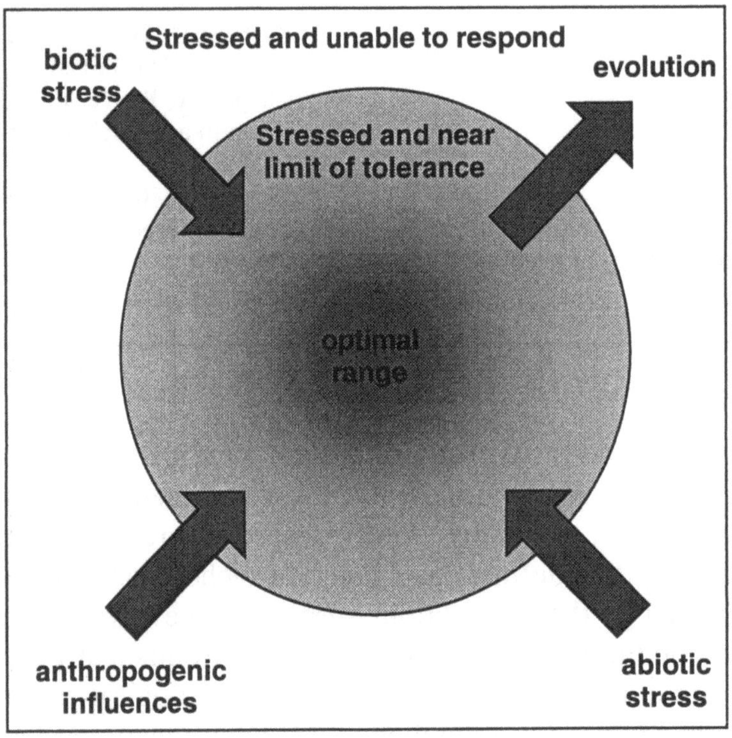

Figure 2. Species range depends upon interactions with the environment. The "normal range" encompasses those environmental conditions in which individuals of a species perform near their optimum in terms of growth and fecundity. Some environmental factor or stress will always limit the size of the range and as extremes are neared plant performance will deteriorate until a point is reached at which the plant no longer survives. In this diagrammatic representation, the nearer an individual is to the centre, the more likely that conditions are optimal. The range of a species may increase through evolutionary adaptation.

THE COMPLEXITY OF PLANT RESPONSES TO THE ENVIRONMENT

Underlying plant adaptation is a complex web of biochemistry, which is immensely flexible, and in different species, and in response to different environments, may be adapted and optimised to support life in almost any environment on the planet. The basic inputs of water, light, carbon dioxide and mineral nutrients are utilised in a wide range of extreme ratios. At the most fundamental level it is the control of expression of the genes encoding pathways and structural components, which determines this flexibility or plasticity. The plasticity is encoded by fixed genetic components, uniquely evolved in different species, with patterns of expression programmed to respond to changing environments.

Inevitably, responses to the environment are complex, involving many interacting branches of metabolism. Expression patterns of large numbers of genes are co-ordinately modified in programmed patterns, and with the flexibility to exploit or adapt to a range of environmental conditions. Many techniques are available to globally assess these patterns of expression and some are outlined below and in the following chapters. Generally from a bewildering complexity, molecular methods seek a reductionism goal, and usually for experimental purposes, specific environmental influences are isolated. In terms of evaluating response to environment, firstly a large-scale description is sought, and thereafter components are elucidated in isolation, eventually to the level of just a few genes. Then the whole process begins in reverse, to set expression of specific genes in context of wider environmental interactions.

What is the complexity of responses?

What is the sequence of events in response to environmental change? Firstly any change must be perceived and transmitted to invoke cellular responses. For most stresses and environmental signals, this is clearly the greatest area of ignorance. Many changes in the environment are perceived by receptors, often located on the plasma membrane. Examples might be G-protein coupled receptors, phytochrome receptor systems and metabolite sensing pathways. Signal transduction events involving constitutively expressed proteins, perhaps linked to cellular fluctuations in ion concentrations or other signal molecules, transduce this signal to an intracellular site for action. This would classically be a change in gene expression of components of the appropriate response pathway. Perhaps the signaling results directly in increased or decreased expression of an enzyme involved in the protection of the cell against a damaging environmental influence. Alternatively the first changes in expression may be transcription factors which in turn can cascade the response and induce or repress multiple genes.

It is clear that the number of genes showing variation in expression will be dependent upon the specific stresses or combinations of stresses. Typically in many natural environments, plant individuals will have to cope with multiple stresses simultaneously and responses will be a compromise between optimal adaptation to any individual stress. In a hot arid desert environment for example, a plant will have to cope with drought stress, an inability to mobilise nutrients, oxidative stress and probably temperature extremes.

All environments are dynamic, with constantly changing abiotic and biotic factors; processes of adaptation must be active, constantly. A few environmental factors are relatively constant and will exert their influence uniformly; such examples would be altitude, available CO_2, temperature in the tropics or poles. In these cases a steady-state adaptation is appropriate, which may not be very plastic. Factors limiting growth and proliferation are more likely to be transient factors, such as cloud cover limiting light availability, seasonal variations in temperature or cyclical availability of nutrients brought about by resource utilisation, exhaustion, senescence and recycling. Experimental procedures adopted by physiologists and molecular biologists often mimic these types of transients. In the case of nutrient availability, following a period of exposure to adequate nutrition, the supply of selected nutrients is restricted and effects on physiology and/or gene expression are monitored. There are many examples of this classical approach (Lee, 1982; Hawkesford and Belcher, 1991; Smith *et al.*, 1997).

Dissecting responses to the environment as illustrated by nutrition

The range of plant responses, which facilitate adaptation to the environment, may be illustrated by plant responses to nutrient availability. In the absence of essential nutrients, plants do not grow. With limiting macronutrients, growth will be severely retarded. In the natural environment or in a crop, during the growing season, sooner or later one or another nutrient, most likely a macronutrient will become limiting after an initial phase of growth. An early response is an up-regulation of specific nutrient transporters, as exemplified by the phosphate and sulphate transporters. In parallel, stored reserves are mobilised, firstly free ions in the vacuole, and subsequently recycling by protein degradation. This may lead to early senescence of vegetative tissues, and specific allocation to reproductive tissues and early flowering and seed set. A detailed example of the responses to phosphate deficiency is given in Chapter 11.

Root architecture shows considerable variation in form as an adaptation to environment (Epstein, 1973; McCully, 1999 and Figure 3). Root profiles may be adapted to exploit surface water and nutrients or much deeper resources. With nutrient deficiency, there is often a shift in root to shoot resource allocation, which favours root proliferation. Irrespective of overall nutrient supply, localised patches of nutrients, including, nitrate, phosphate and ammonium (Robinson, 1994) have been shown to stimulate localised root proliferation. Control of this phenomenon has been linked to expression of specific genes (Zhang and Forde, 1998). Clearly substantial changes in gene expression are required to bring about many of these responses.

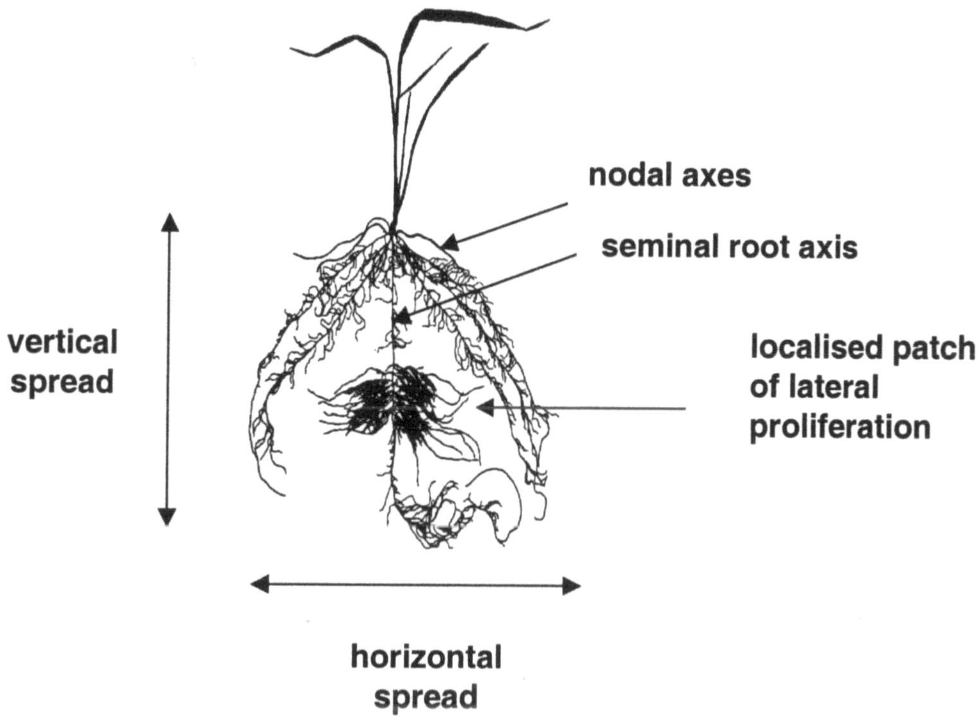

Figure 3. Root morphology is under genetic control, and include the root type (tap-root or fibrous, as shown here), vertical and horizontal spread, numbers of axes, degree of lateral branching and abundance of root hairs. Superimposed upon this are the influences of environmental factors such as water availability, soil compaction and nutrient supply. Shown here is a localised proliferation of lateral roots induced by a nitrate supply (Drew and Saker, 1978). Other specialised root structures such as cluster roots can be induced by phosphate deficiency (Marschner, 1995).

An example of changes in gene expression in response to environment is illustrated in Fig 4. In this case *in vitro* translation products of mRNA isolated from roots of plants with 4 days of a specific nutrient limitation (S) are compared with roots grown on a complete nutrient medium. The translation products are resolved on a 2-dimensional polyacrylamide gel electrophoresis (2D-PAGE) system, a technique able to resolve many hundreds of individual translation products with ease. It is clear that many of the translation products occur with approximately equal abundance in both samples. A number of translation products occur with less abundance or even complete absence (indicated by their presence in Figure 4A) in the nutrient deprived material, indicating a general down-regulation of expression of many genes.

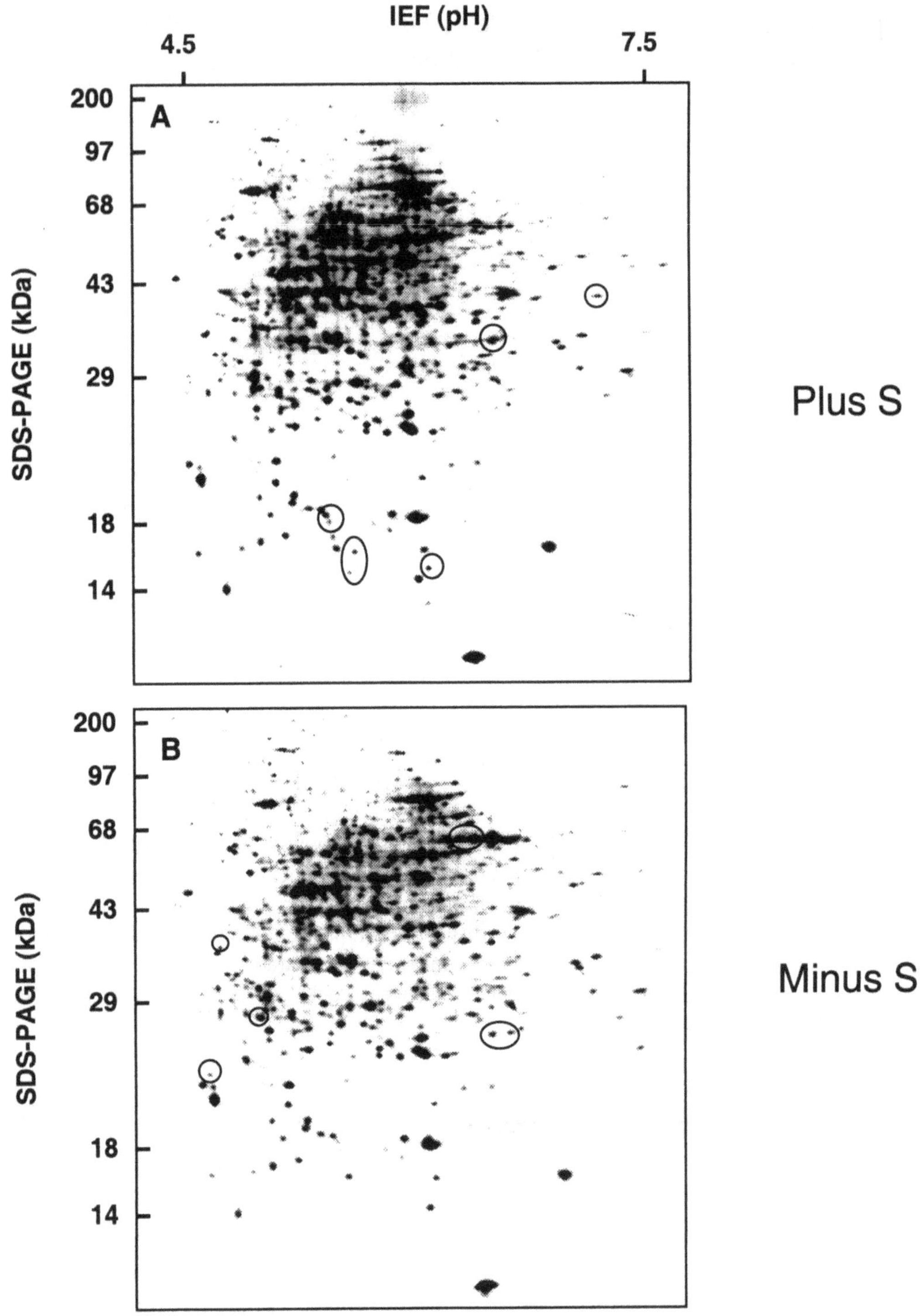

Figure 4. In vitro translation products of roots grown on an adequate *sulfur* supply (A) compared to root subject to 4 d *sulfur*-starvation (B). mRNA was isolated from cultured tomato roots (Hawkesford and Belcher, 1991) and in translation products produced using a wheat germ translation system including radio-labelled leucine. Products were separated by iso-electric focussing and SDS-PAGE as described in Hawkesford and Belcher, 1991. Major products showing higher abundance in either condition are circled.

This may reflect a decrease in cell metabolism or cell division and hence a general decrease in gene expression. As starvation continues this response is likely to become more acute. A few specific translation products show specific expression in the nutrient starved material (indicated in Figure 4B). Such translation products are likely to be proteins involved in specific adaptive responses aimed at alleviating the effect of the nutrient stress, particularly by the capacity for acquisition of the deficient nutrient. Numerous techniques exploit these differences in gene expression and specifically mRNA abundance, including subtractive library cloning, differential display, cDNA-AFLPs and micro-array techniques (see Chapters 3 and 4).

MOLECULAR APPROACHES

Perhaps the most significant recent advance in molecular biology has been the elucidation of details of the genomes of many organisms, including several plant species. The next major challenge, in "the post-genomic era", is the understanding of the roles of individual components and the co-ordination of their expression which is necessary for complex biological traits. A brief overview of the approaches available for the elucidation of gene expression and function is presented below, and further details are available in the succeeding chapters and in the references therein.

Genomes to expression analyses

Several complete bacterial genomes have been sequenced and the first plant genome, *Arabidopsis*, has been completed (see articles in Nature issue 6814, 14th December 2000 and Science, issue of 15th December, 2000). Projects are underway for genome sequencing of rice and tomato, and large EST (expressed sequence tag) libraries are available for many more plant species. Given that most plant species probably share the majority of the same genes, and differ only in minor details or in expression patterns, within a few years the basic information of the genetic make-up of all species will be available. It is estimated that the *Arabidopsis* genome comprises 20-30,000 genes (see Chapter 4), many of which have as yet no identified function.

Knowledge of the base sequence of the chromosomes is a long way from understanding what determines a species and the way it grows in a specific environment. A broad approach, which is beginning to make this connection, is transcriptome analysis. This is being performed by using cDNA micro-arrays (see Chapter 4) and the results are multiple and overlapping populations of transcripts which are co-ordinately expressed depending on developmental and/or environmental cues. A parallel analysis is that of the proteome. In this case proteins are extracted from appropriate plant tissues and analysed by high-resolution techniques, usually 2D-PAGE comprising iso-electric focussing and SDS-PAGE. Proteomic analysis will give a quite different picture to any of the transcriptome analyses and the most abundant protein will dominate the analyses, for example Rubisco subunits in leaf extracts. An improvement can be made by either being extremely selective in tissue selection or by sub-fractionation of the tissue prior to analysis. For example highly purified plasma membrane fractions were prepared from root tissues subjected to nutrient stresses, in order to identify up-regulated plasma membrane transporters (Hawkesford and Belcher, 1991). This approach can easily resolve several hundred of the most abundant proteins and can detect a few nanograms of an individual protein. Recovery of protein spots is now possible, and whilst direct sequencing can be performed on the more abundant proteins, mass spectroscopy techniques such as MALDI-TOF and Q-TOF (Thiellement *et al.*, 1999; Yates, 2000) are able to analyse the amino acid compositions of digested fragments of even low quantities of material. The amino acid compositions can be compared to databases of protein sequences, and thus "spots" on gels can be identified as specific proteins or enzymes, or at least identified with known open reading frames of un-assigned proteins from one of the genome or EST projects. These analyses are a substantial advancement for proteomic analysis.

A direct analysis of the transcriptome is possible by *in vitro* translation of isolated mRNA. The transcriptome analysed may be tissue and/or treatment specific, and for example stressed and non-stressed samples may be compared. *In vitro* translation systems are commercially available and translation reactions include radio-labelled amino acids to facilitate sensitive detection of protein products. The products are resolved by gel electrophoresis, preferably utilising the 2D-PAGE technique (see above and Figure 4). Although proteins are being detected, the abundance will depend upon the transcript abundance and not the steady state protein abundance of the originating tissue. An advantage of this technique is that with appropriate choice of conditions, there is likely to be up-regulation of expression of the specific genes of interest, enabling their easy detection.

The identification of specific gene transcription related to defined developmental and environmental cues is often attempted utilising differential expression techniques (Chapter 3). The chief objective of such an approach is to identify the genes which are expressed uniquely under certain defined conditions, in response to a stimuli or in particular ecotypes. Multiple genes are usually identified including many false positives (see discussion in Chapter 3). Some of these are known genes with known function and in this case, evaluating mechanisms of environmental adaptation are considerably more feasible than if only "unknown" genes are isolated. Various modelling procedures can help in the description of "unknown" gene products and correlation with phenotype is also helpful. With either identified or unidentified genes, the most common next step is expression analysis.

Whether unidentified genes (see above) or known genes are been considered, an important step is to establish roles in adaptation to environment and to undertake an expression analysis. This is usually in the form of Northern analysis in the first instance, and subsequently may take the form of protein analysis by Western blotting. Further elaboration on expression analysis may be undertaken utilising *in situ* techniques, *in situ*-PCR or immuno-localisation to identify specific cellular localisation of expression or even sub-cellular localisation of gene products. The aim of expression analyses may be to aid in identification of a gene, or at least to correlate expression with a known physiological condition, or it may be to demonstrate how a "known" gene is involved in a specific adaptive response. Northern blotting involves the gel electrophoretic separation of total extracted RNA, or enriched/purified mRNA, transfer to a membrane matrix and then probing with specific gene probes. Several samples can be run simultaneously on a gel and compared. Signals can be quantitated, and usually this is achieved by densitometry of visual signals from direct labelling of the probe on the membrane or following auto-radiographic detection of a radio-labelled probe on photographic film. Utilisation of constitutively expressed genes, or rRNA content is often used to normalise the individual samples to obtain quantification of the relative abundance of specific mRNAs.

The sensitivity of classical techniques determining mRNA levels, such as Northern blot analysis and RNase protection assay is often limited, although the time consuming and often-difficult enrichment of mRNA from the bulk of the rRNA can enhance sensitivity considerably. An alternative is to use PCR techniques. PCR offers the possibility of detecting a wide range of expression levels, especially of weakly expressed genes or small samples including single cell expression analysis. Additionally an examination of multiple genes from the same sample is possible. Using sequence information of the target gene, RT-PCR utilising specific primers will amplify the rarest of messages: this will give a positive indication of expression. For quantification, various approaches are possible. The development of competitive (Gilliland *et al.*, 1990) RT-PCR has allowed more accurate quantitative comparison and can be used to quantify expression but requires appropriate internal controls. The quantification depends upon the comparison of the intensity of ethidium bromide stained gels or densitometric image quantification of the bands. A quantitative RT-PCR method has been developed recently (Heid *et al.*, 1996; Gibson *et al.*, 1996), in which release of a fluorescent reporter dye from a hybridisation probe in real time, during the PCR, is proportional to the accumulation of the PCR product. This method is both more accurate and less labour intensive than the manual quantitative PCR methods described above. Several commercially available real-time PCR instruments are available.

Localisation of transcripts within tissues and at the single cell level is possible by *in situ* hybridisation (Johansen, 1997; Koltai and Bird, 2000). In this technique mRNAs are detected by hybridisation using specific gene probes in a manner similar to Northern blots. Detection is achieved using radio-labelled probes or with suitably conjugated nucleotides which can be subsequently detected by fluorescent techniques or secondary labelling reagents. Sensitivity can be further improved with PCR amplification approaches and possibilities exist for multiply and simultaneously labelling of individual tissue sections using for example different fluorescent probes.

The abundance of specific proteins can be evaluated in multiple samples by SDS-PAGE combined with immunological detection using Western blot procedures. Specific localisation to the sub-cellular level can be determined by combining microscopy and immuno-staining. An antibody to the target protein is a pre-requisite for such studies and the lack of available purified target protein for immunisation for antibody production has limited applicability in the past. The advent of cloned genes and procedures for 'over-expression' have facilitated the production of antibodies to difficult to isolate plant proteins. Over-expression is readily achieved in a number of systems including bacterial, yeast and bacculovirus (see for example, Gould 1994). Isolation of the target protein is often simple given the artificial over-abundance but is rendered more facile by utilising vectors which result in the production of the target protein fused to peptide fragments which act as affinity tags. These fragments may be sugar-binding proteins or metal-chelating poly-histadine peptides which will bind to affinity columns with respectively bound sugar or nickel, and may easily be cleaved with proteases if required.

Plant transformation

Transformation systems are available for a wide range of plants and are probably adaptable to any plant species (Birch, 1997; Hansen and Wright, 1999). A direct route to improving plant performance is via genetic manipulation, either up- or down-regulating critical steps which influence viability in any given environment. It is probable that multiple transformations will be required for the manipulation of complex traits and the development of new crops able to withstand previously hostile environments, is not a trivial undertaking. As discussed above, genomic approaches and associated techniques provide the genes, which may be over-expressed or down-regulated by anti-sensing or co-suppression. A major concern is the appropriate control of the introduced transgene. Most genes show differential expression tissue specificity, developmental regulation or responsiveness to specific environmental conditions. Inappropriate expression of a gene may serve to confuse interpretation of results or may even be detrimental to the plant.

To manipulate expression of a gene, an appropriate promoter which is active in the right place at the right time is required. It is probably in this area that the technology lags furthest behind. To even describe the activity of a promoter, much less understand how the activity is modulated, requires an immense amount of expression analysis at the molecular level: interacting factors, transcription factors and cis elements, all need to be catalogued and their interactions mapped. In addition, expression must be combined with eco-physiological data, and the full potential of transgenic manipulation awaits this detailed knowledge.

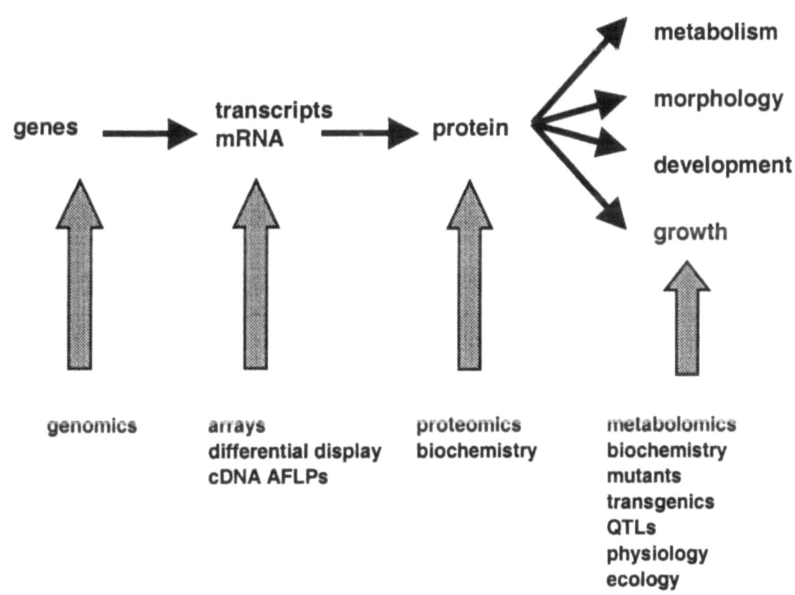

Figure 5. An array of approaches may be used to examine plant adaptation to the environment. A continuum exists from genomic studies and analyses of single genes, through studies of gene transcripts, mRNAs, proteins and metabolism, ultimately leading to whole plant physiology and ecology. AFLP = amplified restriction fragment polymorphism; QTLs = quantitative trait loci.

At present transgenic approaches are particularly useful for assessing the importance of specific steps in a pathway or for associating a phenotype with unknown genes. For these studies the choice of promoter may be less critical, however the resulting phenotypes should be evaluated with care.

Development of safe and acceptable genetically modified crops will require a much more considered approach.

OUTLOOK

Many tools exist to study the sequence of events, relating gene expression to plant adaptation to the environment (see Figure 5), and some of these are highlighted in this volume. Molecular tools will complement classical eco-physiological approaches to give a new level of understanding. Critically, within a few years most plant genes will be isolated and identified, and determining how they interact with one another and with the environment will be the next challenge.

Acknowledgements

Work in the laboratory of the author is sponsored by grants from the BBSRC, Home-Grown Cereals Authority and by Framework V of the EU. IACR receives grant-aided support from the Biotechnology and Biological Sciences Research Council of the UK.

REFERENCES

Birch, R. G. 1997. Plant transformation: problems and strategies for practical application. *Ann Rev. Plant Physiol. Plant Mol. Biol.* 48, 297-326.

Drew, M.C. and Saker, L.R. 1975. Nutrient supply and the growth of the seminal root system in barley. II. Localised compensatory increases in lateral root growth and rates of nitrate uptake when nitrate supply is restricted to only part of the root system. *J. Exp. Bot.* 29, 22-33.

Epstein, E. 1973. Roots. *Scientific American* 228, 48-58.

Gibson, U.E., Heid C.A. and Williams P.M. 1996. A novel method for real time quantitative RT-PCR. *Genome Res.* 6, 995-1001.

Gilliland, G., Perrin, S., Blanchard, K. and Bunn, H.F. 1990. Analysis of cytokine mRNA and DNA: detection and quantitation by competitive polymerase chain reaction. *Proc. Natl. Acad. Sci. USA* 87, 2725-2729.

Gould G.W. ed. 1994. *Membrane Protein Expression Systems: a User's Guide.* Portland press, London.

Hansen, G. and Wright, M.S. 1999. Recent advances in the transformation of plants. *Trends Plant Sci.* 4, 226-231.

Hawkesford, M. J. and Belcher, A. R. 1991. Differential protein-synthesis in response to sulfate and phosphate deprivation - identification of possible components of plasma-membrane transport-systems in cultured tomato roots. *Planta* 185, 323-329.

Heid C.A., Stevens, J., Livak, K.J. and Williams, P.M. 1996. Real time quantitative PCR. *Genome Res.* 6, 986-994

Howarth, C. J. and Ougham, H. J. 1993. Gene-expression under temperature stress. *New Phytol.* 125, 1-26.

Ingram, J. and Bartels, D. 1996. The molecular-basis of dehydration tolerance in plants. *Ann. Rev. Plant Physiol. Plant Mol. Biol.* 47, 377-403.

Johansen, B. 1997. *In situ* PCR on plant material with sub-cellular resolution. *Anal. Bot.* 80, 697-700.

Koltai, H. and Bird, D.M. 2000. High throughput cellular localization of specific plant mRNAs by liquid-phase *in situ* reverse transcription-polymerase chain reaction of tissue sections. *Plant Physiol.* 123, 1203-1212.

Lee, R. B. 1982. Selectivity and kinetics of ion uptake by barley plants following nutrient deficiency. *Anal. Bot.* 50, 429-49.

Lerner H.R. ed. 1999. *Plant Responses to Environmental Stresses, from Phytohormones to Genome Reorganization.* Marcel Dekker, New York.

Marschner H. 1995. *Mineral Nutrition of HigherPlants*, 2nd edition. Academic Press, London.

McCully, M.E. 1999. Roots in soil: unearting the complexities of roots and their rhizospherers. *Ann. Rev. Plant Physiol. Plant Mol. Biol.* 50, 695-718.

Noctor, G. and Foyer, C. H. 1998. Ascorbate and glutathione: keeping active oxygen under control. *Ann. Rev. Plant Physiol. Plant Mol. Biol.* 49, 249-79.

Pearce, R. S. 1999. Molecular analysis of acclimation to cold. *Plant Growth Regulation* 29, 47-76.

Robinson, D. 1994. The responses of plants to non-uniform supplies of nutrients. *New Phytol.* 127, 635-674.

Smallwood, M.F., Calvert, C.M. and Bowles, D.J. eds. 1999. *Plant Responses to Environmental Stress.* Bios Scientific publishers, Oxford.

Smith, H. 2000. Phytochromes and light signal perception by plants – an emerging synthesis. *Nature* 407,585-591.

Smith, F. W., Hawkesford, M. J., Ealing, P. M., Clarkson, D. T., VandenBerg, P. J., Belcher, A. R. and Warrilow, A. G. S. 1997. Regulation of expression of a cDNA from barley roots encoding a high affinity sulphate transporter. *Plant Journal* 12, 875-84.

Thiellement, H., Bahrman, N., Damerval, C., Plomion, C., Rossignol, M., Santoni, V., de Vienne, D. and Zivy, M. 1999. Proteomics for genetic and physiological studies in plants. *Electrophoresis* 20, 2013-26.

Via, S., Gomulkiewicz, R., Scheiner, S.M., Schlichting, C.D. and Van Tiederen, P.H. 1995. Adaptive phenotypic plasticity: consensus and controversy. *Trends Ecol. Evol.* 10, 212-216

Vierling, E. and Kimpel, J.A. 1992. Plant responses to environmental stress. *Current Opin. Biotechnol.* 3, 164-170.

Wilkinson, R.E ed. 2000. *Plant-Environment Interactions.* 2nd edition. Marcel Dekker, New York.

Yates, J. R., 3rd. 2000. Mass spectrometry. From genomics to proteomics. *Trends Genet* 16, 5-8.

Yco, A. 1998. Molccular biology of salt tolerancc in thc contcxt of whole-plant physiology. *J. Exp. Bot.* 49, 915-929.

Zhang, H. and Forde, B. G. 1998. An *Arabidopsis* MADS box gene that controls nutrient-induced changes in root architecture. *Science* 279, 407-409.

[illegible]

Chapter 2

GENETIC DISSECTION OF PLANT STRESS RESPONSES

Andy Pereira

Plant Research International, Business unit Genomics, 6700AA Wageningen, The Netherlands.
A.Pereira@plant.wag-ur.nl

INTRODUCTION

Natural environmental changes around a plant's life cycle and habitat require rapid transitory as well as seasonal phased responses. This innate adaptability is a natural defense mechanism due to the sessile habit that can not be supplemented by active defense or avoidance reactions. The resultant plastic response reactions of plants comprise some of the most sophisticated adaptations, unique in the biological world.

Adverse environmental factors can impose stress on plants that then are restricted in expressing their full genetic potential for growth and reproduction. These physicochemical stress factors can cause depreciation in crop yield up to 70% when compared to the yield under favorable conditions (Boyer, 1982). The major crop losses are due to stresses imposed by drought, excessive water, salinity or cold. Stability of crops to changes in environmental factors are therefore one of the most valued traits for breeding. Traditional breeding is thwarted by the complexity of stress tolerance traits, low genetic variance of yield components and the lack of efficient selection techniques. Stress tolerance components or secondary traits might be useful to follow in breeding by marker assisted selection. The identification of stress tolerance components using mapped loci or gene sequences will provide candidate genes to understand the stress reactions as well as suggest ways to modify plants to make them more resistant.

Amidst the complexities of environmental stress reactions in crop plants the use of the simple model *Arabidopsis*, offers an opportunity for the precise genetic analysis of stress reaction pathways common to most plants.

M.J. Hawkesford and P. Buchner (eds.),
Molecular Analysis of Plant Adaptation to the Environment, 17–42.

The relevance of the *Arabidopsis* model is evident in recent examples of improving drought, salt and freezing tolerance (Jaglo-Ottosen *et al.*, 1998; Kasuga *et al.*, 1999) using genes identified in *Arabidopsis*. In this review the depth of the molecular genetic analyses in *Arabidopsis* will be covered to reach general mechanisms and conclusions for other plants.

The environmental factors that impose stress on plants can be broadly defined as: osmotic stress in the form of dehydration (drought) and salinity; temperature stress including chilling, freezing and high temperature (heat); biotic stresses conferred by pathogens or pests (including wounding) as well as extremes in illumination or atmospheric conditions. In controlled laboratory studies a single stress factor can be imposed and the plant reactions precisely monitored at the level of transcripts, proteins, morphology and growth parameters. In nature most environmental stresses do not occur alone, for example osmotic stress due to drought in summer is accompanied by high temperature stress. Cold stress can also cause osmotic stress as it reduces water absorption and transport. The study of the interrelationships between different stresses is therefore relevant for applications in improving stress tolerance under field conditions.

The early responses of plants to stress are the sensing and subsequent signal transduction leading to stress-responsive gene expression. In response to osmotic stress elicited by water deficit or high salt the expression of a set of genes is altered (Zhu *et al.*, 1997), some of which are also induced by low temperature stress (Thomashow, 1998). These genes have been systematically termed *rd* (responsive to dehydration), *erd* (early responsive to dehydration), *cor* (cold-regulated), *lti* (low-temperature induced) and *kin* (cold-inducible). There is an overlap between the responsiveness of many of these genes, most dehydration-inducible genes also respond to cold stress and vice versa. Thus many of the identified genes have multiple names depending on their discovery. The analysis of stress responsive genes and promoters has enabled the identification of a cis-acting dehydration-responsive element (DRE) (Yamaguchi-Shinozaki and Shinozaki, 1994) and a similar low-temperature C-repeat (CRT) element (Baker *et al.*, 1994).

Osmotic and cold stress influence ABA levels and many osmotic stress and cold-stress responsive genes can also be induced by ABA (Zhu *et al.*, 1997). The analysis of stress responsive gene expression studied in ABA-deficit (*aba*) and ABA-insensitive (*abi*) mutants has revealed the presence of ABA-dependent and ABA-independent signal transduction pathways in response to different stresses (Shinozaki and Yamaguchi-Shinozaki, 2000). At least four independent pathways function under drought, two ABA-independent and two ABA-dependent. Two additional ABA-independent pathways are involved in low-temperature responsive gene expression. There is a common signal transduction pathway between dehydration and cold stress involving the DRE/CRT cis-acting elements, and two additional signal transduction pathways function solely in dehydration or cold response. These pathways are proposed to not necessarily act in a parallel manner but interact and converge to activate stress genes (Ishitani *et al.*, 1997).

Plants respond to dehydration and low temperature by accumulating various proteins and smaller molecules including sugars, proline and glycine betaine. Analysis of the drought-stress inducible genes additionally reveal the plethora of responses that include the induction of transcription factors (MYC, MYB), phospholipase C, protein kinases (MAPK, CDPK), proteinases, water channel proteins, chaperones, detoxification enzymes (GST, sEH), protection factors of macromolecules (LEA proteins) and osmoprotectant synthases (for proline, betaine, sugar). Cold acclimation involves complex physiological changes resulting in reduction of growth, water content, transient increase in ABA, changes in membrane composition and also accumulation of osmolytes like proline, betaine and soluble sugars. The overlap in the induced proteins and other molecules in the different stress responses emphasizes the interrelationship of signaling pathways and also reveals the complexity confronted by analyses of the product of the stress response. An alternative gene-target or gene-product analysis is potentially more productive, an option offered by gene mutation analysis.

MUTAGENESIS SYSTEMS

Mutants can help genetically dissect complex biological processes into discrete steps and identify the genes and their relative interactions. While some processes are more amenable to this approach like the visually evident developmental mutants of plants, other conditional processes need specific screening approaches to uncover the mutations and genes involved. Mutants can be obtained that manifest themselves at the morphological, biochemical or physiological levels and the specific phenotypes provide information on the interaction between different processes. Chemical and physical mutagens generally provide loss of function recessive mutations and can efficiently help define genes involved in specific pathways or processes. The interaction between genes in a particular pathway can be studied by combining mutants for the trait and testing for interaction between individual mutants. A number of biochemical processes like pigmentation has been thoroughly dissected in a number of plants (Mol *et al.*, 1998) in which both the enzymes and regulatory genes have been identified. While dispensable pathways can be analysed completely by mutant analysis, essential pathways need other elaborate methods to complement this approach.

Classical mutagenesis

Chemical and physical mutagenesistechniques and uses in the genetic model *Arabidopsis* have been extensively reviewed (Rédei and Koncz 1992), suggesting many practical and theoretical considerations that are also useful for other plants. Commonly used chemical mutagens include ethylmethane sulphonate (EMS), 1-methyl-3-nitro-1-nitrosoguanidine, 1-ethyl-1-nitroso-urea and 1-methyl-1-nitrosourea. EMS that primarily yields point mutations is most commonly used and at a 0.2-0.3 % treatment for 15 hr *Arabidopsis* exhibits an overall mutation frequency of 0.59. Physical mutagens like x-rays, γ-rays, thermal neutrons and fast neutrons have been successfully used. Hard x-rays, γ-rays and fast neutrons are not as destructive as soft x-rays and thermal neutrons that give more aberrations. The generation of small deletions has become a desired tool, as deletion mutations are assayable at high throughput by PCR reactions and also help in subtractive cloning. Diepoxybutane (Reardon *et al.*, 1987) and chlorambucil (Russel *et al.*, 1989) have been reported to induce small deletions in animal systems. Recently in *Caenorhabditis elegans* a deletion library was constructed using chemicals like EMS, ethylnitrosourea, diepoxyoctane and UV activated trimethyl-psoralen (Liu *et al.*, 1999). In *Arabidopsis* specific doses of γ-radiation were optimized (Cecchini *et al.*, 1998) to generate deletions of about 5 kb that would also be applicable to subtractive hybridization cloning. A HTP method termed TILLING, for the identification of point mutations in plant pools using denaturing HPLC, has recently been described in *Arabidopsis* as a tool for functional genomics (McCallum *et al.*, 2000).

The genetic analysis of mutants requires good model systems that are convenient to handle and are genetically simple with minimal gene redundancy. Primarily *Arabidopsis* but also maize, tomato, petunia, rice and barley are good systems to isolate mutants using chemical or physical agents and map the corresponding loci on chromosomes. These mutants can then be precisely mapped on the dense molecular maps available and lead to gene isolation by map-based cloning procedures from large BAC/YAC libraries. The availability of dense genetic, physical maps and BAC contigs in *Arabidopsis*, rice and tomato can facilitate BAC landing and identification of the corresponding gene without much chromosome walking (Tanksley *et al.*, 1995). Once the gene has been identified complementation of the mutant by transformation is the most straightforward way of establishing the gene-phenotype causal relationship.

Two features determine the ease of gene identification by map based cloning approaches; the availability of linked markers and transformation protocols. Markers can be generated at high efficiency in any plant by methods like AFLP (Vos *et al.*, 1995) using bulked segregation analysis. In *Arabidopsis* the generation of SNPs and HTP (high-throughput) screens in segregating populations (Cho *et al.*, 1999) can aid in determining the precise map position within a few weeks. Together with efficient transformation systems like the floral dip method in *Arabidopsis* (Clough and Bent, 1998) gene-function relationships with mutants can be established in a semi HTP manner. Consequently, with the help of the genome sequence, the genes corresponding to all the known mutants can be identified in a cost-effective way.

Insertional knockout mutagenesis

Insertional mutagenesis systems like transposable elements and *Agrobacterium tumefaciens* mediated T-DNA (transfer DNA) insertions can directly lead to gene identification, offering a strong advantage over chemical/physical mutagenic treatments. Insertion sequences often cause knockout mutations by blocking gene expression and might display a mutant phenotype. The mutant gene that is tagged by the insertion sequence can then be isolated by recovering DNA flanking the insert and subsequently lead to the isolation of the wild-type gene sequence.

The major advance in the use of T-DNA tagging was made in *Arabidopsis*, where non-tissue culture transformation approaches like seed transformation (Feldmann, 1991), vacuum infiltration (Bechtold and Pelletier, 1998) or floral-dip (Clough and Bent, 1998) methods yielded a higher proportion of tagged mutants in comparison to somaclonal mutants obtained using tissue culture approaches. For forward genetics, T-DNA insertional mutagenesis is practical on a large scale for *Arabidopsis* resulting in a large number of tagged genes. Molecular analyses of tagged mutants (Azpiroz-Leehan and Feldmann, 1997) have shown that the position of insertions are distributed quite randomly on all chromosomes as well as within genes, insertions in introns and promoters revealing mutant phenotypes. A number of T-DNA insertion mutagenesis screens have been undertaken for different stress parameters (Quesada et al 2000; Tokuhisa et al 1997; Tsugane et al 1999; Zhu et al 1998)

Transposon tagging has been an effective strategy to identify mutants and isolate genes in plants like maize with its well-studied endogenous transposons (Walbot, 1992). The maize transposon systems *Ac-Ds* and *En-I(Spm)* transpose when introduced into numerous heterologous hosts and after modification in vitro offer several advantages for transposon mutagenesis over the endogenous systems (Pereira, 1998). Two component systems comprising of a mobile transposon component (*Ds* or *I/dSpm*) and the corresponding stable transposase (*Ac* or *En/Spm*) source have been developed for effective tagging strategies (Bancroft *et al.*, 1992; Aarts *et al.*, 1993). The mobile transposon components are constructed to contain various marker or selectable genes like antibiotic or herbicide resistance genes and to monitor their excisions often inserted in other assayable/selectable marker genes (Baker *et al.*, 1987; Jones *et al.*, 1989). To increase or control transposition the transposase is put under the regulation of heterologous promoters (Swinburne *et al.*, 1992), or segregated out in progeny to yield stable transposon inserts. Random tagging in *Arabidopsis* (Aarts *et al.*, 1993; Bancroft *et al.*, 1993; Long *et al.*, 1993) and petunia (Chuck *et al.*, 1993) yielded the first tagged genes, in which plants containing transposed elements were selfed and screened for obvious mutant phenotypes. The frequency of random inserts displaying a mutant phenotype is about 1-5%, depending on the screening strategy used, similar to that observed for T-DNA inserts. Transposons display preferential transpositions to closely linked sites (Jones *et al.*, 1990; Bancroft and Dean, 1993), that can be used to efficiently isolate mutants for genes located near the original position. In heterologous systems this is achieved by mapping a large number of transformed T-DNA inserts that serve as jumping pads for local mutagenesis.

Gene detection methods

Classical genetic analyses of biological processes are based on loss-of-function or knockout mutations. The recovery of loss-of-function mutant phenotypes is limited by several factors. a) Genetic redundancy: in which a homologous gene can substitute the function of a mutated gene, is prevalent even in the simple genome of *Arabidopsis* where large chromosomal segments are duplicated (Lin *et al.*, 1999; Mayer *et al.*, 1999) comprising redundant genes. Sequential disruption of homologous and redundant genes in an individual genotype might therefore ultimately reveal a mutant phenotype. b) Conditional and subtle mutants: require an appropriate screening system for their detection, where phenotypes are revealed only when challenged with appropriate environmental cues like pathogens or abiotic stress. c) Lethal pleiotropic mutants: their role in a process of interest like abiotic stress is difficult to judge as the major obvious phenotype overshadows what might be the primary function in a plant. The recovery of lethal mutants is low in *Arabidopsis* when compared to other organisms (Miklos and Rubin, 1996), suggesting that a large proportion of gametophytic knockout mutants might be difficult to recover. All these reasons suggest that other gene detection systems rather than standard knockouts are required for gene-function identification.

Gene detection strategies (Skarnes, 1990), make use of inserts containing reporter gene constructs, whose expression is dependent on transcriptional regulatory sequences of the adjacent host gene. In this way genes (adjacent or knockouts) can be identified by their expression pattern, even though they might not directly display an obvious mutant phenotype. The identification of very specific expression patterns, e.g. in the meristem or localized after pathogen induction, can suggest the possible phenotype to screen for. The entrapment inserts allow for the selection of inserts in specific classes of genes, based on their expression pattern. Subsequent production of double or multiple mutants in a pathway or between partially redundant genes, as indicated by the expression pattern, might finally reveal mutant phenotypes.

Enhancer detection constructs contain a reporter gene like ß-glucuronidase (GUS) with a weak or minimal promoter, e.g. with a TATA box, situated near the border of the insert. When integrated in the vicinity of an enhancer sequence in the genome, capable of orientation independent transcriptional activation from a distance, the reporter gene can display the expression pattern of the adjacent chromosomal gene.

"Gene trap" type inserts are designed to create fusion transcripts with the target gene (Skarnes, 1990), mostly employing the NPTII and GUS reporter genes (Topping and Lindsey, 1995). Promoter trap types consist of a promoterless reporter gene and are expressed when inserted downstream of the chromosomal gene promoter. The exon trap type is more versatile as it enables reporter gene fusions to be created at various locations within a gene. By introduction of splice acceptor sites upstream of the reporter gene, transcriptional fusions are created even for insertions in introns and thus increases the recover of inserts expressing the reporter gene.

Analysis of expression patterns in T-DNA based constructs showed that 25% promoter traps and 50% enhancer traps displayed reporter gene expression in *Arabidopsis* or larger genomes like tobacco (Koncz *et al.*, 1989; Kertbundit *et al.,* 1991; Topping and Lindsey, 1995), suggesting that T-DNA inserted preferentially in transcriptionally active regions. Large T-DNA collections have been produced in *Arabidopsis* that can be used for forward as well as reverse genetics strategies (Table 2). Promoter trap tagged mutants have been isolated in *Arabidopsis*, exhibiting increased sensitivity to freezing stress (Isaksson *et al.,* 1997). From a screen of 2500 lines two tagged lines displayed low-temperature enhanced expression that was also ABA responsive.

For transposons similar strategies were employed for the *Ac-Ds* transposons engineered as promoter, enhancer or gene/exon trap systems. Using a novel method for efficient selection of stable transpositions about 50% enhancer trap (*DsE*) and 25% gene trap (*DsG*) inserts displayed an expression pattern (Sundaresan *et al.,* 1995). A number of starter *DsG* and *DsE* lines have since then been used by different collaborators to produce large populations of stable transposed elements, many of which have been screened for their expression pattern (Martienssen, 1998; Parinov *et al.*, 2000). A similar strategy for the selection of stable transposed gene trap *DsG* elements has been initiated in rice (Chin *et al.,* 1999) using greenhouse selectable marker systems.

Mis-expression mutants

Classical loss-of-function mutations are limited by redundant gene functions, conditional and subtle phenotypes. Gain-of-function mutants can be achieved by employing insertion sequences that carry a strong enhancer element near the border, thus over-expressing or mis-expressing the adjacent gene. Over/mis-expression mutants are advantageous where positive selection is possible for dominant mutants, or in processes or genes where simple knockout mutants reveal no mutant phenotype.

Gain of function "Activation Tagging" was successful in *Arabidopsis* using a T-DNA vector with multiple CaMV enhancer sequences near the border, in selecting transformants for cytokinin independent regeneration (Kakimoto, 1996) or an early flowering mutant caused by activation of a gene *FT* (Kardailsky *et al.*, 1999), which displays a late flowering knockout mutant phenotype (Kobayashi *et al.*, 1999). Further dissection of genetic pathways using second-step mutants, was demonstrated by the selection of a suppressor of a phytochrome *phyB* mutant with activation tagging (Neff *et al.*, 1999). An extensive screening of more than 25,000 T-DNA activation tags (Weigel *et al.*, 2000), revealed a 1/1000 frequency of confirmed dominant mutants. This frequency is surprisingly much lower than that observed for recessive knockout mutant screens but anyway provides new variants for gene function analysis.

A transposon construct variant containing a CaMV 35S promoter transcribing outward from the *Ds* transposon end, yielded the semi-dominant over-expression mutants *TINY* (Wilson *et al.*, 1996) and *SHI* a dwarf mutant. In this system, dominant mutations might be caused by insertions in the transcription unit over/mis-expressing the gene but also by producing an anti-sense transcript causing suppression. The populations of T-DNA and transposon mis-expression inserts that have been developed are attractive to use in positive selection schemes for gain-of-function stress resistant mutants and thus identify novel genes that can confer stress tolerance in other plants.

IDENTIFICATION OF STRESS RESPONSE MUTANTS

Direct selection for trait mutants

Mutants with altered responses to specific stress parameters can be obtained by directly selecting under the stress conditions. Stress factors that have been used to isolate mutants include osmotic stress using salt response, chilling or cold tolerance, high temperature or thermotolerance. The major limiting factor involved is the selection scheme for selecting mutants that have to be done at the single plant level. A number of the different screens undertaken are summarized in Table 1 and a discussion of the results is described below.

Table 1. Insertion Mutagenesis Systems for Reverse Genetics in *Arabidopsis*.

Stress factor	Selective agent	Type mutagen	# plants screened	# mutants	Mutant name	Reference
Chilling	Low temp	EMS	20,000 M2 plants	21	*chs*	Schneider *et al.*, 1995
Chilling	5°C	EMS, T-DNA	1711 M2 plants 8000 lines	51 21	*pfc, ncr* *sop*	Tokhuisa *et al.*, 1997
Freezing	4°C, -8°C	EMS	1804 M2 lines	13	*sfr*	Warren *et al.*, 1996
Freezing	-8°C	EMS	800,000 M2 plants	26		Xin and Browse, 1998
Thermo-tolerance	45°C	EMS	17,000 M2 plants	4	*hot*	Hong and Vierling, 2000
Thermo-tolerance	50°C	EMS	180,000 M2 plants	36	AtTS	Burke *et al.*, 2000
Salinity	Germination 250 mM Nacl	EMS	500,000 M2 plants	3	*rs*	Saleki *et al.*, 1993
Salinity	Germination 250 mM Nacl	γ-rays	130,000 M2 plants	12	*rss*	Werner and Finkelstein, 1995
Salinity	50 mM NaCl growth sensitivity	EMS Fast neutrons T-DNA	14,500 lines 40,000 lines 11,000 lines	25 13 4	*sos*	Wu *et al.*, 1996 Zhu *et al.*, 1998
Salinity	Germination 250 mM NaCl	EMS Fast neutrons T-DNA Ac-Ds	12,000 lines 47,952 lines 6,480 lines 62,500 F5	5 36 12 0	*sañ*	Quesada *et al.*, 2000
Salinity	Growth 200 mM NaCl	EMS T-DNA	17,400 lines 4,900 lines	2 0	*pst*	Tsugane *et al.*, 1999
Cold, drought, salinity, ABA	RD29A-LUC expression	EMS	300,000 M2 plants	103	*cos, los* *hos*	Ishitani *et al.*, 1997
Hypoxia	Allyl alcohol	EMS	12,000 lines	27	*aar*	Conley *et al.*, 1999
Genotoxic	UV-C roots	EMS	11,000 M2 plants	19	*uvs*	Albinsky *et al.*, 1999

Temperature stress

Chilling tolerant mutants were first obtained in *Arabidopsis* (Hugly *et al.*, 1990; Schneider, Hugly and Somerville 1995), based on a screen of a population of EMS mutagenised plants, where mutants were defined as those that display damage to chilling but appear wild-type at normal growth conditions. About 20 mutants were uncovered and one mutant *chs1* (chilling-sensitive) was analysed further, showing low-temperature induced chlorosis indicating a lesion at chloroplast level. This was demonstrated to be due to a reduced accumulation of proteins localized to the chloroplast (Schneider, *et al.,* 1995). This analysis validated the screen and the relevance of protein and membrane structure in chilling tolerance. The genes involved in the chilling tolerance were not identified at that time, but paved the way for other analyses in this direction.

A later screen for chilling tolerant mutants used EMS and T-DNA mutagenised *Arabidopsis* populations (Tokhuisa *et al.,* 1997). The mutant screen was designed to identify genes that were unnecessary at 22°C but essential for proper growth at 5°C. Using EMS mutagenesis a surprisingly high mutant frequency of 3% was obtained suggesting the requirement of many genes in the process of chilling sensitivity. The diverse chilling sensitive phenotypes identified like chlorosis, reduced growth, necrosis and death suggest the involvement of diverse processes like organelle biogenesis, cell metabolism and cell and organ development. Some of the mutants obtained seem to be tagged by the T-DNA and thus offer the advantage of direct gene isolation.

Freezing tolerance in temperate plants can be acquired by a period of low but non-freezing period. This cold acclimation is associated with complex changes in gene expression, but not all the cold induced genes are required for development of freezing tolerance. To isolate mutants affecting the development of freezing tolerance in *Arabidopsis* (Warren *et al.,* 1996), seedlings were grown under short day length for 5 weeks, transferred to a 4°C acclimation for 2 weeks and then frozen at -8°C for 24 hr. A set of seedlings from 1804 EMS mutagenised M3 lines was screened, yielding 13 potential mutants that were sensitive (and some lethal) to the freezing, but could be recovered from the parental lines. After retesting and complementation analysis, 7 different mutants displayed specific deficiencies in freezing tolerance and were designated *sfr* (sensitive to freezing). Out of these mutants the *sfr6* mutant was found to be deficient in cold-induced gene expression of a set of genes that contain the C repeat/dehydration-responsive element (CRT/DRE) motif in their promoters (Knight *et al.,* 1999). The *sfr6* mutant failed to provide osmotic stress and ABA induction of specific inducible genes and displayed pleiotropic effects on pigmentation and fertility, suggesting a broader role of *SFR6* in signal transduction.

To identify mutants that are constitutively freezing tolerant in the absence of a low temperature acclimation treatment, a freezing tolerance assay was developed (Xin and Browse, 1998). EMS mutagenised M2 seedlings were grown for 10 days on Petri dishes on plant nutrient solid media, then transferred to -8°C for 3 hr before transfer to 4°C and the survivors transferred to soil. From a screen of 800,000 M2 plants, 26 mutant lines were heritable, representing 6 separate loci. One mutant designated *eskimo1* (*esk1*) was analysed further, showing 50% survival at -10.6°C and the tolerance exhibited also in soil grown plants. In continuous light, the *esk1* mutants were darker green and more compact than wild-type, a characteristic of wild-type *Arabidopsis* grown under low temperatures (4°C) of cold acclimation. Interestingly this constitutive freezing tolerance seems unrelated to *COR* gene expression, as they showed cold induction in the *esk1* mutant, and not constitutive expression. The *esk1* and a few other mutants showed high levels of proline and *P5CS* an enzyme in proline biosynthesis, strongly suggesting that proline does play an important role in plant freezing tolerance. The *ESK1* protein is suggested to function normally as a negative regulator to repress *P5CS* expression and other responses in the absence of stress. A model for cold acclimation was proposed in which parallel or branched signaling pathways activate distinct suites of cold-acclimation responses.

At the other temperature extreme, mutants defective in acquisition of tolerance to high temperature were obtained (Hong and Vierling, 2000). A screen based on hypocotyl elongation was developed in which 2.5 day old seedlings were pre-treated at 38°C for 90 min followed by 2 h at 45°C, and after further 2.5 days those seedlings with no further growth were rescued as putative mutants. This screen identified 4 separate genetic loci *hot*1-4 (sensitive to hot temperatures) that are unable to acquire thermotolerance to high temperature stress. By candidate gene identification, the *hot1* mutant was determined to be caused by mutation in the heat shock protein 101 (Hsp 101) gene in an ATP binding domain, and was proven by mutant complementation with a wild-type gene transformation. A similar phenotype seems to be also caused by antisense suppression of this gene. The *hot1* mutant exhibits a thermotolerance defect in the hypocotyl elongation assay, as well as at 10-day old seedling stage and at the seed level, validating the mutant screen as a means of identifying genes involved in plant thermal tolerance. Recently another screen for thermotolerance acquisition (Burke *et al.*, 2000), using a 4 hr pre-incubation at 38°C and a 50°C challenge for 30 min, resulted in the identification of about 36 mutants at a frequency of about 1 mutant per 5000 M2 seedlings tested. The analysis of a mutant showed that there was a reduction in the level of a 27-kD protein.

Salt stress

Salinity tolerance has been a subject of numerous selection schemes in tissue culture (Rhodes *et al.*, 1986; Kirti *et al.*, 1991; Sumaryati *et al.*, 1992) as well mutants in crops like soybean (Abel, 1969), barley (Kueh and Bright, 1982) or tobacco (Sumaryati *et al.*, 1992). A more systematic analysis of salt tolerance was initiated in *Arabidopsis* by selection of salt-resistant (RS) germination mutants (Saleki *et al.*, 1993). Three different RS mutants were obtained by screening about 500,000 M2 seed and selection at 250 mM NaCl concentration. A more stringent selection for germination at 250 mM NaCl in minimal nutrient medium resulted in the selection of 6 recessive mutant alleles of a single gene (Werner and Finkelstein, 1995). The mutants termed *rss* (reduced salt sensitivity) displayed reduced sensitivity to Na^+ and Rb^+ as well as a lesser extent to K^+ and Cs^+ and accumulate lower proline than wild type. The mutants are not affected in ABA sensitivity and did not display improved growth when transferred to high salt after germination.

Another approach to understand salinity responses in *Arabidopsis* was taken in the identification of salt-hypersensitive *sos* (salt overly sensitive) mutants (Wu *et al.*, 1996). A root bending assay was used, where mutagenised seedlings grown first under the absence of salt were transferred to media containing 50 mM NaCl. The seedlings were placed on the plates that were oriented vertically so that the roots were upside down and mutants were scored as those that did not display root bending due to gravitropism, indicative of growth inhibition under NaCl stress. Wild-type seedlings could show a gravitropic root-bending response up to 150 mM NaCl on plates. From a number of different screens (Wu *et al.*, 1996; Liu and Zhu, 1997; Zhu, *et al.*, 1998) of 267,000 seedlings arising from 14,500 EMS, 4000 fast neutron and 11,000 T-DNA mutagenised lines, 3 different *sos* mutants were obtained. From the 3 mutagenic treatments 25, 13 and 4 mutants respectively were obtained (shown by allelism testing to be *sos1*, *sos2* or *sos3*) showing that the fast neutron screen was the most effective and additionally yielded all 3 mutant loci. The *sos* mutants are hypersensitive to inhibition by high Na^+ or Li^+ and are defective in the regulation of intracellular Na^+ and K^+ homeostasis, accumulating more Na^+ and less K^+ in response to high Na^+ challenge.

The *SOS3* gene was first isolated by map based cloning and displays homology to the calcineurin B subunit of yeast and neuronal calcium sensors of animals (Liu and Zhu, 1998). The *SOS2* gene was then isolated by positional cloning, and shown to encode a protein kinase (Liu *et al.*, 2000). Analysis of double mutants of *sos2* and *sos3* revealed no additive phenotype suggesting that these genes act in the same pathway. This was supported by biochemical evidence as the SOS2 kinase was shown to physically interact and be activated by the calcium-binding SOS3 protein (Halfter *et al.*, 2000). The recently isolated *SOS1* gene is predicted to encode a protein with N-terminal part of 12 transmembrane domains having homology to plasma membrane Na^+/H^+ antiporters from microbes and animals, and a hydrophilic cytoplasmic tail in the C-terminal part (Shi *et al.*, 2000). Sequence analyses of the numerous *sos1* mutant alleles reveal the importance of several regions in the transmembrane and tail parts for salt tolerance. The *SOS1* gene is expected to function in exporting Na^+ from the cytosol to the extracellular space, preventing Na^+ accumulation in the cytoplasm. The induction of the

SOS1 gene by salt stress is inhibited in the *sos2* or *sos3* mutants, indicating that it is controlled by the *SOS2/SOS3* regulatory pathway.

A very extensive screen for resistance at germination to 250 mM NaCl was carried out on a population of 66,432 families mutagenised by EMS, fast neutrons or T-DNA insertions (Quesada *et al.*, 2000). Although a total of 578 mutants were obtained, 53 survived the recovery process to set seed in the greenhouse. From 16 strong phenotypes studied further, genetic analysis distinguished 4 complementation groups and the mutants were named *saň* (*SALOBREŇO*) and numbered 1-4. These mutants were NaCl and mannitol tolerant but KCl and Na_2SO_4 sensitive at germination. Another weaker mutant, *saň5*, was NaCl, KCl and mannitol tolerant but ABA-insensitive at germination. Genetic analysis revealed this to be a null allele of the *ABI4* gene that encodes an APETALA2 DNA binding domain.

In addition to selection for resistance to salt at seed germination as described above, salinity tolerance at seedling growth stage was used as criteria to identify salt stress resistance genes (Tsugane *et al.*, 1999). *Arabidopsis* seedlings mutagenised by EMS or T-DNA insertions were germinated under low salt and then transferred after a week to 200 mM NaCl in solid medium for 2 weeks. Normal seedlings bleached while two EMS-derived salt tolerant mutant lines designated *pst* (photoautotrophic salt tolerance) were obtained from screening 22,300 families (17,400 EMS and 4,900 T-DNA). The mutant *pst1* was analysed further and was shown to be incompletely penetrant as only about 20% seedlings survived at concentrations between 100 and 300 mM NaCl at which wild-type seedlings died. In contrast to the previously reported mutants described above, the *pst1* mutant was not resistant to germination under salt stress and did not differ from wildtype in uptake of Na^+, K^+, Ca^{2+} or Mg^{2+} after NaCl challenge. Moderate light intensity increased the effect of salt stress suggesting the role of photosynthesis and active oxygen species. This was confirmed by the demonstration that the *pst1* seedlings were more tolerant to methyl viologen that generates superoxide radicals during photosynthesis. The increased activity of the oxygen scavenging enzymes, superoxide dismutase (SOD) and ascorbate peroxidase (APX), in the *pst1* mutant under salt stress indicates that active oxygen species produced by salt stress was detoxified more efficiently and conferred tolerance. The recessive nature of this mutant suggests that this mechanism for salt-stress tolerance is blocked in wild-type *Arabidopsis* and is probably under inducible control.

Indirect selection for mutants

The complexity of plant responses to environmental stresses makes it difficult to identify mutants in stress signal transduction solely by morphological or physiological responses. To identify genes that cannot be addressed by direct selection for defects in stress sensitivity, indirect responses can be selected for using novel selection schemes. The most common response to different stress factors is the alteration in the expression of a set of genes. Mutants can be selected that are deficient in the expression of specific stress regulated transcripts. Another indirect screen can be related to the secondary or associated stress responses.

Expression of stress responses

A convenient assay was devised with a chimaeric gene construct RD29A-LUC containing a firefly luciferase reporter driven by the DRE/C-repeat and ABRE-containing RD29A promoter (Ishitani *et al.*, 1997). The RD29A-LUC transgenic plants emit bioluminescence in response to cold, osmotic stress or exogenous application of ABA. Homozygous reporter plants were mutagenised by EMS and the M2 seedlings systematically screened using a high-throughput luminescence imaging system for mutants with altered bioluminescence. The expression mutants of osmotically responsive genes were designated as *cos* (constitutive), *los* (low) and *hos* (high). Seedlings grown on MS agar plates for a week were sprayed with luceferin and imaged before stress identifying putative *cos* mutants. The seedlings were then treated at 0°C for 48 hr and imaged again to identify *los* and *hos* mutants. After 2 days the seedlings were sprayed with ABA and a third luminescence imaging again identified potential ABA responsive *los* and *hos* mutants.

From about 300,000 M2 seedlings 3000 were selected with altered basal, cold and ABA responses. From those surviving to maturity, 833 mutants were confirmed in progeny as *cos*, *los* or *hos* mutants and about a hundred strong phenotypes analysed further. The mutants were challenged by cold, NaCl and ABA and defined by their response to these treatments as either responsive to only one, two or all three factors. These mutants were distributed among the classes with about 9% *cos,* 44% *los* and 47% *hos* mutants, the majority of the *los* and *hos* mutants being responsive to all the tested stress factors. Unique response mutants were identified in NaCl response only for 3 *los* and 1 *hos* type, for cold response with 3 *los* and 9 *hos* types. The mutant classes helped define ABA-dependent and ABA-independent stress signaling pathways that act in a parallel manner and cross-talk between the stress response pathways to activate stress regulated expression.

The *los1* mutant that shows a reduced response specifically to low-temperature stress (Xiong *et al.,* 1999a), displays no difference to wild-type in response to NaCl, ABA or NaCl combined with ABA. Unlike wild-type plants, heat shock at 30°C did not enhance the effect of NaCl or ABA while the synergism between ABA and NaCl was observed at room temperature as well as 30°C. When treated at 0°C for 48 hr no induction was evident in *los1* even in combination with NaCl, ABA or both together.

Another cold-response mutant *hos1* displayed superinduction of the reporter *RD29A* and other cold-responsive genes *COR47*, *COR15A*, *KIN1* and *ADH* (Ishitani *et al.*, 1998).Though these genes are also induced by ABA or osmotic stress in wild-type, the *hos1* response is restricted to cold stress. The *hos1* mutant appeared less cold resistant than wild type though it acquired the same level of freezing tolerance after cold acclimation. A pleiotropic phenotype exhibited was early flowering suggesting constitutive vernalisation. Based on the interaction to the various factors, the *HOS1* gene is proposed to negatively regulate low-temperature-induced expression of many stress-responsive genes but is also a positive factor for specific gene regulation by osmotic stress or ABA. An analysis of the *hos2* mutant (Lee *et al.*, 1999) showed enhancement of RD29A and other stress genes under low temperature treatment but not by ABA or osmotic stress. The *hos2* mutant plants have a lower capacity of developing freezing tolerance after acclimation by low non-freezing temperatures. This indicates that *HOS2* is a negative regulator of low temperature signal transduction important for plant cold acclimation. One other *hos* mutant (number 693) was reported to show an enhanced response of 30, 7 and 15 times higher to the inducers cold, NaCl and ABA respectively (Xiong *et al.*, 1999a).

The *hos5* mutant was identified by increased expression in response to osmotic stress and ABA and unaffected by cold (Xiong *et al.*, 1999b). Double mutant analysis of *hos5* with the ABA-deficient *aba1* as well as the ABA insensitive *abi1* mutants showed that the osmotic stress superinduction was independent of ABA deficiency or insensitivity. In addition the *hos5* mutant did not affect stomatal control and only slightly influenced growth regulation and proline accumulation by ABA. This analysis reveals the *hos5* mutation as a negative regulator of osmotic stress-responsive gene expression shared by ABA dependent and independent pathways.

Associated stress responses

Stress responses induce changes in reactive oxygen species, ABA, proline and other metabolites or signaling molecules. The role of ABA in stress signaling is well documented by a set of stress induced genes in an ABA dependent pathway. Mutants in ABA responses (Koornneef *et al.*, 1984), ABA-deficient and ABA-insensitive are therefore relevant for stress response studies. The ABA deficient mutants *aba1*, *aba2* and *aba3* mutants that have reduced seed dormancy also display a wilty phenotype due to excessive water loss. In addition the *aba3* mutant is also tolerant to 200 mM NaCl. The ABA insensitive *abi1* and *abi2* mutant plants wilt readily, the *abi1* mutations also characterised to exhibit reduced seed dormancy and abnormal drought rhizogenesis. The *ABI1* and *ABI2* genes encode protein phosphatases involved in ABA signaling and required for ABA responsiveness in seeds and vegetative tissues (Leung *et al.*, 1997).

The expression of the alcohol dehydrogenase gene (*ADH*) is induced under oxygen deprivation (anoxia) and decreased availability (hypoxia) including other environmental stresses in *Arabidopsis*. A negative selection approach was taken using the ADH enzyme as a selectable marker in which EMS mutagenised M2 seed were treated to allyl alcohol that allows selection of mutants in ADH expression (Conley *et al.*, 1999). From 150,000 M2 seeds 40 resistant mutants designated *aar* (allyl alcohol resistant) were tested on a *ADH* promoter-GUS reporter to identify those mutants that were only affected in the *ADH* gene expression (expressing the *ADH*-GUS reporter). The remaining 27 mutants, not expressing the ADH-GUS reporter were expected to have mutations in signaling pathways leading to *ADH* expression. Two mutants *aar1* and *aar2* were analysed in detail and shown to be defective in response to cold and osmotic stress as well as in the anoxic and hypoxic induction of *ADH* and other glycolytic genes in mature plants. These mutants are proposed to be involved in a late step of the signaling pathways that lead to increased expression of the *ADH* gene and glycolytic genes.

In attempts at a genetic dissection of plant DNA repair components mutants have been isolated that are hypersensitive to UV-light, γ- or X-ray radiation. The characterised mutants can be grouped into a class that is defective in the production of UV-absorbent flavonoid compounds (Landry *et al.*, 1995; Li *et al.*, 1993) and a class defective in DNA repair (Britt *et al.*, 1993). A mutant *uvs66* was isolated that is highly sensitive to UV-light and DNA damaging chemicals but not impaired in basic DNA repair processes (Albinsky *et al.*, 1999). This mutant exhibits a hypersensitivity to ABA and NaCl, providing a surprising link between genotoxic stress and ABA/salinity signaling pathways.

REVERSE GENETICS STRATEGIES TO ANALYSE STRESS RESPONSES

The reactions to different stress factors have identified a multitude of transcripts that are induced by specific and multiple stress factors. Their role in the plant can be analysed by either generating transformed plants over-expressing the specific gene or by isolating and analysing mutants. These reverse genetics strategies to identify the functions of genes are emerging as important tools since the added genome sequence information is providing more genes with unknown function. A large proportion of the genome and of individual genes in *Arabidopsis* is duplicated, so multiple mutants for the redundant genes will have to be generated and analysed to uncover their functions. Added to this is the analysis of micro-array experiments where a large number of new genes induced by various stress factors can be identified and mutants of genes with similar expression patterns tested for interaction.

Site selected insertional mutagenesis

A reverse genetics technique called "site selected insertion mutagenesis" was developed using transposons in Drosophila (Ballinger and Benzer, 1989; Kaiser and Goodwin, 1990) to inactivate genes that had been sequenced, but whose function was unknown. Insertion mutants of the entire genome are first generated and then individuals with an insertion in a gene of interest can be identified by PCR. In maize and petunia the multiple endogenous transposon copies provide genome saturation and site selected insertional mutagenesis has been shown to work to identify knockout mutations in specific genes (Das and Martienssen, 1995; Koes *et al.,* 1995). As an alternative to insertions, deletion mutants generated by chemical or physical mutagens can be employed in a "high-throughput (HTP)" screen to identify mutants of specific genes described by the genome sequence (Liu *et al.,* 1999).

To be able to recover an insert in any gene of interest, a population of plants carrying inserts in most genes is required. Using transgenic systems large populations of inserts can be generated by T-DNA or transposon inserts. These can be knockout insertions as well as the variety of gene detection and mis-expression inserts described above that will help gene function analysis. Currently two strategies for genome saturation with inserts are being pursued; (i) single/few stable elements, similar to the T-DNA population structure, and (ii) multiple elements from active transposing populations, similar to the endogenous transposon systems. In *Arabidopsis* about 110,000 inserts would give an insert every kb, with about a 99% chance to mutate an *Arabidopsis* gene of 5 kb (Krysan *et al.,* 1999). Larger genomes being experimented like rice, tomato and maize would require sizeable larger number of inserts unless the frequency of insertions in genes is higher.

The selection of defined gene mutations is done by PCR using a pair of primers, one of which anneals to the insertion sequence and the other to the specific target gene. If an insert close to the gene primer is present in the population then a PCR product specific for the target gene can be observed. These sensitive PCR screens can be scaled up to HTP so that by specific pooling strategies the individuals containing the insert can be directly identified from a large population. Progeny of the individual containing the insert can then be analysed to characterise the mutant phenotype.

In *Arabidopsis* a number of T-DNA populations have been generated (Table 2) that have been used to isolate and characterise mutants providing new insights into gene function and validating the strategy (Hirsch *et al.,*

1998; Gaymard *et al.,* 1998). At present a collection of 60,480 transformants are publicly available for screening by PCR through the *Arabidopsis* Knockout Facility at the University of Wisconsin (Krysan *et al.,* 1999). The pooling strategy and organized screening done on available DNA involves a primary PCR screen on 30 DNA pools and after identification of a positive primary pool, a secondary PCR screen. This identifies a pool that has to be sown out by the experimenter, DNA isolated and the individual containing the insert identified.

There are various *En-I* (*Spm-dSpm*) transposon systems available in *Arabidopsis* (Table 2) that are also accessible to PCR screening methods (Wisman *et al.,* 1998; Speulman *et al.,* 1999; Tissier, *et al.,* 1999). Two of the populations contain multiple actively transposing elements per line that have been accumulated by propagation of the lines for a number of generations (Wisman *et al.,* 1998; Speulman *et al.,* 1999). A large stable *dSpm* insert population has been recovered by an ingenious system (Tissier *et al.,* 1999), using positive and negative herbicide selectable markers in the greenhouse.

Table 2. Insertion Mutagenesis Systems for Reverse Genetics in *Arabidopsis.*

Tag	Screenable marker	# inserts per plant	Insert Population	Construct Type	Reverse Genetics	Reference
T-DNA	NPTII	~ 1.5	5,300 lines	KO	X	McKinney *et al.*, 1995
T-DNA	NPTII	Low	9,100 lines	KO	X	Krysan *et al.*, 1996
T-DNA	NPTII	~ 1.5	6,000 lines	KO	X	Winkler *et al.*, 1998
T-DNA	PT- NPTII	Low	Cell culture	PT	X	Mathur *et al.*, 1998
T-DNA	ET-GUS	Low	11,370 lines	ET		Campisi *et al.*, 1999
T-DNA	NPTII	Low	60,480 lines	KO	X	Krysan *et al.*, 1999
T-DNA		1-low	9,264 lines	KO	X	Meissner *et al.*, 1999
T-DNA	BAR	1-low	25,000 lines	AT		Weigel *et al.*, 2000
DsG/DsE	NPTII	1-low	2,000 lines	ET/GT	X	Martienssen, 1998
En-1	-	~ 6	8,000 lines	KO	X	Wisman *et al.*, 1998
I/dSpm	-	~ 20	2,592 lines	KO	X	Speulman *et al.*, 1999
DSpm	BAR	1-low	48,000 lines	KO	X	Tissier *et al.*, 1999
DsG	NPTII	1-low	931 lines	GT/ET	X	Parinov *et al.*, 1999

KO Knockout, PT Promoter trap, ET Enhancer trap, GT Gene trap, AT Activation tag.

With more than 80% of the *Arabidopsis* genome sequence now available it is efficient to identify inserts in genes by sequencing the DNA flanking different inserts and comparing them to the genome sequence (Ito *et al.*, 1999; Parinov *et al.*, 1999; Speulman *et al.*, 1999; Tissier *et al.*, 1999). The DNA flanking an insert can be isolated by a variety of ways (Maes *et al.*, 1999), sequenced and compared to the genome sequence. After sequencing a large number of these insertion sites, inserts in genes of interest can be identified and their mutant phenotype analysed. These efforts are very valuable as an accumulating resource towards insertions in all genes. The value of sequencing transposon-flanking DNA has also been recognised in the larger genome rice where active *Ac* transposon populations have been used for the identification of inserts in genes by sequencing in parallel to PCR based screening (Enoki *et al.*, 1999).

Gene silencing

The use of transgenes for homology dependent gene silencing (HDGS) offers a new tool to shut down the function of endogenous genes, consequently generating mutants without mutation. Two types of silencing have been described (Kooter *et al.*, 1999); transcriptional gene silencing (TGS) involving promoter homology/methylation and post-transcriptional gene silencing (PTGS) requiring homology between interacting genes in the transcribed regions. To obtain gene silencing most experiments typically use a complete or partial cDNA clone placed downstream of a strong promoter in sense or anti-sense orientation. In about 10-50% of the transformants a dominant mutant phenotype can be observed for genes with a visual effect, e.g., enzymes in anthocyanin colouration (Meyer and Saedler, 1996). PTGS has been suggested to be probably due to the involvement of a form of aberrant or double stranded RNA and constructs containing repeats that specifically confer aberrant RNAs have been shown to be particularly useful for silencing (Hamilton *et al.*, 1998; Waterhouse *et al.*, 1998). By the transformation of specific gene constructs that confer aberrant or double stranded RNA quite predictable silencing of homologous genes in the genome is likely. This approach of HDGS is also applicable to create mutations for multigene families, for example many of the stress-induced transcripts, where clues to their biological role may be obtained.

An efficient method for silencing that can be conducted in a HTP way is virus-induced gene silencing (VIGS). This employs virus vectors carrying fragments from plant host genes, that can be introduced by transformation or viral infection into the host and can result in the suppression of endogenous gene expression (Baulcombe, 1999). For a HTP based forward genetics screen libraries of gene fragments (e.g. cDNA) can be cloned in vectors developed from the genomes of TMV, PVX or TGMV that are used for plant infection with pools or individual clones. Even members of multigene families that have sufficient DNA homology within the fragments used can be effective in gene suppression.

FUNCTIONAL GENOMIC PERSPECTIVES

Gene discovery through genome sequence information is an enormous growing resource that is expected to generate a new genomic revolution. With all the genes of the two model plants *Arabidopsis* and rice in public databases, the discovery of the gene functions is going to take central stage. It is evident that a large proportion of these plant genes will be affected by environmental challenges. Genome-wide micro-array analysis of gene transcript expression patterns will soon be able to reveal every gene that is perturbed in expression by any particular stress (see Chapter 4). Coupled to transcript information, changes reflected in the proteome and metabolome will help identify the biological and chemical reactions taking place. The vast amount of gene expression information has to be understood in the context of signal perception, transduction and gene regulatory circuits in response to external influences. This will provide a genomic view of biological processes like environmental stress response but not necessarily provide more answers to the functions of genes involved. Reductionism methods like the use of mutants will help dissect the complex process as the relevance of a specific gene activity to the biological process can be made. The use of mutants for expression analysis will provide another dimension, where regulatory genes can be identified and the cross talk between complete regulatory and signaling pathways deciphered. The mutants in turn will have to be tested in combination, to obtain functional knockouts for redundant processes, and analysed in a much more HTP manner that done so far. The outcome will be an understanding in the genomic biology of stress responses that will provide us with a better view of this complex process and help in crop improvement.

Acknowledgements

I would like to thank Nayelli Marsch Martinez and Pirjo Mäkelä for a critical appraisal of this manuscript.

REFERENCES

Aarts, M.G.M., Dirkse, W., Stiekema, W.J. and Pereira, A. 1993. Transposon tagging of a male sterility gene in *Arabidopsis*. *Nature* 363, 715-717.

Abel, G.H. 1969. Inheritance of the capacity for chloride inclusion and exclusion by soybean. *Crop Sci.* 9, 697-698.

Albinsky, D., Masson, J.E., Bogucki, A., Afsar, K., Vaas, I., Nagy, F. and Paszkowski, J. 1999. Plant responses to genotoxic stress are linked to an ABA/salinity signaling pathway. *Plant J.* 17, 73-82.

Azpiroz-Leehan, R. and Feldmann, K.A. 1997. T-DNA insertion mutagenesis in *Arabidopsis*: going back and forth. *Trends Genetics* 13, 152-156.

Baker, B., Coupland, G., Fedoroff, N., Starlinger, P. and Schell, J. 1987. Phenotypic assay for excision of the maize controlling element *Ac* in tobacco. *EMBO J.* 6, 1547-1554.

Baker, S.S., Wilhelm, K.S. and Thomashow, M.F. 1994. The 5'-region of *Arabidopsis thaliana cor15a* has *cis*-acting elements that confer cold-, drought- and ABA-regulated gene expression. *Plant Mol. Biol.* 24, 701-713.

Ballinger, D.G. and Benzer, S. 1989. Targeted gene mutations in *Drosophila*. *Proc. Natl. Acad. Sci. USA* 86, 9402-9406.

Bancroft, I. and Dean, C. 1993. Transposition pattern of the maize element *Ds* in *Arabidopsis thaliana*. *Genetics* 134, 1221-1229.

Bancroft, I., Jones, J.D.G. and Dean, C. 1993. Heterologous transposon tagging of the *DRL1* locus in *Arabidopsis*. *Plant Cell* 5, 631-638.

Bancroft, I., Bhatt, A.M., Sjodin, C., Scofield, S., Jones, J.D.G. and Dean, C. 1992. Development of an efficient two-element transposon tagging system in *Arabidopsis thaliana*. *Mol. Gen. Genet.* 233, 449-461.

Baulcombe, D.C. 1999. Fast forward genetics based on virus-induced gene silencing. *Curr. Opin. Plant Biol.* 2, 109-113.

Bechtold, N. and Pelletier, G. 1998. *In planta Agrobacterium*-mediated transformation of adult *Arabidopsis thaliana* plants by vacuum infiltration. *Methods Mol. Biol.* 82, 259-266.

Boyer, J.S. 1982. Plant productivity and environment. *Science* 218, 443-448.

Britt, A.B., Chen, J-J., Wykoff, D. and Mitchell, D. 1993. A UV-sensitive mutant of *Arabidopsis* defective in the repair of pyrimidine-pyrimidinone (6-4) dimers. *Science* 261, 1571-1574.

Burke, J.J., O'Mahony, P.J. and Oliver, M.J. 2000. Isolation of *Arabidopsis* mutants lacking components of acquired thermotolerance. *Plant Physiol.* 123, 575-587.

Campisi, L., Yang, Y., Yi, Y., Heilig, E., Herman, B., Cassista, A.J., Allen, D.W., Xiang, H. and Jack, T. 1999. Generation of enhancer trap lines in *Arabidopsis* and characterization of expression patterns in the inflorescence. *Plant J.* 17, 699-707.

Cecchini, E., Mulligan, B.J., Covey, S.N. and Milner, J.J. 1998. Characterization of gamma irradiation-induced deletion mutations at a selectable locus in *Arabidopsis*. *Mut. Res. Fund. Mol. Mech. Mutag.* 401, 199-206.

Chin, H.G., Choe, M.S., Lee, S-H., Park, S.H., Park, S.H., Koo, J.C., Kim, N.Y., Lee, J.J., Oh, B.G., Yi, G.H., Kim, S.C., Choi, H.C., Cho, M.J. and Han, C-D. 1999. Molecular analysis of rice plants harboring an *Ac/Ds* transposable element-mediated gene trapping system. *Plant J.* **19**, 615-623.

Cho, R.J., Mindrinos, M., Richards, D.R., Sapolsky, R.J., Anderson, M., Drenkard, E., Dewdney, J., Reuber, T.L., Stammers, M,, Federspiel, N., Theologis, A., Yang, W.H., Hubbell E, Au M, Chung EY, Lashkari D, Lemieux B, Dean, C., Lipshutz, R.J., Ausubel, F.M., Davis, R.W. and Oefner, P.J. 1999. Genome-wide mapping with biallelic markers in *Arabidopsis thaliana*. *Nat. Genet.* 23, 203-207.

Chuck, G., Robbins, T., Nijjar, C., Ralston, E., Courtney-Gutterson, N. and Dooner, H.K. 1993. Tagging and cloning of a petunia flower color gene with the maize transposable element *Activator*. *Plant Cell*, 5, 371.

Clough, S.J. and Bent, A.F. 1998. Floral dip: a simplified method for Agrobacterium-mediated transformation of *Arabidopsis thaliana*. *Plant J.* 16, 735-743.

Conley, T.R., Peng, H.P., Shih, M.C. 1999. Mutations affecting induction of glycolytic and fermentative genes during germination and environmental stresses in *Arabidopsis. Plant Physiol.* 119, 599-608.

Das, L. and Martienssen, R. 1995. Site-selected transposon mutagenesis at the *hcf106* locus in maize. *Plant Cell* 7, 287-294.

Enoki, H., Izawa, T., Kawahara, M., Komatsu, M., Koh, S., Kyozuka, J. and Shimamoto, K. 1999. *Ac* as a tool for the functional genomics of rice. *Plant J.* 19, 605-613.

Feldmann, K.A. 1991. T-DNA insertion mutagenesis in *Arabidopsis*: mutational spectrum. *Plant J.* 1, 71-82.

Gaymard, F., Pilot, G., Lacombe, B., Bouchez, D., Bruneau, D., Boucherez, J., Michaux-Ferriere, N., Thibaud, J.B. and Sentenac, H. 1998. Identification and disruption of a plant Shaker-like outward channel involved in K^+ release into the xylem sap. *Cell* 94, 647-655.

Halfter, U., Ishitani, M. and Zhu, J-K. 2000. The *Arabidopsis* SOS2 protein kinase physically interacts with and is activated by the calcium-binding protein SOS3. *Proc. Natl. Acad. Sci. USA* 97, 3735-3740.

Hamilton, A.J., Brown, S., Yuanhai, H., Ishizuka, M., Lowe, A., Solis, A-G.A. and Grierson, D. 1998. A transgene with repeated DNA causes high frequency, post-transcriptional suppression of ACC-oxidase gene expression in tomato. *Plant J.* 15, 737-746.

Hirsch, R.E., Lewis, B.D., Spalding, E.P. and Sussman, M.R. 1998. A role for AKT1 potassium channel in plant nutrition. *Science* 280, 918-921.

Hong, S-W. and Vierling, E. 2000. Mutants of *Arabidopsis* thaliana defective in the acquisition of tolerance to high temperature stress. *Proc. Natl. Acad. Sci. USA* 97, 4392-4397.

Hugly, S., McCourt, P., Browse, J., Patterson, G.W. and Somerville, C. 1990. A chilling sensitive mutant of *Arabidopsis* with altered steryl-ester metabolism. *Plant Physiol.* 93, 1053-1062.

Isaksson, J., Mantyla, E., Welin, B., Lindsey, K. and Mandal, A. 1997. Isolation and characterization of promoter trap tagged mutants of *Arabidopsis thaliana* exhibiting an increased sensitivity to freezing stress. *Acta Agron. Hungar.* 45, 405-411.

Ishitani, M., Xiong, L., Stevenson, B., Zhu, J-K. 1997. Genetic analysis of osmotic and cold stress signal transduction in *Arabidopsis*: interactions and convergence of abscisic acid-dependent and abscisic acid-independent pathways. *Plant Cell* 9, 1935-1949.

Ishitani, M., Xiong, L., Lee, H., Stevenson, B. and Zhu, J-K. 1998. *HOS1*, a genetic locus involved in cold-responsive gene expression in *Arabidopsis. Plant Cell* 10, 1151-1161.

Ito, T., Seki, M., Hayashida, N., Shibata, D. and Shinozaki, K. 1999. Regional insertional mutagenesis of genes on *Arabidopsis thaliana* chromosome V using the *Ac/Ds* transposon in combination with a cDNA scanning method. *Plant J.* 17, 433-444.

Jaglo-Ottosen, K.R., Gilmour, S.J., Zarka, D.G., Schabenberger, O. and Thomashow, M.F. 1998. *Arabidopsis CBF1* overexpression induces *cor* genes and enhances freezing tolerance. *Science* 280, 104-106.

Jones, J.D.G., Carland, F.M., Maliga, P. and Dooner, H.K. 1989. Visual detection of transposition of the maize element *Activator* (*Ac*) in tobacco seedlings. *Science* 244, 204-207.

Jones J.D.G., Carland, F.M., Lin, E., Ralston, E. and Dooner, H.K. 1990. Preferential transposition of the maize element *Activator* to linked chromosomal locations in tobacco. *Plant Cell* 2, 701-707

Kaiser, K. and Goodwin, S.F. 1990. Site-selected transposon mutagenesis of *Drosophila. Proc. Natl. Acad. Sci. USA* 87, 1686-1690.

Kakimoto, T. 1996. CKI1, a histidine kinase homolog implicated in cytokinin signal transduction. *Science* 274, 982-985.

Kardailsky, I., Shukla, V.K., Ahn, J.H., Dagenais, N., Christensen, S.K., Nguyen, J.T., Chory, J., Harrison, M.J. and Weigel, D. 1999. Activation tagging of the floral inducer *FT. Science* 286, 1962-1965.

Kasuga, M., Liu, Q., Miura, S., Yamaguchi-Shinozaki, K. and Shinozaki ,K. 1999. Improving plant drought, salt and freezing tolerance by gene transfer of a single stress-inducible transcription factor. *Nat. Biotech.* 17, 287-291.

Kertbundit, S., De Greve, H., De Boeck, F., van Montague, M. and Hernalsteens, J.P. 1991. In vivo random ß-glucuronidase gene fusions in *Arabidopsis thaliana*. *Proc. Natl. Acad. Sci. USA* 88, 5212-5216.

Kirti, P.B., Hadi, S., Kumar, P.A. and Chopra, V.L. 1991. Production of sodium-chloride-tolerant *Brassica juncea* plants by in vitro selection at the somatic embryo level. *Theor. Appl. Genet* 83, 233-237.

Knight, H., Veale, E.L., Warren, G.J. and Knight, M.R. 1999. The *sfr6* mutation in *Arabidopsis* suppresses low-temperature induction of genes dependent on the CRT/DRE sequence motif. *Plant Cell* 21, 875-886.

Kobayashi, Y., Kaya, H., Goto, K., Iwabuchi, M. and Araki, T. 1999. A pair of related genes with antagonsitic roles in mediating flowering signals. *Science* 286, 960-1962.

Koes, R., Souer, E., van Houwelingen, A., Mur, L., Spelt, C., Quattrocchio, F., Wing, J., Oppedijk, B., Ahmed, S., Maes, T., Gerats, T., Hoogeveen, P., Meesters, M., Kloos, D. and Mol, J.N.M. 1995. Targeted gene inactivation in petunia by PCR-based selection of transposon insertion mutants. *Proc. Natl. Acad. Sci. USA* 92, 8149-8153.

Koncz, C., Martini, N., Mayerhofer, R., Koncz-Kalman, Zs., Körber, H., Redei, G.P. and Schell, J. 1989. High-frequency T-DNA-mediated gene tagging in plants. *Proc. Natl. Acad. Sci. USA* 86, 8467-8471.

Kooter, J.M., Matzke, M.A. and Meyer, P. 1999. Listening to the silent genes: transgene silencing, gene regulation and pathogen control. *Trends Plant Sci.* 4, 340-347.

Koornneef, M., Reuling, G. and Karssen, C.M. 1984. The isolation and characterization of abscisic acid-insensitive mutants of *Arabidopsis thaliana. Physiol. Plant.* 61, 377-383.

Krysan, P.J., Young, J.F., Tax, F. and Sussman, M.R. 1996. Identification of transferred DNA insertions within *Arabidopsis* genes involved in signal transduction and ion transport. *Proc. Natl. Acad. Sci. USA* 93, 8145-8150.

Krysan, P.J., Young, J.C. and Sussman, M.R. 1999. T-DNA as an insertional mutagen in *Arabidopsis. Plant Cell* 11, 2283-2290.

Kueh, J.S.H. and Bright, S.W.J. 1982. Biochemical and genetic analysis of three proline accumulating barley mutants. *Plant Sci. Lett.* 27, 233-241.

Landry, L.G., Chapple, C.C. and Last, R.L. 1995. *Arabidopsis* mutants lacking phenolic sunscreens exhibit enhanced ultraviolet-B injury and oxidative damage. *Plant Physiol.* 109, 1159-1166.

Lee, H., Xiong, L., Ishitani, M., Stevenson, B. and Zhu, J-K. 1999. Cold-regulated expression and freezing tolerance in an *Arabidopsis thaliana* mutant. *Plant J.* 17, 301-308.

Leung, J., Merlot, S., Giraudat, J. 1997. The *Arabidopsis* ABSCISIC ACID-INSENSITIVE2 (ABI2) and ABI1 genes encode homologous protein phosphatases 2C involved in abscisic acid signal transduction. *Plant Cell* 9, 759-71.

Li, J., Ou-Lee, T.M., Raba, R., Amundson, R.G. and Last, R.L. 1993. *Arabidopsis* flavonoid mutants are hypersensitive to UV-B irradiation. *Plant Cell* 5, 171-179.

Lin, X., Kaul, S., Rounsley, S., Shea, T.P., Benito, M.I., Town, C.D., Fujii, C.Y., Mason, T., Bowman, C.L., Barnstead, M., Feldblyum, T.V., Buell, C.R., Ketchum, K.A., Lee, J., Ronning, C.M., Koo, H.L., Moffat, K.S., Cronin, L.A., Shen, M., Pai, G., van Aken, S., Umayam, L., Tallon L.J., Gill, J.E., Adams, M.D., Carrera, A.J., Creasy, T.H., Goodman, H.M., Somerville, C.R., Copenhaver, G.P., Preuss, D., Nierman, W.C., White, O., Eisen, J.A., Salzberg, S.L., Fraser, C. and Venter, J.C., et al. 1999. Sequence and analysis of chromosome 2 of the plant *Arabidopsis thaliana*. *Nature* 402, 761-768.

Liu, J. and Zhu, J-K. 1997. An *Arabidopsis* mutant that requires increased calcium for potassium nutrition and salt tolerance. *Proc. Natl. Acad. Sci. USA* 94, 14960-14964.

Liu, J. and Zhu, J-K. 1998. A calcium sensor homolog required for plant salt tolerance. *Science* 280, 1943-1945.

Liu, J., Ishitani, M., Halfter, U., Kim, C-S. and Zhu, J-K. 2000. The *Arabidopsis thaliana SOS2* gene encodes a protein kinase that is required for salt tolerance. *Proc. Natl. Acad. Sci. USA* 97, 3730-3734.

Liu, L.X., Spoerke, J.M., Mulligan, E.L., Chen, J., Reardon, B., Westlund, B., Sun, L., Abel, K., Armstrong, B., Hardiman, G., King, J., McCague, L., Basson, M., Clover, R. and Johnson, C.D. 1999. High-throughput isolation of *Caenorhabditis elegans* deletion mutants. *Genome Res.* 9, 859-867.

Long, D., Martin, M., Sundberg, E., Swinburne, J., Puangsomlee, P. and Coupland, G. 1993. The maize transposable element system *Ac/Ds* as a mutagen in *Arabidopsis*: identification of an *albino* mutation induced by *Ds* insertion. *Proc. Natl. Acad. Sci. USA* 90, 10370-10374.

Maes, T., de Keukeleire, P. and Gerats, T. 1999. Plant tagnology. *Trends Plant Sci* 4, 90-96.

Martienssen, R.A. 1998. Functional genomics: probing plant gene function and expression with transposons. *Proc. Natl. Acad. Sci. USA* 95, 2021-2026.

Mathur, J., Szabados, L., Schaefer, S., Grunenberg, B., Lossow, A., Jonas-Straube, E., Schell, J., Koncz, C. and Koncz-Kálmán, Z. 1998. Gene identification with sequenced T-DNA tags generated by transformation of *Arabidopsis* cell suspension. *Plant J.* 13, 707-716.

Mayer, K. *et al.,* 1999. Sequence and analysis of chromosome 4 of the plant *Arabidopsis thaliana. Nature* 402, 769-777.

McKinney, E.C., Ali, N., Traut, A., Feldmann, K.A., Belostotsky, D.A., McDowell, J.M. and Meagher, R.B. 1995. Sequence-based identification of T-DNA insertion mutations in *Arabidopsis* actin mutants *act2-1* and *act4-1*. *Plant J.* 8, 613-622.

Meissner, R.C. Jin, H., Cominelli, E., Denekamp, M., Fuertes, A., Greco, R., Kranz, H.D., Penfield, S., Petroni, K., Urzainqui, A., Martin, C., Paz-Ares, J., Smeekens, S., Tonelli, C., Weisshaar, B., Baumann, E., Klimyuk, V., Marillonnet, S., Patel, K., Speulman, E.,Tissier, A.F., Bouchez, D., Jones, J.J., Pereira, A., Wisman, E., *et al.,* 1999. Function search in a large transcription factor gene family in *Arabidopsis*: assessing the potential of reverse genetics to identify insertional mutations in R2R3 *MYB* genes. *Plant Cell* 11, 1827-1840.

Meyer, P. and Saedler, H. 1996. Homology dependent gene silencing in plants. *Ann. Rev. Plant Physiol. Plant Mol. Biol.* 47, 23-48.

Miklos, G.L.G. and Rubin, G.M. 1996. The role of the genome project in determining gene function: insights from model organisms. *Cell* 86, 521-529.

Mol, J., Grotewold, E. and Koes, R. 1998. How genes paint flowers and seeds. *Trends. Plant Sci.* 3, 212-217.

Neff, M.M., Nguyen, S.M., Malancharuvil, E.J., Fujioka, S., Noguchi, T., Seto, H., Tsubuki, M., Honda, T., Takatsuto, S., Yoshida, S. and Chory, J. 1999. *BAS1*: A gene regulating brassinosteroid levels and light responsiveness in *Arabidopsis*. *Proc. Natl. Acad. Sci. USA* 96, 15316-15323.

Parinov, S., Sevugan, M., Ye, D., Yang, W-C., Kumaran, M. and Sundaresan, V. 1999. Analysis of flanking sequences from dissociation insertion lines: a database for reverse genetics in *Arabidopsis. Plant Cell* 11, 2263-2270.

Pereira, A. 1998. "Heterologous transposon tagging systems". In *Transgenic Plant Research*. eds. K. Lindsey, and J.L. Harwood, pp. 91-108. Academic Publishers, UK.

Quesada, V., Ponce, M.R. and Micol, J.L. 2000. Genetic analysis of salt-tolerant mutants in *Arabidopsis thaliana. Genetics* 154, 421-436.

Reardon, J.T., Liljestrand-Golden, C.A., Dusenberry, R.L. and Smith, P.D. 1987. Molecular analysis of diepoxybutane-induced mutations at the *rosy* locus of *Drosophila melanogaster. Genetics* 115, 323-331.

Rédei, G.P. and Koncz, C. 1992. "Classical mutagenesis". In *Methods in Arabidopsis Research*, eds. C. Koncz, N-H. Chua and J. Schell, pp.16-82. World Scientific Publishing Co., Singapore.

Rhodes, D., Handa, S. and Bressan, R.A. 1986. Metabolic changes associated with adaptation of plant cells to water stress. *Plant Physiol.* 82, 890-903.

Russel, L.B., Hunsicker, P.R., Cacheiro, N.L.A., Bangham, J.W., Russel, W.L. and Shelby M.D. 1989. Chlorambucil effectively induces deletion mutations in mouse germ cells. *Proc. Natl. Acad. Sci. USA* 86, 3704-3708.

Saleki, R., Young, P.G. and Lefebvre, D.D. 1993. Mutants of *Arabidopsis thaliana* capable of germination under saline conditions. *Plant Physiol.* 101, 839-845.

Schneider, J.C., Hugly, S. and Somerville, C.R. 1995. Chilling sensitive mutants of *Arabidopsis*. *Plant Mol. Biol. Rep.* 13, 11-17.

Schneider, J.C., Nielsen, E. and Somerville, C. 1995 A chilling-sensitive mutant of *Arabidopsis* is deficient in chloroplast protein accumulation at low temperature. *Plant Cell Environ.* 18, 23-31.

Shi, H., Ishitani, M., Kim, C. and Zhu, J-K. 2000. The *Arabidopsis thaliana* salt tolerance gene *SOS1* encodes a putative Na^{+}/H^{+} antiporter. *Proc. Natl. Acad. Sci. USA* 97, 6896-6901.

Shinozaki, K. and Yamaguchi-Shinozaki, K. 2000. Molecular responses to dehydration and low temperature: differences and cross-talk between two stress signaling pathways. *Curr. Opin. Plant Biol.* 3, 217-223.

Skarnes, W.C. 1990. Entrapment vectors: a new tool for mammalian genetics. *Biotechnology* 8, 827-831.

Speulman, E., Metz, P.L.J., van Arkel, G., te Lintel Hekkert, B., Stiekema, W.J. and Pereira, A. 1999. A two-component *enhancer-inhibitor* transposon mutagenesis system for functional analysis of the *Arabidopsis* genome. *Plant Cell* 11, 1853-1866.

Sumaryati, S., Negrutiu, I. and Jacobs, M. 1992. Characterization and regeneration of salt- and water-stress mutants from protoplasts culture of *Nicotiana plumbaginifolia* (Viviani). *Theor. Appl. Genet.* 83, 613-619.

Sundaresan,V., Springer, P., Volpe, T., Haward, S., Jones, J.D.G., Dean, C., Ma, H. and Martienssen, R. 1995. Patterns of gene action in plant development revealed by enhancer trap and gene trap transposable elements. *Genes Dev.* 9, 1797-1810.

Swinburne, J., Balcells, L., Scofield, S.R., Jones, J.D.G. and Coupland, G. 1992. Elevated levels of *Activator* transposase mRNA are associated with high frequencies of *Dissociation* excision in *Arabidopsis*. *Plant Cell* 4, 583-595.

Tanksley, S.D., Ganal, M.W. and Martin, G.B. 1995. Chromosome landing: a paradigm for map-based cloning in plants with large genomes. *Trends Genet.* 11, 63-68.

Thomashow, M.F. (1998) Role of cold-responsive genes in plant freezing tolerance. *Plant Physiol.* 118, 1-7.

Tissier, A.F., Marillonnet, S., Klimyuk, V., Patel, K., Torres, M.A., Murphy, G. and Jones, J.D.G. 1999. Multiple independent defective *suppressor-mutator* transposon insertions in *Arabidopsis*: a tool for functional genomics. *Plant Cell* 11, 1841-1852.

Tokuhisa, J.G., Feldmann, K.A., LaBrie, S.T. and Browse, J. 1997. Mutational analysis of chilling tolerance in plants. *Plant Cell Environ.* 20, 1391-1400.

Topping, J.F. and Lindsey, K. 1995. Insertional mutagenesis and promoter trapping in plants for the isolation of genes and the study of development. *Transgenic Res.* 4, 291-305.

Tsugane, K., Kobayashi, K., Niwa, Y., Ohba, Y., Wada, K. and Kobayashi, H. 1999 A recessive *Arabidopsis* mutant that grows photoautotrophically under salt stress shows enhanced active oxygen detoxification. *Plant Cell* 11, 1195-1206.

Vos, P., Hogers, R., Bleeker, M., Reijans, M., van de Lee, T., Homes, M., Frijters, A., Pot, J., Peleman, J., Kuiper, M. and Zabeau, M. 1995. AFLP: a new technique for DNA fingerprinting. *Nucleic Acids Res.* 23, 4407-4414.

Walbot, V. 1992. Strategies for mutagenesis and gene cloning using transposon tagging and T-DNA insertional mutagenesis. *Annu. Rev. Plant Physiol. Plant Mol. Biol.* 43, 49-82.

Warren, G., McKown, R., Marin, A.L. and Teutonico, R. 1996. Isolation of mutations affecting the development of freezing tolerance in *Arabidopsis thaliana* (L.) Heynh. *Plant Physiol.* 111, 1011-1019.

Waterhouse, P.M., Graham, M.W. and Wang, M-B. 1998. Virus resistance and gene silencing in plants can be induced by simultaneous expression of sense and antisense RNA. *Proc. Natl. Acad. Sci. USA* 95, 13959-13964.

Weigel, D. *et al.*, 2000. Activation tagging in *Arabidopsis*. *Plant Physiol.* 122, 1003-1013.

Werner, J. and Finkelstein, R.R. 1995. *Arabidopsis* mutants with reduced response to NaCl and osmotic stress. *Physiol. Plant.* 93, 659-666.

Wilson, K., Long, D., Swinburne, J. and Coupland, G. 1996. A *dissociation* insertion causes a semidominant mutation that increases expression of *tiny*, an *Arabidopsis* gene related to *APETALA2*. *Plant Cell* 8, 659-671.

Winkler, R.G., Frank, M.R., Galbraith, D.W., Feyereisen, R. and Feldmann, K.A. 1998. Systematic reverse genetics of transfer-DNA-tagged lines of *Arabidopsis*. *Plant Physiol.* 118, 743-750.

Wisman, E., Hartmann, U., Sagasser, M., Baumann, E., Palme, K., Hahlbrock, K., Saedler, H. and Weisshaar, B. 1998. Knock-out mutants from an *En-1* mutagenized *Arabidopsis thaliana* population generate phenylpropanoid biosynthesis phenotypes. *Proc. Natl. Acad. Sci. USA* 95, 12432-12437.

Wu, S-J., Ding, L. and Zhu, J-K. 1996. *SOS1*, a genetic locus essential for salt tolerance and potassium acquisition. *Plant Cell* 8, 617-627.

Xiong, L., Ishitani, M. and Zhu, J-K. 1999a. Interaction of osmotic stress, temperature, and abscisic acid in the regulation of gene expression in *Arabidopsis*. *Plant Physiol.* 119, 205-212.

Xiong, L., Ishitani, M., Lee, H. and Zhu, J-K. 1999b. *HOS5* – a negative regulator of osmotic stress-induced gene expression in *Arabidopsis thaliana*. *Plant J.* 19, 569-578.

Xin, Z. and Browse, J. 1998. *eskimo1* mutants of *Arabidopsis* are constitutively freezing-tolerant. *Proc. Natl. Acad. Sci. USA* 95, 7799-7804.

Yamaguchi-Shinozaki, K. and Shinozaki, K. 1994. A novel cis-acting element in an *Arabidopsis* gene is involved in responsiveness to drought, low-temperature, or high-salt stress. *Plant Cell* 6, 251-264.

Zhu, J-K., Hasegawa, P.M. and Bressan, R.A. 1997. Molecular aspects of osmotic stress in plants. *Crit. Rev. Plant Sci.* 16, 253-277.

Zhu, J-K., Liu, J. and Xiong, L. 1998. Genetic analysis of salt tolerance in *Arabidopsis*: evidence for the critical role of potassium nutrition. *Plant Cell* 10, 1181-1191.

Chapter 3

DIFFERENTIAL CLONING

Peter Buchner

Agriculture and Environment Division, IACR-Rothamsted, Harpenden, Hertfordshire AL5 2JQ, UK.
peter.buchner@bbsrc.ac.uk

INTRODUCTION

Plants are subject to different environmental situations in which the plant is able to react and adapt by numerous physiological and metabolic changes. These changes are mostly accompanied and/or induced by alterations of gene expression. Altered gene expression is likely to be critical for the establishment of the resistant or susceptible state or may reflect early attempts by the plant at defence against changed environmental situations. The identification of differentially expressed genes under different environmental conditions has been used as an experimental approach to understand not only the gene function but also the molecular mechanisms underlying the biological process of plant adaptation to environmental stress. Considerable progress has been made in identifying genes responses to different environmental stresses and their importance for tolerance to abiotic stress (Vierling *ct al.*, 1992; Tabaeizadeh, 1998; Martin; 1999).

Differentially expressed genes are usually identified by comparing steady-state mRNA concentrations. Several techniques have been developed which provide the researcher with the ability to detect variations in gene expression between different RNA populations of plant material, e.g. subjected to environmental stress situations as compared to a control situation. These techniques are all very powerful, allowing the analysis of changes of expression of mRNAs utilising selective detection or enrichment without any prior knowledge of the sequence information of the specific genes in question. The most commonly used methods to clone differentially expressed genes are differential colony hybridisation, subtractive

M.J. Hawkesford and P. Buchner (eds.),
Molecular Analysis of Plant Adaptation to the Environment, 43–60.

hybridisation, representational difference analysis (RDA), mRNA differential display, and AFLP based RNA fingerprinting. Each of these will be discussed in turn.

DIFFERENTIAL SCREENING OF cDNA LIBRARIES

The technique of differential screening of cDNA libraries is widely used to identify differential gene expression between a particular tissue in its control state and the same tissue in an altered state. The simplest technique, differential colony/plaque hybridisation is based on the generation of cDNA libraries containing the gene sequences of interest. Colonies are transferred to bacterial plates, in an orderly array, and replica plated onto duplicate membrane filters (Cochran et al. 1987). In the case of phages duplicate plaque lifts were performed according to standard procedures (Sambrook *et al.,* 1989). Each membrane filter is then hybridised to two different ^{32}P-labelled cDNA probes made from poly(A)$^+$mRNA of the control tissue and from tissue during or after the change being studied. The major disadvantages of these regular screening procedures are that they are very time-consuming. Often thousands of cDNAs in the form of colonies or plaques have to be screened. An additional limiting factor is often the low quantities of mRNA isolated from the target tissue resulting in inadequate amounts of poly(A)$^+$mRNA for the cDNA library synthesis and/or probe labelling. The development of cDNA libraries from small numbers of cells or single cells using PCR-techniques have made the production of cDNA libraries feasible from defined tissue regions (Belyavski *et al.,* 1989; Domec *et al.,* 1990; Jepson *et al.,* 1991; Jena *et al.,* 1996; Brandt *et al.,* 1999). The problem of surrounding noise signals masking the signals of differentially expressed genes can be avoided by comparative hybridisation. This method is based on competition between unlabelled cDNA from the mRNA of one sample and labelled cDNA from another sample. By manipulating the amount of competing unlabeled cDNA, background signal from non-regulated genes can be increased or reduced. This enables a contrasted signal from differentially regulated genes to be identified in a quantitative manner (Zhong *et al.,* 2000).

More recently the application of PCR to differential screening of cDNA libraries has improved the efficiency of this technique. The PCR-assisted differential screening method (Figure 1) was successful used to identify differentially expressed mRNA in association to symbiosis related fungal transcripts in the developing *Eucalyptus globulus-Pisolithus tinctorius* mycorrhizas (Tagu *et al.,* 1993) and to identify genes that are differentially expressed in cotton plant after *Verticillium dahliae* infection (Hill *et al.,* 1999).

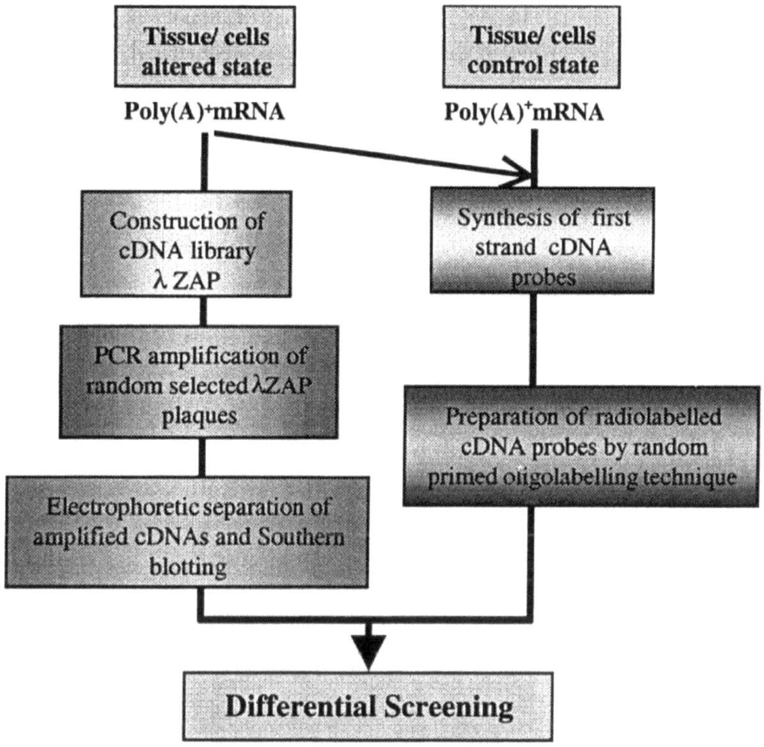

Figure 1. Flow diagram for the isolation of up- and down-regulated genes by PCR-assisted differential screening.

SUBTRACTIVE HYBRIDISATION

The subtractive hybridisation technique leads to an enrichment of cDNAs corresponding to differentially expressed mRNAs. The original subtractive cloning method (Figure 2) is based on the hybridisation of first strand cDNAs from target cells/tissues as so called "tracer" to a vast molar excess of "driver" mRNA (control cell/tissue). The single stranded cDNAs that do not form hybrids during two rounds of hybridisation are then separated from the double-stranded nucleic acid hybrids. The resulting subtracted cDNA is then used to construct a cDNA library or as a labelled probe to screen libraries.

The efficiency of subtraction, integrity of the residual single-stranded cDNA and the efficiency of recovery of the subtracted cDNA are the three important factors affecting quality of subtractive hybridisation. The chromatographic purification of the subtracted cDNA on hydroxyapatite used in the early protocols (Sambrook *et al.*, 1989) has considerable disadvantages. The cDNA recovered after two rounds of hybridisation and separation is limited in quantity and concentration. Moreover the necessary use of a buffer with high phosphate ion concentration and high temperature during chromatography increases the probability of degradation.

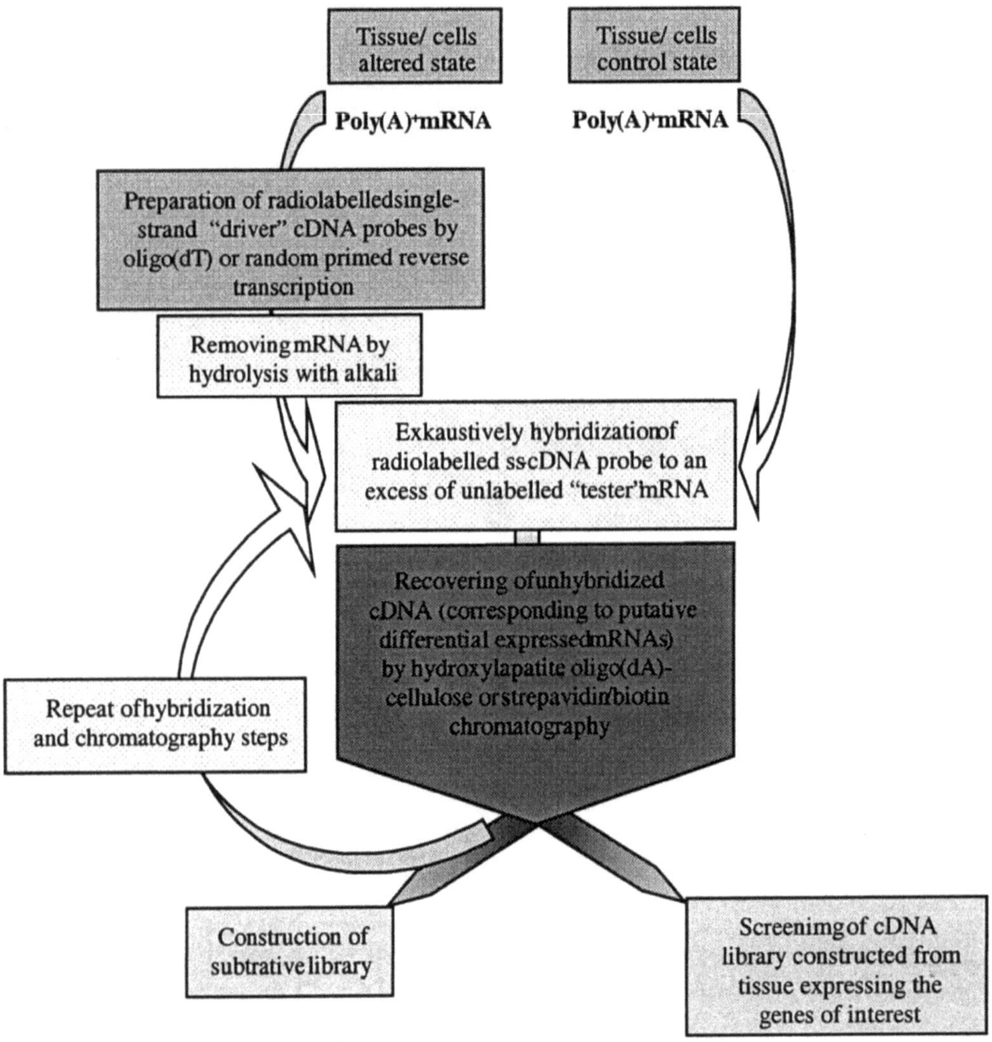

Figure 2. Flow diagram for the isolation of up-regulated genes by subtractive hybridisation.

Owing to such disadvantages, new strategies were developed to optimise the efficiency of subtraction. Synthesising cDNA on a solid support such oligo(dT)30 primers covalently linked to Latex particles (Hara *et al.*, 1991) or paramagnetic oligo(dT) beads (Heinrich *et al.*, 1997) before hybridisation or the elimination of cDNA/RNA hybrids after hybridisation (Meszaros and Morton, 1996) improved the efficiency of subtraction and the enrichment of cDNA clones corresponding to low abundance mRNAs. An improved cDNA enrichment was also obtained by the use of biotinylated driver mRNA and the simple removing of the cDNA/mRNA hybrids with several rounds of streptavidin/phenol subtraction (Sive and St. John, 1988; Gastel and Sutter,

1996) or by binding to streptavidin beads (Duguid *et al.*, 1988; Lopez-Fernandez and del Mazo, 1993). Methods using a physical separation of single-stranded or double stranded cDNA are usually inefficient for obtaining low abundance transcripts and require large amounts of poly(A)$^+$mRNA and multiple labour intensive subtraction steps.

A profound improvement of subtractive hybridisation was obtained by coupling it to PCR-amplification to increase the amount of starting cDNA to be subtracted (Wang and Brown, 1991) and/or to enhance the amount of the subtracted cDNA product (Timblin *et al.*, 1990; Meszaros and Morton, 1996; Coche, 1997). The adapter ligated PCR amplification of cDNA can be used for generating a representative cDNA library from prohibitively low amounts of mRNA template (Akowitz and Manuelidis, 1988). A similar approach has been applied to amplify subtracted cDNA (Timblin *et al.*, 1990). After separation, the RNA:cDNA duplex was converted into double-stranded cDNA. Following generation of phosphorylated blunt end termini, a double stranded "amplification linker" oligonucleotide, with a 5'-phosphate at the blunt end and a non-phosphorylated 5'overhang, is attached to both ends of the cDNA. After purification the adapter ligated cDNAs are subjected to PCR-amplification with adapter specific oligonucleotide primers.

Subtractions can be performed effectively with double-stranded driver cDNA produced by exponential PCR (Wang and Brown, 1991). This PCR procedure to amplify the starting material in a subtraction experiment is based on the preparation of tester and driver double stranded cDNA fragments suitable for PCR. Aliquots of both cDNAs are separately digested completely with four-base blunt end cutting restriction enzymes Alu I and AluI/Rsa I, and then ligated with the same or two different double-stranded phosphorylated oligonucledotide linkers (adapters) with one blunt end and one 4-base 3'protruding end. The linker has an EcoRI site near the blunt end. After separation of the unligated linker, the cDNA-fragments are amplified by PCR. The amplified – and + cDNA fragments are the starting material for subtractive hybridisation. Including the photobiotinylation and separation of the driver DNAs magnetic beads this method allows several repeated rounds of enrichment and amplification of the remaining tracer DNA (Wang and Brown, 1991, see also Fig. 3).

A normalisation procedure to equalise the abundance of all cDNAs in a population can increase the efficiency of subtractive hybridisation procedures. Highly redundant sequences can interfere with subtractive hybridisation and lead to an unacceptably high level of false positives. Normalisation not only increases the discrimination power of differential cloning strategies, but also provides access to functionally important classes of lowly expressed genes (Bento Soare *et al.*, 1994; Coche and Dewez, 1994). cDNA libraries are generally normalised by submitting thermally denatured cDNA to a self-reassociation reaction and physically separating the abundant, re-annealed sequences from the rare single stranded species by hydroxyapatite chromatography. The selection of rare single stranded cDNA's by Oligo-d$(T)_{25}$ coupled magnetic beads overcomes of single and double stranded DNA separation (Coche and Dewez, 1994).

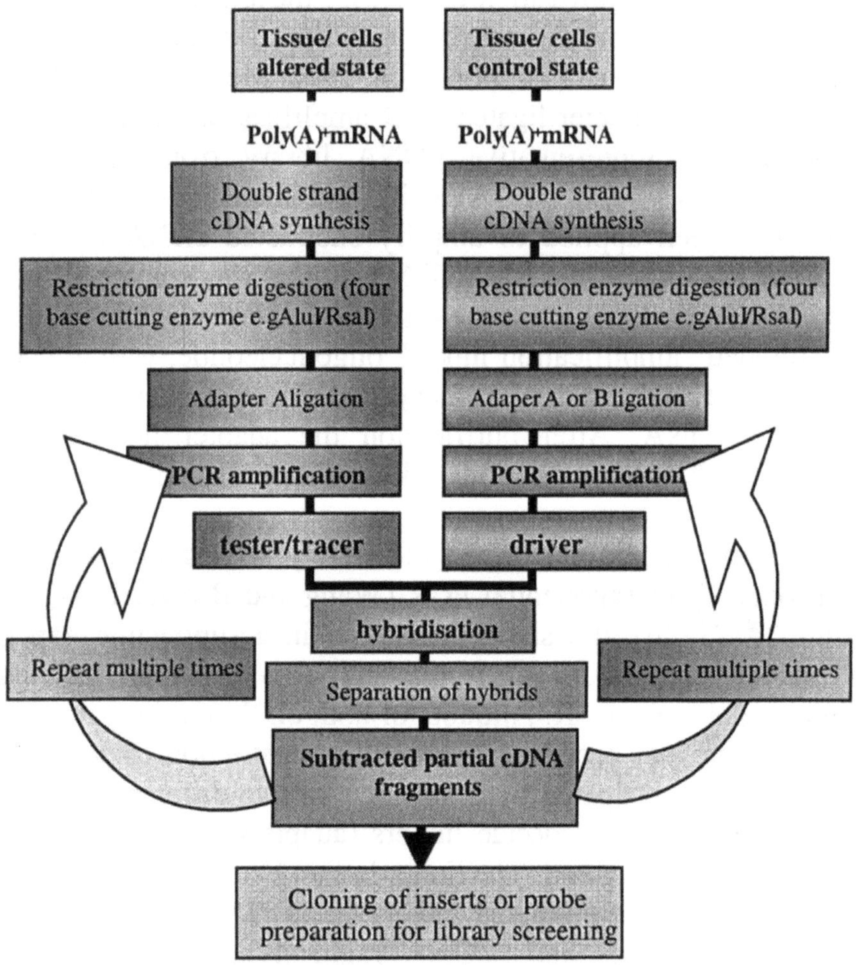

Figure 3. Flow diagram of the isolation of up-regulated genes by PCR based subtractive hybridisation.

The PCR-based cDNA subtraction method, termed suppression subtractive hybridisation (the principle is illustrated in Fig. 4), overcomes the problem of differences in mRNA abundance by incorporation of a hybridisation step that normalises sequence abundance during the course of subtraction by standard hybridisation kinetics.

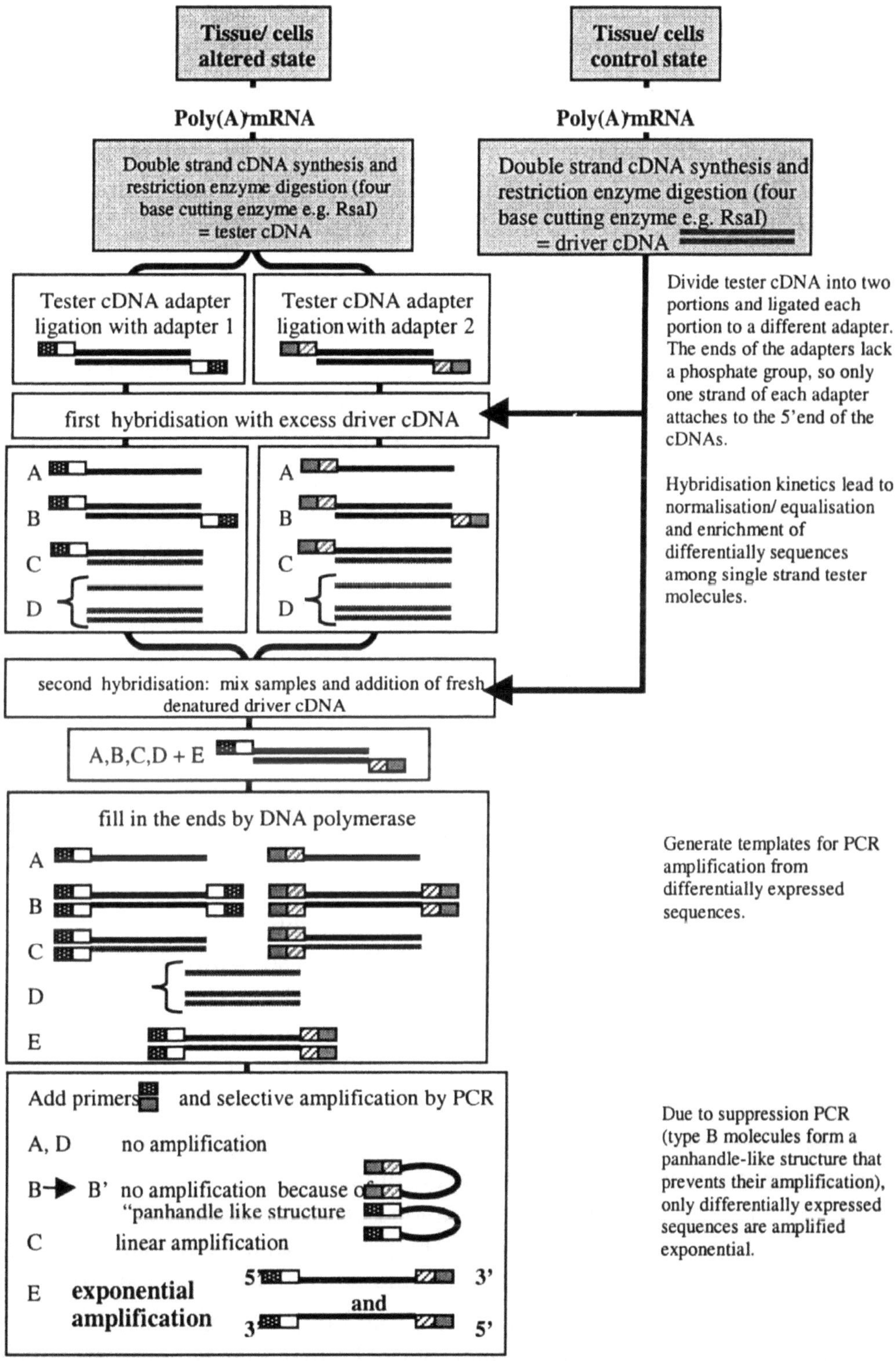

Figure 4. Flow diagram of the suppression subtractive hybridisation (SSH) method. The SSH technology is part of Clontech laboratories (Palo Alto, CA, USA) intelectual property.

This procedure eliminates any intermediate step for the physical separation of single and double stranded cDNAs, requires only one subtractive round, and can achieve a very high enrichment for differentially expressed cDNAs.

The suppression subtractive hybridisation is used to selectively amplify target cDNA fragments (differentially expressed) and simultaneously suppress nontarget DNA amplification (Diatchenko *et al.,* 1996). The method is based on a suppression effect in which long terminal repeats when attached to DNA fragments, can selectively suppress amplification of undesirable sequences in PCR reactions (Siebert *et al.,* 1995). An adapter primer is used which is shorter in length than the adapter sequences and is capable of hybridising to the outer primer-binding site. If any PCR products are generated containing the double stranded adapter sequences on both ends, because of the long inverted repeats on the ends, the individual DNA strands will form panhandle-like structures which cannot serve as template for exponential PCR. This is the suppression effect (Siebert *et al.,* 1995). Only DNA strands with different adapter sites at their ends are exponentially amplified using PCR (Figure 4). The use of this technique has allowed the identification of clones associated with Cys nematode resistance in soybean (Tomkins *et al.,* 1999).

cDNA-REPRESENTATIONAL DIFFERENCE ANALYSIS (RDA)

The cDNA-RDA is also a process of subtraction coupled to PCR amplification. It was originally developed for use with genomic DNA to isolate the differences between two complex genomes (Lisitsyn *et al.,* 1993) and was adapted subsequently for the analysis of differences in expressed mRNA populations (Hubank and Schatz, 1994). Based on successive rounds of subtractive hybridisation followed by PCR, RDA enriches for, and isolates, differentially expressed mRNAs while simultaneously removing all non-differentially expressed transcripts (Figure 5).

In the RDA procedure, double-stranded cDNA is created from two cell or tissue populations of interest. The cDNA is digested by a four base cutting restriction enzyme and linkers are ligated to the ends of the cDNA fragments. A linker combination is used and consists of a desired PCR 24-mer primer along with an oligonucleotide 12-mer that is complementary to both the 24-mer and the restriction site used based on the primer bridging method (Jones and Winistorfer, 1993; Pastorian *et al.,* 2000). Slow annealing creates a temporary bridge formed by the 12-mer that positions the 24-mer adjacent to the digested cDNA fragment, allowing for the direct ligation of the 24-mer PCR primer to the 5' ends of the cDNA fragment. The cDNA pools are amplified by PCR and the cDNA pool from which unique clones are desired is designated "tester", and the cDNA pool that is used to subtract away shared sequences is designated as the "driver". Following initial PCR, the linkers are removed from both cDNA pools by restriction enzyme digestion, and unique linkers are ligated to the tester cDNA. The tester cDNA is then hybridised to an excess of the driver cDNA. cDNAs common to both populations are removed as driver cDNAs reanneal to form tester/driver cDNA hybrids that are not exponentially amplified in the subsequent rounds of PCR. In every round of subtractive hybridisation, the amount of driver cDNA remains constant.

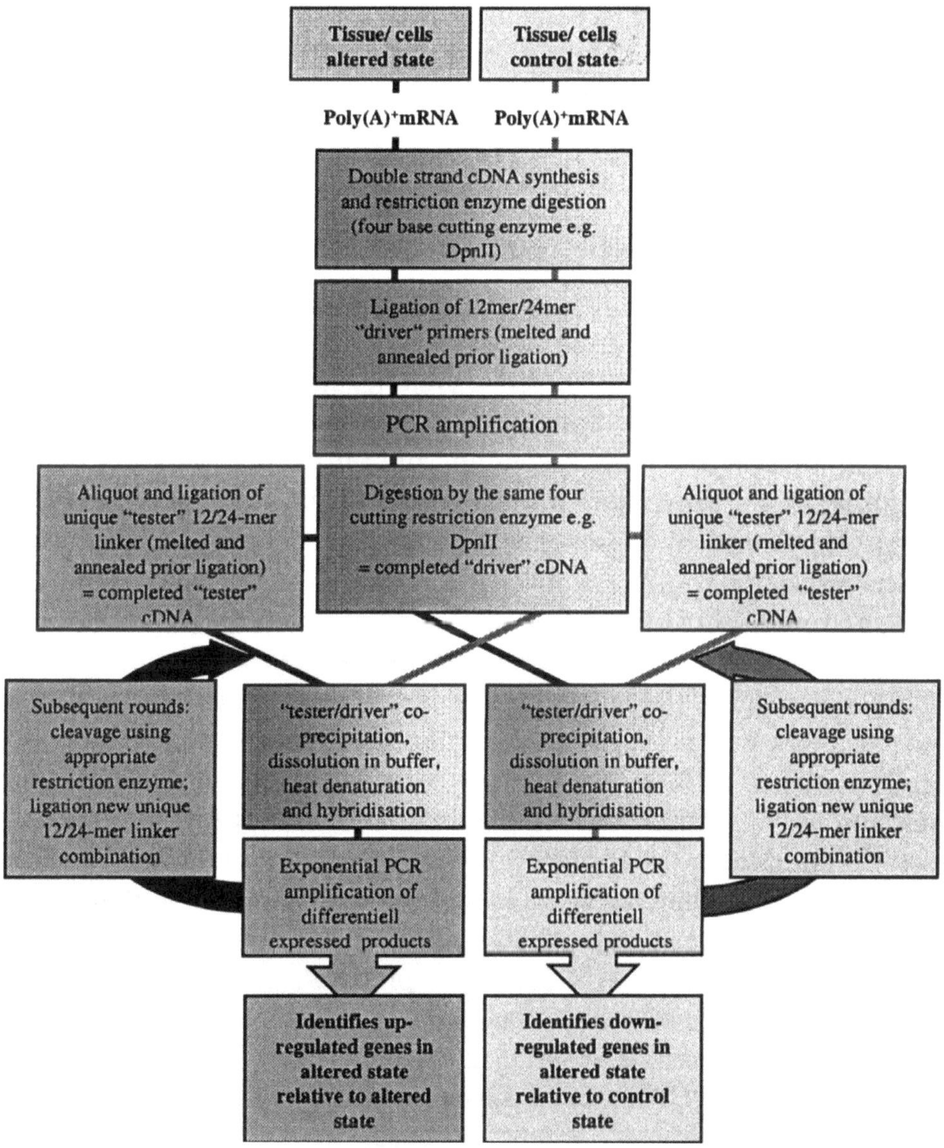

Figure 5. Flow diagram of the isolation of up- and down regulated genes by the representational difference analysis (RDA) method.

The amount of non-differentially expressed tester (the product of the previous hybridisation/ amplification step) will diminish round by round to yield only the differentially expressed sequences (Pastorian *et al.*, 2000). The flexibility of RDA is based on the possibility that the cDNA population can be fractionated by a number of restriction enzymes with short recognition sequences (i.e. four base cutters) to produce unique subsets of cDNAs. An additional RDA in the procedure, using a different four-cutting restriction enzyme, increases the number of usable cDNA fragments, which improves the chance of successfully cloning differentially expressed transcripts. Like any technique, the RDA is not always comprehensive. The cDNAs identified are typically those, that differ significantly in expression level between the two cDNA populations while subtle quantitative differences are often missed. Nevertheless the relative abundance of false positives is quite low and the cDNA-RDA is effective, sensitive and rapid.

DIFFERENTIAL DISPLAY

The most commonly used method to date to identify differentially expressed genes in plants under different environmental conditions is differential display. This approach combines both the power of PCR amplification and the high resolution of denaturing polyacrylamide gel electrophoresis for separation of the amplified cDNA products. The classical principle, developed by Liang and Pardee, 1992, is the reverse transcription and systematically amplification of the 3'termini of mRNAs using a set of anchored primers consisting of oligo-dT and an arbitrary primer (10 bases is a common length). The PCR products that represent a subpopulation of mRNAs defined by the given primer set are separated on a denaturing polyacrylamide gel. By changing primer combinations, most of the RNA species in target cells or tissue may be represented. Differences in gene expression are visualised by comparison of the banding pattern across gel lanes; the presence or absence of bands on the gel, and quantitative differences in gene expression are identified by differences in the intensity of bands. These bands can be excised and the cDNA fragment extracted. The fragment can be used directly for probe preparation to confirm differential expression by Northern blot analysis or to screen a library. Direct sequencing is also possible, or after reamplification using the original primer combination, the cDNA can be subcloned for further uses (Figure 6).

Differential display is popular because it allows the simultaneous comparison of up- and down regulated genes with small amounts of starting material. Another advantage is the sensitivity for detecting low copy number transcripts. However the major criticism has been the high number of false positives reported in some studies (Liang *et al.*, 1993; Li *et al.*, 1994). In addition, the amplified cDNA fragment corresponds to the 3'noncoding region of the mRNAs and this renders the identification of the genes difficult by database comparison. Isolation of the full-length cDNA is therefore necessary for this purpose. Many approaches have attempted to improve the method. The modification of downstream primers, such as the use of single-base anchored primers (i.e. $T_{11}V$) (Liang *et al.*, 1994), and the use of longer

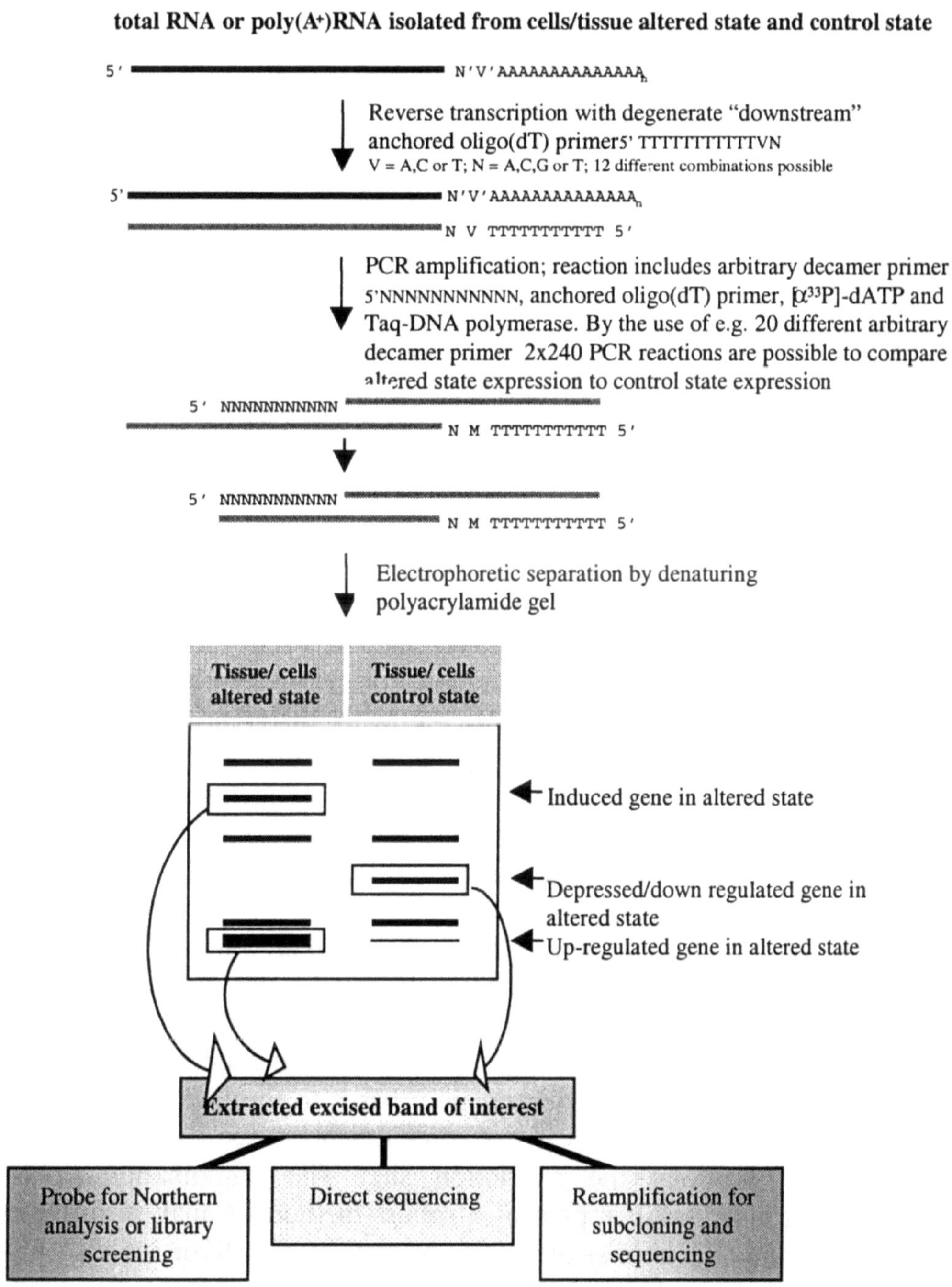

Figure 6. Schematic diagram of the differential display

primers for both up- and down-stream primers (Zhao *et al.*, 1995; Diatchenko *et al.*, 1996) reduced the incidence of false positives and enabled direct sequencing of the PCR products. Another reason for false positives is the contamination of the excised target cDNA with non-differentially expressed cDNA sequences. This can be overcome by using a single-stranded conformational polymorphism gel or restriction analysis (Mathieu-Daude *et al.*, 1996; Zhao *et al.*, 1996). If care is be taken in several steps of the differential display procedure to avoid false positives, the possibility to false positive is reduced to approximately 5 % (Martin *et al.*, 1998). Further strategies for differential display have been developed. The amplification of adapter ligated cDNA after specific restriction using several combinations of adapter specific primer and FITC-labelled oligodT primers. The selected cDNA fragments are displayed on a polyacrylamid gel (Kohroki *et al.*, 1999). By combining of the differential display method with long-distance PCR cDNAs from 150 bp to 2 kb can be successfully amplified and comparatively displayed (Jurecic *et al.*, 1998). The regular occurrence of published differentially expressed genes from plants in relation to different environmental conditions indicates, that the differential display represents an effective method for identifying differentially expressed genes.

AFLP BASED RNA FINGERPRINTING (cDNA-AFLP)

RNA fingerprinting, an alternative method to differential display, is based on the use of arbitrary primers both for cDNA synthesis and for PCR amplification (Welsh *et al.*, 1992). As with differential display, RNA fingerprinting requires low temperature annealing during PCR amplification in order to achieve visible products. AFLP is a rapid technique for displaying large numbers of DNA polymorphisms. It is used extensively for genetic mapping and fingerprinting in plants. The use of highly stringent PCR conditions, facilitated by the ligation of double-stranded adapters to the ends of restriction fragments which serve as primer binding sites during amplification, avoids problems which may encountered with the reproducibility and optimisation of reaction conditions when using arbitrarily primed PCR (Vos *et al.*, 1995). Selective fragment amplification can be performed by adding one or more bases to the PCR primers, which allows successfully extension if the complementary sequence is present in the fragment flanking the restriction site. The combination of RNA fingerprinting and AFLP represent a robust method to analyse differential gene expression (Bachem *et al.*, 1996; Money *et al.*, 1996; see also Figure 7). As with genomic AFLP, two restriction enzymes are generally used for the preparation of the cDNA based PCR template. The choice of the restriction enzyme pair depends on the frequency of cutting.

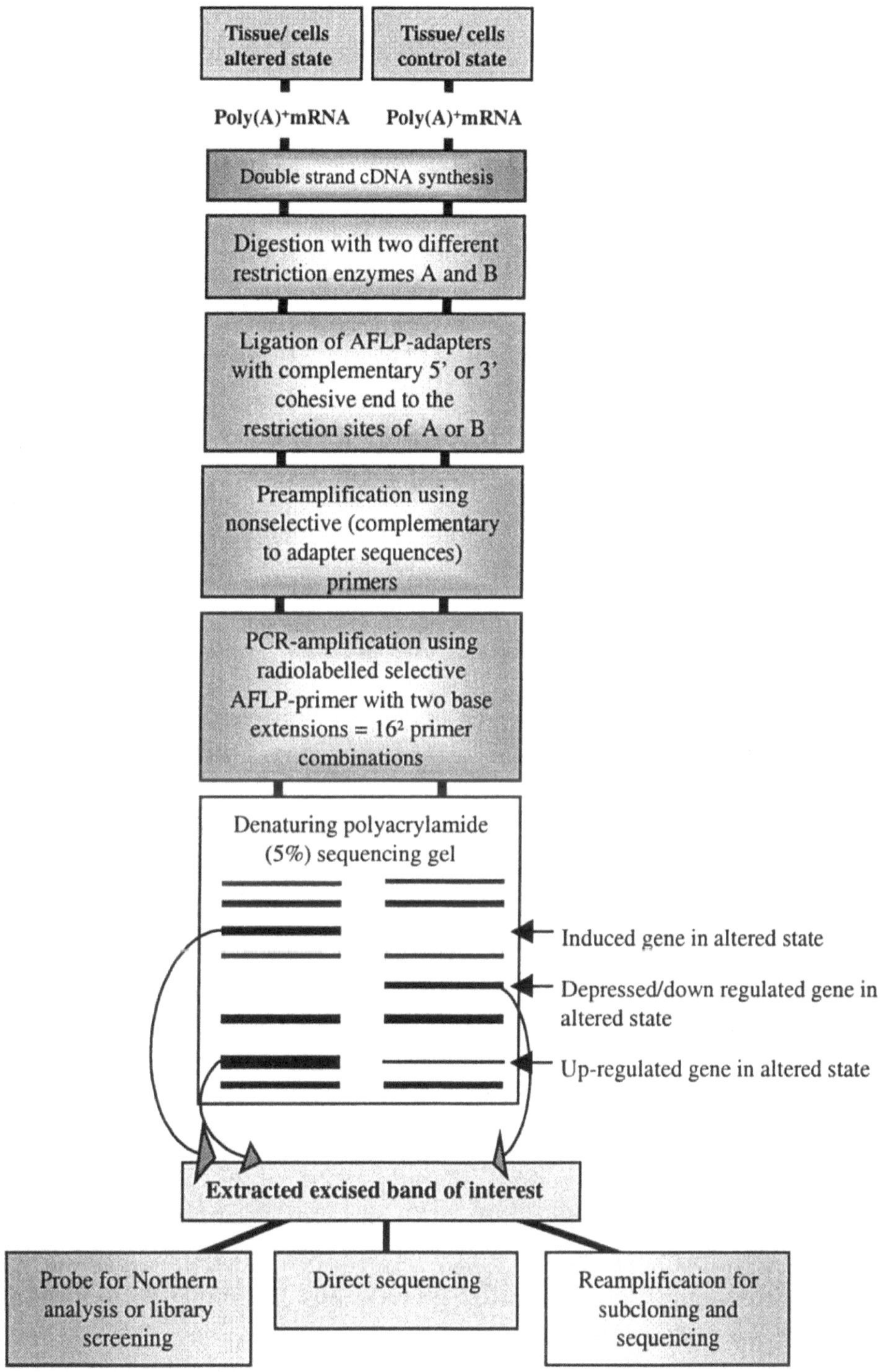

Figure 7. Flow diagram of the cDNA-AFLP

The ideal choice of the restriction enzyme pair are a rare cutter which should cut every cDNA species and a frequent cutter which subsequently cuts the cDNA on one or both sides finally yielding fragment sizes between 100 and 1000 bp (Bachem *et al.*, 1996). For cDNA AFLP the lower complexity allows the use of two selectable bases for each primer.

The total combination of 256 possible primers may enable the detection of all mRNA species in target cells or tissue. After denaturing polyacrylamide gel electrophoresis, target bands can be excised and the cDNA fragment extracted for further amplification and cloning. As with differential display, the verification of the excised target band by contamination of non-target band in the post-amplication procedure may be a problem. In cDNA-AFLP it is possible to almost eliminate all non-target bands from PCR by increasing the length of the selective extensions (Bachem *et al.*, 1996). The cDNA-ALFP technique has been successfully used to identify ACRE genes involved in the stress response caused by fungal infection and also other stresses such as wounding (Durrant *et al.*, 2000).

CONCLUSION

A number of techniques have been developed which provide the researcher with the ability to study differential gene expression in any tissue or cell to identify the gene(s) involved in a given process. The continuous improvement and optimization of the techniques, and also development of new approaches, enables the researcher to determine which is the most suitable for any given experiment. Several factors influence the choice of a specific technique. The amount of starting material available; ability to identify the entire spectrum of abundant and rare mRNAs; the level or degree of differential expression detectable; the ability to compare multiple samples simultaneously; and the amount of false positives compared to the amount of differentially expressed genes isolated. Differential expression analysis can be labour- and time-consuming, requiring considerable effort and expense. The handling and overall speed of a technique is also an important consideration. The majority of the work remains after the cloning process in the form of sequencing, expression analysis, subcloning, the isolation of full-length clones of the gene(s) and the analysis of the function of the unknown gene products.

The traditional approaches of differential screening and subtractive hybridisation require large amounts of RNA making them unsuitable if the starting material is limited. The use of PCR techniques for cDNA library construction and amplification make these methods accessible for the analysis of rare material. The PCR based differential screening methods have drastically reduced the need for large amounts of starting material and differential display can be performed with sub-microgram amounts of total

RNA (Wan *et al.,* 1996). Although full-length library based subtraction can yield full-length clones, it is difficult to perform enough rounds of subtraction to isolate differences from complex tissues. The PCR-based techniques allow multiple rounds of subtractions and isolation of differences from complex tissues, but yield only small cDNA fragments. It is time-consuming to determine how many different clones are represented by the fragments derived from a PCR-based subtraction. Differential display or cDNA-ALFP allow very fast simultaneous comparisons of multiple samples. But like the PCR based subtraction, verification and subsequent isolation of the corresponding full-length clones is necessary. In summary a combination of methods is often useful: differential screening is often performed in combination with subtractive hybridization (Hortensteiner *et al.,* 2000; Matsuyama *et al.,* 1999; Jin *et al.,* 1997). This has advantages in terms of reproducibility and enhanced sensitivity (Pardinas *et al.,* 1998). New technical developments in the analysis of differential gene expression are likely to be automated techniques. The TOGA (total gene expression analysis) technology is conceptually related to differential display and RNA arbitrarily primed (RAP) methods. The TOGA method utilizes exact primer matching and, therefore, systematically attributes each cDNA with a single, defined feature based on the combination of nucleotide sequence and nucleotide length, two parameters derived from the primary sequence of the mRNA. Four bases parsing sequences of the primer are used as a part of the primer binding sites in 256 PCR based assays, performed robotically on tissue extracts to determine simultaneously the presence and relative concentration of nearly every mRNA in the extracts. Visualization of the electrophoretically separated fluorescent assay products from different extracts via a Netscape-browser-based graphical user interface allows the status of each mRNA to be compared among samples and its identity to be matched with sequences of known mRNAs compiled in databases (Sutcliffe *et al.,* 1999). DNA chips constitute the most promising and revolutionary technique to study differential gene expression (see chapter 4). The basic idea is simple, consisting of the arrangement of known large sets of different sequences (ESTs or oligonucleotides) in an ordered array on a membrane or small glass surface. Samples of interest e.g. two mRNA populations are labeled and allowed to hybridize to the array. An image of the array is acquired and the representation of individual RNA species in the sample is reflected by the amount of hybridization to complementary DNAs immobilized in known positions on the array (Eisen and Brown, 1999). The robot-based production of micro-arrays and the image of the hybridized arrays by laser confocal scanning analysis facilitates the survey of the differential expression of thousands of genes under many conditions.

Acknowledgements

I would like to thank Dr. Malcolm Hawkesford and Dr. Jonathan Howarth for critical reading and discussion.

REFERENCES

Akowitz, A. and Manuelidis, L. 1989. A novel cDNA/PCR strategy for efficient cloning of small amounts of undefined RNA. *Gene* 81, 295-306.

Bachem, C.W.B., van der Hoeven, R.S., de Buijn, S.M., Vreugdenhil, D., Zabeau, M and Vidder, R.G.F 1996. Visualization of differential gene expression using a novel method of RNA fingerprinting based on AFLP: analysis of gene expression during potato tuber development. *Plant J.* 9, 745-753.

Belyavsky, A., Vinogradova, T. and Rajewski, K. 1989. PCR-based cDNA library construction: general cDNA libraries at the level of few cells. *Nucleic Acid Res.* 17, 2919-2932.

Bento Soares, M., de Fatima Bonaldo, M., Jelen,P., Su, L., Lawton, L. and Eftratiadis, A. 1994. Construction and characterization of normalized cDNA library. *Proc. Natl. Acad. Sci. USA* 91, 9228-9232.

Brandt S., Kehr J., Walz C., Imlau A., Willmitzer L. and Fisahn J. 1999. Technical Advance: a rapid method for detection of plant gene transcripts from single epidermal, mesophyll and companion cells of intact leaves. *Plant J.* 20, 245-250

Coche, T. and Dewez, M. 1994. Reducing bias in cDNA sequences representation by molecular selection. *Nucleic Acids Res.* 22, 4545-4546.

Coche, T.G. 1997. Subtractive cDNA cloning using magnetic beads and PCR. *Methods Mol. Biol.* 67, 371-387.

Conchran, B.H., Zumstein, P., Zullo, J., Rollins, B., Mercola, M. and Stiles, C.D. 1987. "Differential colony hybridization: molecular cloning from a zero base". In: *Methods in Enzymology* Vol. 147, eds. D. Barnes and D.A. Sirbasku, pp. 64-85. Academic Press, London, San Diego.

Diatchenko, L., Lau Y.F, Campell, A.P., Chenchik, A., Moqadam, F., Huang, B., Lukyanov, S., Lukyanov K., Gurskaya, N., Sverdlov, E.D. and Siebert, P.D. 1996. Suppression subtractive hybridization: A method for generating differentially regulated or tissue-specific probes and libraries. *Proc. Natl. Acad. Sci. USA* 93, 6025-6030.

Diatchenko, L.B., Ledesma, J., Chenchik, A.A and Siebert, P.D. 1996. Combining the technique of RNA fingerprinting and differential display to obtain differentially expressed mRNA. *Biochem. Biophys. Res. Commun.* 219, 824-828.

Duguid, J.R., Rohwer, R.G. and Seed, B. 1988. Isolation of cDNAs of scrapie-modulated RNAs by subtractive hybridization of a cDNA library. *Proc. Natl. Acad. Sci. USA* 85, 5738-42.

Durrant, W.E., Rowland, O., Piedras, P., Hammond-Kosack, K.E. and Jones, J.D.G. 2000. CDNA-AFLP reveals a striking overlap in race-specific resistance and wound response gene expression profiles. *Plant Cell* 12, 963-977.

Eisen M.B. and Brown, P.O. 1999. "DNA arrays for analysis of gene expression". In: *Methods in Enzymology* Vol. 303, ed. S.M. Weissman, pp. 179-205. Academic Press, London, San Diego.

Gastel., J.A. and Sutter, T.R. 1996. A control system for cDNA enrichment reactions. *Biotechniques* 20, 870-875.

Hara, E., Kato, T., Sekiya, S. and Oda, K. 1991. Subtractive cDNA cloning using oligo(dT)30-latex and PCR: isolation of cDNA clones specific to undifferentiated human embryonal carcinoma cells. *Nucleic Acids Res.* 19, 7097-7104.

Heinrich, T., Washer, S., Marshall, J., Jones, M.G. and Potter, R.H. 1997. Subtractive hybridization of cDNA from small amounts of plant tissue. *Mol. Biotech.* 8, 7-12.

Hill, M.K., Lyon, K.J. and Lyon, B.R. 1999. Identification of disease response genes expressed in *Gossypium hirsutum* upon infection with the wilt pathogen *Verticillium dahlia. Plant Mol. Biol.* 40, 289-296.

Hortensteiner, S., Chinner, J., Matile, P., Thomas, H. and Donnison, I.S. 2000. Chlorophyll breakdown in *Chlorella protothecoides*: characterization of degreening and cloning of degreening-related genes. *Plant Mol. Biol.* 42, 439-50.

Hubank, M and Schatz, D.G. 1994. Identifying differences in mRNA expression by representational difference analysis of cDNA. *Nucleic Acids Res.* 22, 5640-5648.

Jena, P.K., Liu A.H., Smith D.S. and Wysocki L.J. 1996. Amplification of genes, single transcripts and cDNA libraries from one cell and direct sequence analysis of amplified products derived from one molecule. *J. Immunol. Methods.* 190, 199-213.

Jin, H., Cheng, X., Diatchenko, L., Siebert, P.D. and Huang, C.C. 1997. Differential screening of a subtracted cDNA library: a method to search for genes preferentially expressed in multiple tissues. *Biotechniques* 23, 1084-1086.

Jones, D.H. and Winistorfer, S.C. 1993. A method for the amplification of unknown flanking DNA: targeted inverted repeat amplification. *Biotechniques* 12, 894-904.

Jurecic, R., Nachtman. R.G., Colicos, S.M and Belmont, J.W. 1998. Identification and cloning of differentially expressed genes by long-distance differential display. *Anal. Biochem.* 259, 235-244.

Kohroki, J., Tsuchiya, M., Fujita, S., Nakanishi, T., Itoh, N., Tanaka, K. 1999. A novel strategy for identifying differential gene expression. An improved method of differential display analysis. *Biochem. Biophys. Res. Comm.* 262, 365-367.

Li., F., Barnathan, E.S. and Kariko, K. 1994. Rapid method for screening and cloning cDNAs generated in differential messenger-RNA display and application of northern blot for affinity capture of cDNAs. *Nucleic Acids Res.* 22, 1764-1765.

Liang, P. and Pardee, A.B. 1992. Differential display of eukaryotic messenger RNA by means of the polymerase chain reaction. *Science* 257, 967-971.

Liang, P., Averboukh, L. and Pardee, A.B. 1993. Distribution and cloning of eukaryotic mRNAs by means of differential display: refinements and optimization. *Nucleic Acids Res.* 21, 3269-3275.

Liang, P., Zhu, W., Zhang, X., Guo, Z., O'Connell, R.P., Averboukh, L., Wang, F. and Pardee, A.B. 1994. Differential display using one-base anchored oligo-dT primers. *Nucleic Acids Res.* 22, 5763-5764.

Lisitsyn, N., Lisitsyn, N. and Wigler, M. 1993. Cloning the differences between two complex genomes. *Science* 259, 946-951.

Lopez-Fernadez, L.A. and del Mazo, J. 1993. Construction of subtractive cDNA libraries from limited amounts of mRNA and multiple cycles of subtraction. *Biotechniques* 15, 654-656, 658-659.

Martin, K.J., Kwam, C.-P., O'Hare, M.J., Pardee, A.B. and Sager, R. 1998. Identification and verification of differential display cDNAs using gene-specific primers and hybridization arrays. *Biotechniques* 24, 1018-1026.

Mathieu-Daude, F., Cheng, R., Welsh, J. and McClelland, M. 1996. Screening of differentially amplified cDNA products from RNA arbitrarily primed PCR fingerprints using single strand conformation polymorphism (SSCP) gels. *Nucleic Acids Res.* 24, 1504-1507.

Matsuyama, T., Yasumura, N., Funakoshi, M., Yamada, Y. and Hashimoto, T. 1999. Maize genes specifically expressed in the outermost cells of root cap. *Plant Cell Physiol.* 40, 469-76

Mertin, G.B. 1992. Functional analysis of plant disease resistance genes and their downstream effectors. *Curr. Opin. Plant Biol.* 2, 272-279.

Meszaros, M. and Morton, D.B. 1996. Subtractive hybridization strategy using parametic oligo(dT) beads and PCR. *Biotechniques* 20, 413-419.

Money, T., Reader, S., Qu, L.J., Dunford, R.P. and Moore, G. 1996. AFLP-based mRNA fingerprinting. *Nucleic Acids Res.* 24, 2616-2617.

Pardinas, J.R., Combates, N.J., Prouty, S.M., Stenn, K.S and Parimoo, S. 1998. Differential subtraction display: an unified approach for the isolation of cDNAs from differentially expressed genes. *Anal. Biochem.* 257, 161-168.

Pastorian, K., Hawel III, L. and Byus, C.V. 2000. Optimization of cDNA representational difference analysis for the identification of differentially expressed mRNAs. *Anal. Biochem.* 283, 89-98.

Sambrook, J., Fritsch, E.F. and Maniatis, T. 1989. *Molecular Cloning: A Laboratory Manual.* Cold Spring Harbor Laboratory Press: Cold Spring Harbor.

Siebert, P.D., Chechnik, A., Kellogg, D.E., Lukyanov, K.A. and Lukyanov, S.A. 1995. An improved PCR method for walking in uncloned genomic DNA. *Nucleic Acids Res.* 23, 1087-1088.

Sive, H.L. and St John, T. 1988. A simple subtractive hybridization technique employing photoactivatable biotin and phenol extraction. *Nucleic Acids Res.* 16, 10937.

Sutcliffe, J.G., Foye, P.E., Erlander, M.G., Hilbush, B.S., Bodzin, L.J., Durham, J.T. and Hasel, K.W. 1999. TOGA: An automated parsing technology for analyzing expression of nearly all genes. *Proc. Natl. Acad. Sci. USA* 97, 1976-1981.

Tabaeizadeh, Z. 1998. Drought-induced responses in plant cells. *Int. Rev. Cytol.* 182, 193-247.

Tagu, D., Python, M., Crétin, C. and Martin, F. 1993. Cloning symbiosis-related cDNAs from eucalypt ectomycorrhiza by PCR-assisted differential screening. *New Phytol.* 125, 339-343.

Timblin, C., Battey, J. and Kuehl, W.M. 1990. Application for PCR technology to subtractive cDNA cloning: identification of genes expressed specifically in murine plasmacytoma cells. *Nucleic Acids Res.* 18, 1587-1595.

Tomkins, J.P., Mahalingam, R., Smith, H., Goicoechea, J.L., Knap, H.T. and Wing, R.A. 1999. A bacterial artificial chromosome library for soybean PI 437654 and identification of clones associated with cyst nematode resistance. *Plant Mol. Biol.* 41, 25-32.

Vierling, E. and Kimpel, J.A. 1992. Plant respondes to environmental stress. *Curr. Opin. Biotechnol.* 3, 164-170.

Vos, P., Hogers, R., Bleeker, M., Reijans, M., van der Lee, T., Hornes, M., Frijters, A., Pot, J., Peleman, J., Kuiper, M. and Zabeau, M.1995. AFLP: a new technique for DNA fingerprinting. *Nucleic Acids Res.* 23, 4407-4414.

Wan, J.S., Sharp, S.J., Poirier, G M.-C., Wagaman, P.C., Chambers, J., Pyati, J., Hom, Y-L., Galindo, J.E., Huvar, A., Peterson, P.A., Jackson, M.R and Erlander, M.G. 1996. Cloning differentially expressed mRNAs. *Nature Biotechnology* 14, 1685-1691.

Wang, Z. and Brown, D.D. 1991. A gene expression screen. *Proc. Natl. Acad. Sci. USA* 88, 11505-11509.

Welsh, J., Chada, K., Dala, S.S, Cheng, R. and McClelland, M. 1992. Arbitrarily primed PCR fingerprinting of RNA. *Nucleic Acids Res.* 20, 4965-4970.

Zhao, S., Ooi, S.L. and Pardee A.B. 1995. New primer strategy improves precision of differential display. *Biotechniques* 18, 842-6, 848, 850.

Zhao, S., Ooi, S.L., Yang, F.C. and Pardee A.B. 1996. Three methods for identification of true positive cloned cDNA fragment in differential display. *Biotechniques.* 20, 400-404.

Zhong, G.Y., Riov, J., Goren, R. and Holland, D. 2000. Competitive hybridization: theory and application in isolation and quantification of differentially regulated genes. *Anal. Biochem.* 282, 129-135.

Chapter 4

APPLICATION OF GENOMICS IN AGRICULTURE

Holger Hesse[1] and Rainer Höfgen[2]

[1]*Freie Universität Berlin, Institut für Biologie, Angewandte Genetik, Albrecht-Thaer-Weg 6, 14195 Berlin, Germany.*
hesse@mpimp-golm.mpg.de
[2]*Max-Planck-Institut für Molekulare Pflanzenphysiologie, Am Mühlenberg 1, 14476 Golm, Germany.*
hoefgen@mpimp-golm.mpg.de

INTRODUCTION

The goals of agricultural plant science are to increase crop productivity and the quality of agricultural products and to protect the environment by maintaining a system of sustainable agriculture that preserves the ecological basis of plant production. These goals have significant economic implications, which are affected by environmental conditions. The inherent diversity among plant species demonstrates clearly that plants are able to adapt to environmental stresses using genetically based programs. Crop improvement, i.e. the optimisation of plant features and performance according to agricultural needs, has been undertaken for hundred's of years: agronomists, breeders, and gardeners have used classical plant breeding methods based on selection of natural variants to improve genetic sources. Methods and techniques developed in molecular biology in recent years, especially reliable transformation systems for essentially all crops and growing numbers of complete genome sequences of higher plants, are providing tools to support plant breeding strategies and allowing scientists to tackle as yet unsolved problems or to speed up breeding programmes. Such tools will extend plant breeding by introducing new, unanticipated traits, in order to develop plants in which both crop productivity and stress tolerance are enhanced. There are two basic types of genomics approaches: structural

M.J. Hawkesford and P. Buchner (eds.),
Molecular Analysis of Plant Adaptation to the Environment, 61–79.

genomic approaches (such as mapping, sequencing, investigations into genome organisation, and comparative genomics) and functional genomic approaches with which the genome is analysed at transcriptome, proteome, or metabolome levels, often using high throughput technology. These approaches provide both tools with which to analyse crop productivity at a molecular level and strategies for improving crop productivity by generating high value output traits for increasing stress tolerance. They all enable scientists to reveal information about trait-conferring genes responsible for crop productivity, quality, and stress tolerance. The discovery of novel genes, the determination of their expression patterns in response to abiotic stress, and an improved understanding of their roles in stress adaptation (obtained through the use of functional genomics) will provide the basis for effective engineering strategies leading to greater stress tolerance. Genomics approaches are generating massive amounts of data, but the success of these approaches will depend upon finding ways to use this information to improve crops.

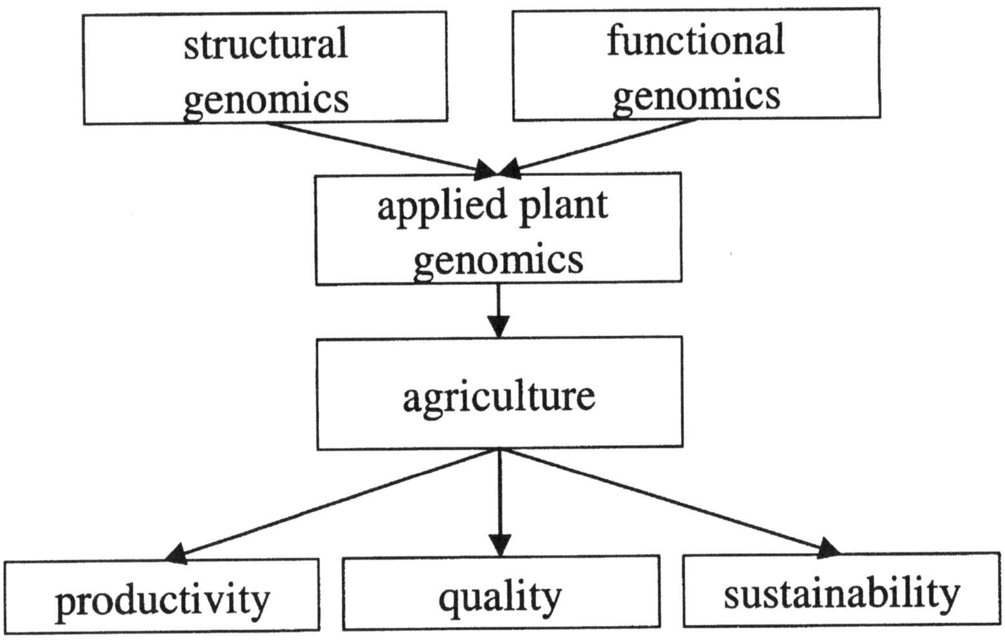

Figure 1. Schematic presentation of the influence of genomics on agriculture.

Agricultural plant science has had two main goals for decades: to increase yield and quality of agricultural products and to improve the protection of crops from diseases or limitations caused by pathogens, insects, nutrients, and stress (Figure 1). Environmental factors that impose stress (e.g. drought, salinity and temperature extremes) place major limits on plant productivity (Boyer, 1982). To overcome these limitations and improve production efficiency in the face of a burgeoning world population, crops with greater stress tolerance must be developed (Ingrim and Bartels, 1996; Holmberg and Bülow, 1998; Kasuga *et al.*, 1999; Kush, 1999). Classical breeding programmes develop new traits by combining different germ plasms in order to exploit natural or artificially induced diversity and, subsequently, to select for desired properties. Attempts to use plant breeding and genetic engineering in order to increase tolerance to conditions such as drought, salinity, and low temperature have had only marginal success for several reasons. These include the genetic complexity of stress responses (Flowers and Yeo, 1995), low genetic variance of yield components under stress conditions, and the lack of efficient selection techniques (Ribaut *et al.*, 1996, 1997; Frova *et al.*, 1999a and 1999b). Furthermore, quantitative trait loci (QTLs) that are linked to tolerance at one stage in development can differ from those at other stages (Foodlad, 1999). Improving this correlation with respect to productivity and protection is one of the remaining challenges in plant breeding. The availability of complete sequence information for crop genomes has been hailed as the driving force behind a new green revolution, which will significantly increase crop yield, processing features, and nutritional quality through the discovery, delivery, and targeted manipulation of new traits in transgenic crop plants (Mazur *et al.*, 1999, DellaPenna, 1999). Progress toward these applied goals is now anticipated through comparative genomic studies of an evolutionary diverse set of model organisms, and through the use of techniques such as high-throughput analysis of expressed sequence tags, analysis of microvariants (SNPs), large-scale parallel analysis of gene expression, targeted or random mutagenesis, and gain-of-function or mutant complementation (Michelmore, 2000). Genomics is a new discipline that extends traditional genetics and molecular biology using multiparallel techniques. In the near future, approaches allowing recombination hotspots to be highlighted will further aid plant-breeding efforts (Figure 2). Germ plasm improvement will continue to depend on non-transgenic methods that use sophisticated assays and molecular genetic markers to identify and exploit relevant genetic combinations, while genomics will accelerate the discovery of genes that confer key traits, enabling their rapid modification.

Transgenics was first used on a large scale to engineer for improved crop protection traits, primarily those conferring insect and herbicide resistance. These modified crops were first to market because the target traits are conferred by single genes. The next generation of improved traits will include disease resistance. Multiple genes whose products cause major changes in cell physiology may confer these traits. Combining multiple traits into a single elite cultivar will be a challenge, especially since these improvements must not be concomitant with decreased crop yields.

Plants have evolved genetically based physiological strategies to react to changing environmental conditions, be they abiotic or biotic. Most plants complete their life cycle in a single location and are therefore plagued by challenges such as nutrient acquisition, pathogen attack, and environmental stresses.

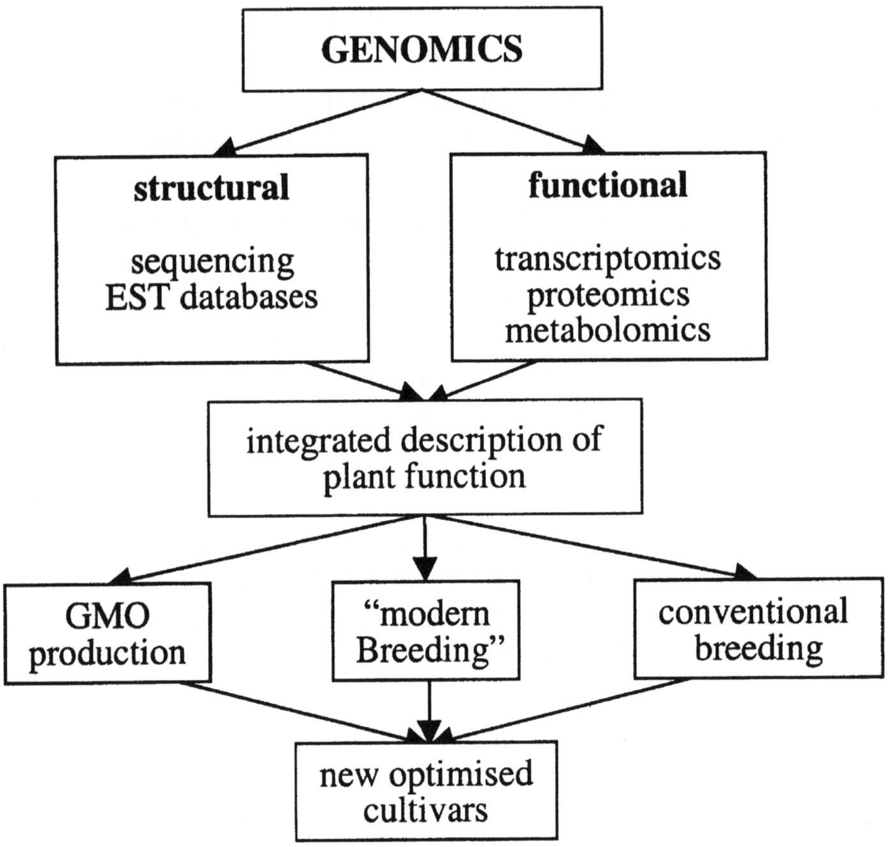

Figure 2. Genomics as a tool to generate new traits in plant breeding programmes.

In response to nutrient limitation, some plants undergo physiological and developmental adaptations to actively scavenge limited nutrients from their environment. These changes include adjustments in root architecture, induction of genes encoding high-affinity transporters (e.g. phosphate, nitrate, and sulfate transporters), rhizosphere acidification, and exudation of organic acids (Schachtmann *et al.,* 1998; Hawkesford, 2000). Stress adaptations are often more physiological than developmental in nature because of the necessity of responding rapidly to environmental changes. Therefore, it is reasonable to assume that stress tolerance is based on multiple, probably even redundant, co-acting gene products and metabolites. Historically, research on the effects of stresses has been largely descriptive by necessity, and relevant genes have been identified either by reference to physiological evidence or by differential screening. A large number of genes with potential roles in stress tolerance have been described and, in consequence, studies examining the control mechanisms of gene expression and analysing putative regulatory pathways have been initiated.

Most transgenic work attempts to unravel gene function and the results obtained are often of high agronomic impact. Conventional plant breeding has already provided an improvement in stress tolerance through the transfer of drought resistance genes from close wild relatives to crop species and the successive selection of the desired trait (Azevedo and Fereres, 1993). However, most such traits are complex, and, since their molecular basis is mostly still not understood, it is mandatory that knowledge of molecular processes continue to expand. This knowledge, when combined with the ever-growing technology of genomic research, will enable improved crop plants to be developed.

STRUCTURAL GENOMICS OR GENE DISCOVERY DRIVEN BY LARGE SCALE DNA SEQUENCING

Arabidopsis, a member of the crucifer family (closely related to brassica species and other food plants such as canola), has become an important model species for the study of almost all aspects of plant biology (Menke *et al.,* 1998) due to its diploid genetics, rapid regeneration time and large seed set. It has a relatively small genome of approximately 120-130 Mb, which contains approximately 10% repetitive DNA and is therefore well suited for map-based cloning strategies (Pruitt and Meyerowitz, 1986). For these reasons a wide range of resources for gene isolation, such as genetic and physical maps, have been developed for *Arabidopsis*.

The definition of the term genomics is rapidly evolving and expanding: genomics makes use of a variety of technologies that allow organisms to be described at DNA, mRNA, protein, and metabolite levels and allow relationships within and among these levels to be deciphered, both during growth and development and in response to various environmental stimuli. Genomics seeks to provide a complete, integrated, and multipurpose set of consistent data harbouring the answers to a multitude of questions. The resulting data can be systematically and repeatedly analysed and mined using bioinformatics to provide the information needed to address various problems and to answer a multitude of questions (many of which have not yet been imagined) using unbiased data analysis programs.

The outputs of plant genome sequencing programmes are typically databases of DNA sequences and associated information (Table 1). An international consortium called the *Arabidopsis* Genome Initiative began sequencing this genome in late 1996 and is expected to complete the project before the end of 2000.

Yet more tools for *Arabidopsis* genomics research include yeast and bacterial artificial chromosome (YAC and BAC, respectively) libraries and a high-resolution physical map covering the entire mappable genome (YAC: Burke *et al.,* 1987; BAC: Shizuya *et al.,* 1992; (Mozo *et al.,* 1998, 1999; www.mpimp-golm.mpg.de/101/mpi_mp_map/bac.html).

Chromosome maps are generated by assembling sequences of overlapping YAC or BAC clones. Such maps are also available for a wide range of other plants, although only for a limited number of agriculturally important species.

Sequence data acquisition was pioneered in bacteria and data from the genomes of several animal models is rapidly accumulating: the entire human genome has now been nearly sequenced. Plant genome sequencing projects are also nearing completion. The focus is now shifting to the examination of the transcribed regions of genomes. To this end, data is generated through the use of expressed sequence tags (ESTs, cDNAs). Sequencing projects reading 200-400 bases per clone of randomly selected cDNAs are used for identification. ESTs provide a cost-effective and rapid approach to describe the genes expressed for example in certain tissues or at various developmental stages of an organism. The short sequence reads are use for an immediate search of sequence databases for similar sequences with proteins of known function from any organism. Thus, the information content of an EST database is primarily derived from links to other databases.

A bonus that comes from generating EST collections, is the rapid discovery of novel genes (genes without known analogues in other plants), and this can be achieved without complete accuracy. When sequence similarities are perceived, the discovery of the function of a given gene motif, or protein domain in one organism provides clues to related sequences in other organisms. The cDNAs used to establish EST arrays can be used as probes to isolate full-length genes. For *Arabidopsis* and rice, large EST collections (http://www.tigr.org/tdb/index.shtml) of approximately 35,000 clones are already available (Newman *et al.,* 1994; Cooke *et al.,* 1996; Yamamoto and Sasaki, 1997), and EST collections for many other plant species are growing, for instance for poplar (Sterky *et al.,* 1998) and soybean (http://129.186.26.94/soybeanest.html). The dbEST section of GenBank (http://www.ncbi.nlm.nih.gov/dbEST/dbEST summary. html) provides information about the various EST collections in use. In addition, several companies have private EST collections, which, in some cases, can be consulted on request for specific academic searches.

Arabidopsis, originally chosen for the first sequencing project because of its small genome size, is now uniquely important for a wide range of studies in plant sciences (Meyerowitz and Somerville, 1994; Meinke *et al.*, 1998). Genomic information can be obtained for *Arabidopsis* at http://sequence-www.stanford.edu/ara/ArabidopsisSeqStanford.html. The first analyses of the *Arabidopsis* genome reveal that it is extremely gene-rich with few repeated elements. On average, there is one gene every 4-5 kb, the average number of introns is five, with a mean size of approximately 160 bp (Bevan *et al.*, 1998; Terryn *et al.*, 1999). The total gene content of *Arabidopsis* is estimated to be approximately 20,000-25,000. About half of the predicted genes can be assigned to a functional category based on their similarity to known genes and protein.

FUNCTIONAL GENOMICS

Information on where and when a certain gene is expressed can indicate the biological function of the encoded protein. Classical approaches such as RNA blot analysis have long allowed expression to be studied at a single gene level. However, the huge amount of data generated by expression profiling projects require multiparallel analyses and enable profiles from samples generated from different tissues to be compared (Kozian and Kirschbaum, 1999).

cDNA libraries are used to fabricate micro-arrays comprised of spots of DNA (ESTs) immobilised onto nylon or glass supports (Baldwin *et al.*, 1999; Brown and Botstein, 1999; Dugan *et al.*, 1999; Granjeaud *et al.*, 1999; Lemieux *et al.*, 1999; Lipshutz *et al.*, 1999; Richmond and Somerville, 2000). Three general types of experiments can be performed with micro-arrays (Figure 3). In marker discovery experiments, the goal is to discover a limited number of highly specific marker genes for a cell type, a developmental stage, or an environmental treatment. In such experiments, the researcher is often interested only in genes that show a dramatic and selective induction or repression of expression. The second type of experiments are exploratory or biological discovery experiments, where expression analysis of a multitude of genes (the micro-array) is exploited to provide an integrative view of a plant's response to a treatment (transcriptomics). Such experiments have the potential to provide new insights into biological processes. The third type of experiments are the gene-function discovery experiments. As micro-array data accumulates in centralised public databases, hypotheses as to the function of novel, uncharacterised genes can be generated by *in silico* analyses of when and where such genes exhibit altered gene expression. These kinds of experiments will be beneficial to plant biology by offering a broader view of how a plant functions as well as providing a detailed characterisation of a multitute of individual genes.

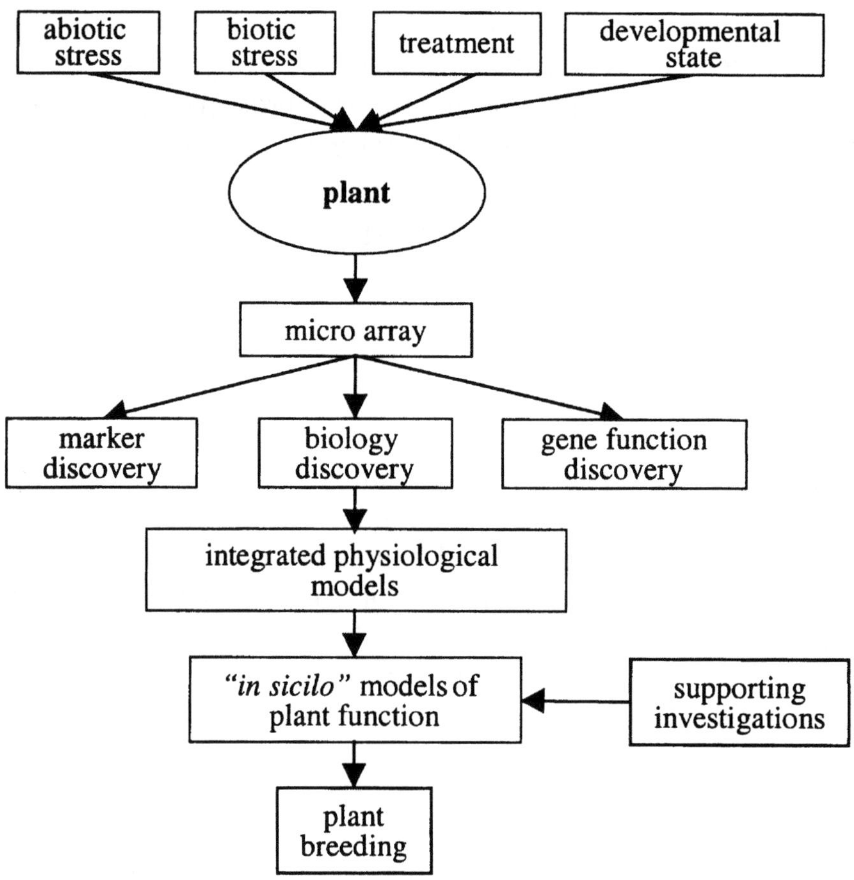

Figure 3. Functional and structural genomics strategy from gene discovery to evaluation of new cultivars.

In principle micro-arrays are a kind of reverse Northern blot in which DNA clones (cDNA clones, PCR-generated fragments, etc.) are spotted in a dense array and hybridised to RNA-derived probes (Figure 4). The hybridisation signal can be quantified automatically (using robotics at the most elaborate) and reflects the abundance of the corresponding mRNA in the total RNA pool. The novelty and benefit of this technique lies in the extreme miniaturisation, allowing large numbers of clones to be simultaneously analysed (Eisen and Brown, 1999; Kehoe *et al.,* 1999; Shena and Davis, 1999). Spatial or temporal expression patterns of all of the arrayed genes may be analysed by probing the array with fluorescent or radio-labelled DNA generated from mRNA. This mRNA can be harvested from various tissues, at multiple developmental stages, or after specific treatments.

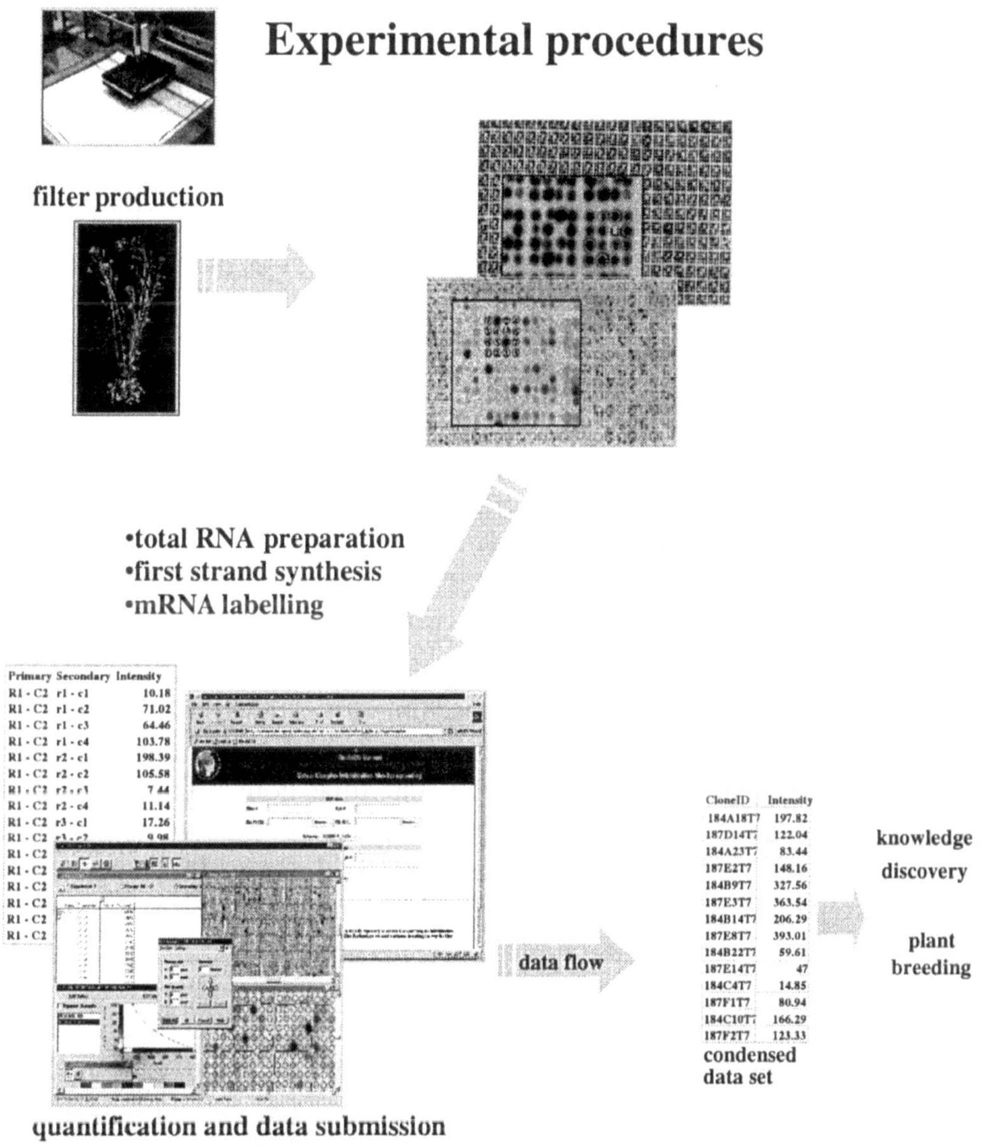

Figure 4. Schematic presentation of a high density gene-expression micro-array experiment. Radioactively labelled probes from test and control tissues are hybridised to arrays and the proportion of each probe bound determined. Every arrayed gene can be analysed quantitatively for expression in the target tissue and changes in expression resulting from experimental treatment (illustration has been provided by Dr. Bernd Essigmann, MPI-MP Golm, Germany).

Three techniques are currently in use: DNA spotted on nitrocellulose and hybridised to radioactive probes (Desprez *et al.,* 1998); DNA spotted on coated glass slides and hybridised to fluorescently labelled probes, which can be determined by scanning confocal fluorescence microscopy and image analysis (Schena *et al.,* 1995; Ruan *et al.,* 1998); and DNA oligonucleotides synthesised *in situ* on solid supports and hybridised to fluorescent probes (developed by Affimetrix (Santa Clara, CA, USA; see Wodicka *et al.,* 1997; see also http://gem.incyte.com/gem/index.shtml). The output indicates whether or not each arrayed gene is expressed in the tested sample, and, when two arrays are compared, whether the level of expression is increased or decreased in response to the experimental treatment. Furthermore, understanding cross-regulation of metabolic pathways in plants (Guyer *et al.,* 1995; Aharoni *et al.,* 2000; Mekhedov *et al.,* 2000) presents a challenge: the goal is to separate potentially informative changes in gene expression from a pathway of interest from unrelated, non-specific responses, i.e. to separate signal from noise (Zhao *et al.,* 1998).

MUTANT ANALYSIS

The role of genes is also sought using large-scale, systematic, genome wide mutation programmes. For a long time, classical mutagenesis, and map-based cloning have been like searching for a needle in a haystack. Indeed, once a certain mutant with an interesting phenotype was obtained by random mutagenesis it was difficult and time consuming to map the position of the gene and walk to or land on it (Tanksley *et al.,* 1995). However, recent techniques, such as amplified fragment length polymorphisms (AFLP; Vos *et al.,* 1995) and the availability of the detailed maps including sequencing data allow mutant genes to be mapped quickly (Vos *et al.,* 1998).

The principle of transcript profiling, a specific AFLP technique, is comparable to that of differential display (see below) and is adapted for mRNA material (Bachem *et al.,* 1996). cDNAs are cut with restriction enzymes and then an adapter is annealed to the resulting fragments. Primers based on this adapter sequence (including a few additional nucleotides to provide further specificity) are then used to amplify the cDNAs. Following this procedure it is possible to obtain information from an estimated 80% of the mRNAs in the original pool. In comparison to transcript profiling, differential display allows transcribed genes to be analysed and compared systematically. In differential display, mRNA is amplified by PCR using random primers (Liang and Pardee, 1992). The PCR products are then subjected to gel electrophoresis and the patterns of the amplified cDNAs are compared. Differentially expressed transcripts can be eluted and cloned. The advantage of this method is its simplicity, the small amount of RNA required, and the possibility of detecting virtually any differentially expressed mRNA. However, there are disadvantages with respect to reproducibility and reliability of the observed differences (RNA blot data of identified genes do not always confirm the data obtained by differential display; see also chapter 3).

ASSIGNING FUNCTION

Since classical mutant identification is a time-consuming effort, an approach has been designed in which naturally occurring DNA can be introduced into the plant genome to interrupt genes at random simultaneously flagging the interrupted sequence with a tag. The phenotype exhibited by the plant may indicate gene function, although further experimental data is required for gene confirmation. Both *Agrobacterium* T-DNA and transposons have been used to this end (Martienssen, 1998; Azpiroz-Leehan and Feldman, 1997). The known sequence of the inserted element enables libraries to be screened using PCR-based strategies, either exploiting known gene sequences to identify plant lines, knock-outs, or applying inverse PCR or plasmid rescue strategies to clone unknown genes.

Transposons have been modified to provide a convenient tool with which to construct large populations of mutated plants within which each individual transgenic line has a single transposon tag to either completely block gene function or to increase gene expression to amplify its function. A knock out or knock in effect depends on the specific genetic components of the system, and on whether the tagged gene is heterozygous or homozygous, which may affect the phenotype. The identity of each tagged gene can be readily determined by sequencing the regions flanking the tag and referring back to the genome sequence. The expression of promoterless reporter genes carried within the tag may also allow genes that only function in specific cell types or at specific developmental stages to be detected, even when there may be no other obvious phenotype (Martienssen, 1998).

Detailed experimental analysis of tagged plant lines may eventually provide enough information so that a function can be unambiguously assigned to the tagged gene. Routine analyses might search for changes in outward appearance and in expression, metabolite, or protein profiles. As data accumulates, it will become progressively easier to assign biological roles to genes. Further indirect data may be derived from the patterns of temporal and spatial gene expression patterns derived from array experiments, whereby the function of a gene may be associated with a specific tissue, developmental process or in response to specific biotic or abiotic stimuli.

It is widely recognised, however, that the assignment of functions to genes will be a significantly more challenging problem than determining their sequences.

BIOINFORMATICS

Bioinformatics uses information technology to tackle the mountains of data generated through genomics approaches. It involves the capture, storage, retrieval, processing, analysis, visualisation, and interpretation of these data. Furthermore, it correlates these data with the currently available wealth of biological information already available on genetics, biochemistry, molecular biology and biological function. Bioinformatics provides the critical tools required to maintain the databases in which these data are stored, and enables information to be mined from them. One such type of tool are algorithms to search databases and recognise similarities between genes and their encoded protein products. Such sequence relationships are found by comparing each newly identified gene with the entire database of pre-existing genes. Sequence similarities often imply related function, and are, as previously discussed, the basis of assigning putative functions to newly identified genes. A significant proportion of the genes identified through plant genome sequencing are expected to be new to science, their DNA or encoded protein sequences providing few clues about their functions. As an effort toward building such knowledge, computer tools have been developed to decipher the gene architecture of the *Arabidopsis* genome and provide annotation of the genomic sequences (Ermolaeva *et al.,* 1998; Bassett *et al.,* 1999; Ewing *et al.,* 1999). However, distinguishing coding from non-coding sequences and setting the proper gene structure remains a problem.

USING GENOMICS TO DISSECT PLANT STRESS RESISTANCE PATHWAYS

The large number of ongoing sequencing projects in a variety of organisms represent a significant development for every field of biology (Table 1).

Analysis of the growing DNA databases reveals a striking degree of inter-kingdom homology at the primary protein sequence level (The *C.*

elegans sequencing consortium, 1998; Sterky *et al.,* 1998; Dellapenna, 1999; Michelmore, 2000). That many of these inter-kingdom orthologs are involved in basic cellular functions (for example protein synthesis, cell division, primary carbon metabolism, and signal transduction) attests to the evolutionary conservation of these processes. Of the remaining sequences in an organism, about half have orthologs of unknown function in other organisms, while the remainder encode novel sequences with no related sequences in the database (The *C. elegans* sequencing consortium, 1998; Sterky *et al.,* 1998). Many unknown and novel sequences are likely to have species-, family- or kingdom-specific functions that have evolved to meet circumstances unique to the particular organism or group of organisms.

Table 1.

Plant genomics, EST or micro-array projects funded in the last years
http://www.tigr.org/tdb/index.shtml
http://129.186.26.94/soybeanest.html
http://www.ncbi.nlm.nih.gov/dbEST/dbEST_summary.html
http://sequence-www.stanford.edu/ara/ArabidopsisSeqStanford.html
http://www.nsf.gov/bio/pubs/awards/genome99.htm
http://www.mpimp-golm.mpg.de/101/mpi_mp_map/bac.html
http://www.ncga.com
http://www.expasy.ch
http://probe.nalusda.gov
http://www.agron.missouri.edu
http://www.iastate.edu/~usda-gem
http://www.reeusda.gov/crgam/nri/programs/rice/rice.htm
http://orsay1.moulon.inra.fr/imgd/

Plant researchers are beginning to use genomic resources and the power of DNA micro-arrays to study all areas of plant biology, including stress metabolism (Desprez *et al.,* 1998; Ruan *et al.,* 1998). These technologies complement, and can be readily integrated with, existing biochemical and genetic approaches to add new dimensions to the elucidation of complex metabolic pathways and physiological reactions in plants and provide significant clues for plant breeding toward stress tolerance.

GENETIC ENGINEERING OF TOLERANCE TRAITS

In contrast to traditional breeding regimes and marker-assisted selection programmes, the direct introduction of a small number of genes by genetic engineering seems to be a more attractive and rapid approach with which to improve stress tolerance. Current engineering strategies, however, rely on the transfer of one or several genes that either encode an enzyme in a biochemical pathway or a protein involved at the endpoint of a signaling pathway (where expression of the transgenes is mainly controlled by constitutive promoters). These gene products protect, either directly or indirectly, against environmental stresses (Bohnert and Sheveleva, 1998; Holmberg and Bülow, 1998; Smirnoff, 1998). The engineered over-expression of biosynthetic enzymes for osmoprotectants (Nuccio *et al.*, 1998; McNeil *et al.*, 1999), scavengers of reactive oxygen species (Noctor and Foyer, 1998; Blaszczyk *et al.*, 1999; Harms *et al.*, 2000), stress-induced proteins such as cold-regulated or late embryogenesis abundant proteins (Malik *et al.*, 1999; Thomashow, 1999), and heavy metal tolerance (Zhu *et al.*, 1999a and b; Pilon-Smith *et al.*, 1999) are among the reported approaches. Identifying genes (proteins) providing increased tolerance against stress will have a significant impact on crop quality and yield for both conventional breeding strategies and genetically modified organisms (GMOs).

The first step towards cataloguing and categorising genetically complex abiotic stress responses is the rapid discovery of genes using large-scale partial sequencing of randomly selected cDNA clones or ESTs. Large-scale EST sequencing initiatives are also well under way for various crop species (Walbot, 1999) including cotton, *Medicago truncatula*, maize, soybean, tomato and sorghum and also for Loblolly pine (http://www.nsf.gov/bio/pubs/awards/genome99.htm). The collections are, however, biased towards high and moderate abundance classes that are derived from specific tissues, organs, or cells, different developmental states, various external stimuli such as heat-shock or nitrogen and sulfur starvation, and treatments with plant growth regulators. In contrast, relatively few studies have focused specifically on ESTs from plants that have been exposed to environmental stresses and this is a significant opportunity for the future.

CONCLUSIONS: THE IMPLICATIONS OF GENOMICS APPLICATIONS FOR INCREASING CROP PLANT STRESS TOLERANCE

The genome sequencing and expression profiling initiatives currently underway promise to allow genes essential for tolerance towards stresses to be identified and characterised. Mining the data generated through these initiatives will supply a systematic agenda for functional analysis with the use of tagged mutant collections, complementation and overexpression tests accompanied by micro-array analyses to reveal hierarchical relationships between specific signaling components and downstream effector genes.

Understanding specific protein-protein interactions will require the construction of protein-linkage maps using yeast two-hybrid technologies. Proteomics approaches will be necessary to clarify the structural predictions of genome sequence information and to assess the protein modifications and protein-ligand interactions that are relevant to stress tolerant phenotypes. Ultimately, understanding the roles of all the gene products participating in stress adaptation or tolerance reactions will provide an integrated understanding of the biochemical and physiological basis of stress responses in plants. Armed with such information from established models, it will be possible to rationally manipulate and optimise tolerance traits for improved crop productivity well into the twenty-first century.

Acknowledgements

We would like to thank Prof. Lothar Willmitzer and Dr. Bernd Essigmann for critical discussion and Megan McKenzie for enthusiastically editing the manuscript.

REFERENCES

Aharoni, A., Keizer, L.C.P., Bouwmeester, H.J., Sun, Z.K., Alvarez-Huerta, M., Verhoeven, H.A., Blaas, J., van Houwelingen, A.M.M.L., De Vos, R.C.H., van der Voet, H., Jansen , R.C., Guis, M., Mol, J., Davis, R.W., Schena, M., van Tunen,A.J. and O'Connell, A.P. 2000. Identification of the SAAT gene involved in strawberry flavor biogenesis by use of DNA micro-arrays. *Plant Cell* 12, 7-661.

Azevedo, E. and Fereres, E. 1993. "Resistance to abiotic stresses". In: *Plant Breeding*, eds. M.D. Hayward, N.O. Bosemark and I. Romagosa, pp. 406-412. Chapman and Hall, London.

Azpiroz-Leehan, R. and Feldmann, K.A. 1997. T-DNA insertion mutagenesis in *Arabidopsis*: going back and forth. *Trends Genet.* 13, 2-156.

Bachem, C.W.B., Vanderhoeven, R.S., Debruijn, S.M., Vreugdenhil, D., Zabeau, M. and Visser, R.G.F. 1996. Visualization of differential gene expression using a novel method of RNA fingerprinting based on AFLP - analysis of gene expression during potato tuber development. *Plant J.* 9, 745-753.

Baldwin, D., Crane, V. and Rice, D. 1999. A comparison of gel-based, nylon filter and micro-array techniques to detect differential RNA expression in plants. *Curr. Opin. Plant Biol.* 2, 96-103.

Bassett, D.E., Eisen, M.B. and Boguski, M.S. 1999. Gene expression informatics - it's all in your mine. *Nature Genet.* 21, 51-55.

Bevan, M., Bancroft, I., Bent, E., Love, K., Goodman, H., Dean, C., Bergkamp, R., Dirkse, W., Vanstaveren, M., Stiekema, W., Drost, L., Ridley, P., Hudson, S.A., Patel, K., Murphy, G., Piffanelli, P., *et al.,* 1998. Analysis of 1.9 Mb of contiguous sequence from chromosome 4 of *Arabidopsis thaliana. Nature* 391, 485-488.

Blaszczyk, A., Brodzik, R. and Sirko, A. 1999. Increased resistance to oxidative stress in transgenic tobacco plants overexpressing bacterial serine acetyltransferase. *Plant J.* 20, 237-243.

Bohnert, H.J. and Sheveleva, E. 1998. Plant stress adaptations - making metabolism move. *Curr. Opin. Plant Biol.* 1, 267-274.

Boyer, J.S. 1982. Plant productivity and environment. *Science* 218, 443-448.

Cooke, R., Raynal, M., Laudie, M., Grellet, F., Delseny, M., Morris, P.C., Guerrier, D., Giraudat, J., Quigley, F., Clabault, G., Li, Y.F., Mache, R., Krivitzky, M., Gy, I.J.J., Kreis, M., Lecharny, A., Parmentier, Y., Marbach, J., Fleck, J., Clement, B., Philipps, G., Herve, C., Bardet, C., Tremousaygue, D., Lescure, B., Lacomme, C., Roby, D., Jourjon, M-F., Chabrier, P ., Charpenteau, J-L., Desprez, T., Amselem, J., Chiapello, H. and Höfte, H. 1996. Further progress towards a catalogue of all *Arabidopsis* genes - analysis of a set of 5000 non-redundant ESTs. *Plant J.* 9, 101-124.

Burke, D.T., Carle, G.F. and Olson, M.V. 1987. Cloning of large segments of exogenous DNA into yeast by means of artificial chromosome vectors. *Science* 236, 806-12.

Brown, P.O. and Botstein, D. 1999. Exploring the new world of the genome with DNA micro-arrays. *Nature Genet.* 21, 33-37.

DellaPenna, D. 1999. Nutritional genomics: manipulating plant micronutrients to improve human health. *Science* 285, 375-379.

Desprez, T., Amselem, J., Caboche, M. and Höfte, H. 1998. Differential gene expression in *Arabidopsis* monitored using cDNA arrays. *Plant J.* 14, 643-652.

Duggan, D.J., Bittner, M., Chen, Y.D., Meltzer, P. and Trent, J.M. 1999. Expression profiling using cDNA micro-arrays. *Nature Genet.* 21, 10-14.

Eisen, M.B. and Brown, P.O. 1999. DNA arrays for analysis of gene expression. *Methods Enzymol.* 303, 179-205.

Ermolaeva, O., Rastogi, M., Pruitt, K.D., Schuler, G.D., Bittner, M.L., Chen, Y.D., Simon, R., Meltzer, P., Trent, J.M. and Boguski, M.S. 1998. Data management and analysis for gene expression arrays. *Nature Genet.* 20, 19-23.

Ewing, R.M., Ben Kahla, A., Poirot, O., Lopez, F., Audic, S. and Claverie, J.M. 1999. Large-scale statistical analyses of rice ESTs reveal correlated patterns of gene expression. *Genome Res.* 9, 950-959.

Flowers, T.J. and Yeo, A.R. 1995. Breeding for salinity resistance in crop plants: where next? *Aust. J. Plant Physiol.* 22, 875-884.

Foodlad, M.R. 1999. Comparison of salt tolerance during seed germination and vegetative growth in tomato by QTL mapping. *Genome* 42, 727-734.

Frova, C., Krajewski, P., di Fonzo, N., Villa, M. and Sari-Gorla, M. 1999. Genetic analysis of drought tolerance in maize by molecular markers. 1. Yield components. *Theor. Appl. Genet.* 99, 280-288.

Frova, C., Caffulli, A. and Pallavera, E. 1999. Mapping quantitative trait loci for tolerance to abiotic stresses in maize. *J. Exp. Zool.* 282, 164-170.

Granjeaud, S., Bertucci, F. and Jordan, B.R. 1999. Expression profiling: DNA arrays in many guises. *Bioessays* 21, 781-790.

Guyer, D., Patton, D. and Ward, E. 1995. Evidence for cross-pathway regulation of metabolic gene expression in plants. *Proc. Natl. Acad. Sci. USA* 92, 4997-5000.

Harms, K., von Ballmoos, P., Brunold, C., Höfgen, R. and Hesse, H. 2000. Expression of a bacterial serine acetyltransferase in transgenic potato plants leads to increased levels of cysteine and glutathione. *Plant J.* 22, 335-343.

Hawkesford, M.J. 2000. Plant responses to sulphur deficiency and the genetic manipulation of sulphate transporters to improve S-utilization efficiency. *J. Exp. Bot.* 51, 131-138.

Holmberg, N. and Bülow, L. 1998. Improving stress tolerance in plants by gene transfer. *Trends Plant Sci.* 3, 61-66.

Ingram, J., and Bartels, D. 1996. The molecular basis of dehydration tolerance in plants. *Ann. Rev. Plant Physiol. Plant Mol. Biol.* 47, 377-403.

Kasuga, M., Liu. Q., Miura, S., Yamaguchi-Shinozaki, K. and Shinozaki, K. 1999. Improving plant drought, salt, and freezing tolerance by gene transfer of a single stress-inducible transcription factor. *Nature Biotech.* 17, 287-291.

Kozian, D.H. and Kirschbaum, B.J. 1999. Comparative gene-expression analysis. *Trends in Biotech.* 17, 73-78.

Kehoe, D.M., Villand, P. and Somerville, S. 1999. DNA micro-arrays for studies of higher plants and other photosynthetic organisms. *Trends Plant Sci.* 4, 38-41.

Kush, G.S. 1999. Green revolution: preparing for the 21st century. *Genome* 42, 646-655.

Liang. P. and Pardee, A,B, 1992. Differential display of eukaryotic messenger RNA by means of the polymerase chain reaction. *Science* 257, 967-71.

Lemieux, B., Aharoni, A. and Schena, M. 1998. Overview of DNA chip technology. *Mol. Breed.* 4, 277-289.

Lipshutz, R.J., Fodor, S.P.A., Gingeras, T.R. and Lockhart, D.J. 1999. High density synthetic oligonucleotide arrays. *Nature Genet.* 21, 20-24.

Malik, M.K., Slovin, J.P., Hwang, C.H. and Zimmerman, J.L. 1999. Modified expression of a carrot small heat shock protein gene, Hsp17.7, results in increased or decreased thermotolerance. *Plant J.* 20, 89-99.

Martienssen, R.A. 1998. Functional genomics - probing plant gene function and expression with transposons. *Proc. Natl. Acad. Sci. USA* 95, 2021-2026.

Mazur, B., Krebbers, E. and Tingey, S. 1999 Gene discovery and product development for grain quality traits. *Science* 285, 372-375.

McNeil, S.D., Nuccio, M.L. and Hanson, A.D. 1999. Betaines and related osmoprotectants. Targets for metabolic engineering of stress resistance. *Plant Physiol.* 120, 945-949.

Meinke, D.W., Cherry, J.M., Dean, C., Rounsley, S.D. and Koornneef, M. 1998. *Arabidopsis thaliana*: a model plant for genome analysis. *Science* 282, 678-682.

Mekhedov, S., de Ilarduyam O.M. and Ohlrogge, J. 2000. Toward a functional catalog of the plant genome. A survey of genes for lipid biosynthesis. *Plant Physiol.* 122, 389-401.

Michelmore, R. 2000. Genomic approaches to plant disease resistance. *Curr. Opin. Plant Biol.* 3, 125-131.

Mozo, T., Fischer, S., Meier-Ewert, S., Lehrach, H. and Altmann, T. 1998. Use of the IGF BAC library for physical mapping of the *Arabidopsis thaliana* genome. *Plant J.* 16, 377-384.

Mozo, T., Dewar, K., Dunn, P., Ecker, J.R., Fischer, S., Kloska, S., Lehrach, H., Marra, M., Martienssen, R., Meier-Ewert, S. and Altmann, T. 1999. A complete BAC-based physical map of the *Arabidopsis thaliana* genome. *Nature Genet.* 22, 271-275.

Newman, T., de Bruijn, F.J., Green, P., Keegstra, K., Kende, H., McIntosh, L., Ohlrogge, J., Raikhel, N., Somerville, S., Thomashow, M., Retzel, E. and Somerville, C. 1994. Genes galore: a summary of methods for accessing results from large-scale partial sequencing of anonymous *Arabidopsis* cDNA clones. *Plant Physiol.* 106, 1241-55.

Noctor, G. and Foyer, C.H. 1998. Ascorbate and glutathione - keeping active oxygen under control. *Annu. Rev. Plant Physiol. Plant Mol. Biol.* 49, 249-279.

Nuccio, M.L., Rhodes, D., McNeil, S.D. and Hanson, A.D. 1999. Metabolic engineering of plants for osmotic stress resistance. *Curr. Opin. Plant. Biol.* 2, 128-34.

Pilon-Smits, E.A.H., Hwang, S.B., Lytle, C.M., Zhu, Y.L., Tai, J.C., Bravo, R.C., Chen, Y.C., Leustek, T. and Terry, N. 1999. Overexpression of ATP sulfurylase in Indian mustard leads to increased selenate uptake, reduction, and tolerance. *Plant Physiol.* 119, 123-132.

Pruitt, R.E. and Meyerowitz, E.M. 1986. Characterization of the genome of *Arabidopsis thaliana*. *J. Mol. Biol.* 187, 169-183.

Ribaut, J.M., Hoisington, D.A., Deutsch, J.A., Jiang, C. and Gonzalez-de-Leon, D. 1996. Identification of quantitative trait loci under drought conditions in tropical maize. 1. Flowering parameters and the anthesis-silking interval. *Theor. Appl. Genet.* 92, 905-914.

Ribaut, J.M., Jiang, C., Gonzalez-de-Leon, D., Edmeades, G.O. and Hoisington, D.A. 1997. Identification of quantitative trait loci under drought conditions in tropical maize. 2. Yield components and marker-assisted selection strategies. *Theor. Appl. Genet.* 94, 887-896.

Ruan, Y., Gilmore, J. and Conner, T. 1998. Towards *Arabidopsis* genome analysis - monitoring expression profiles of 1400 genes using cDNA micro-arrays. *Plant J.* 15, 821-833.

Richmond, T. and Somerville, S. 2000. Chasing the dream: plant EST micro-arrays. *Curr. Opin. Plant Biol.* 3, 108-116.

Schachtmann, D.P., Reid, R.J. and Ayling, S.L. 1998. Phosphorus uptake by plants: from the soil to cells. *Plant Physiol.* 116, 447-453.

Schena, M., Shalon, D., Davis, R.W. and Brown, P.O. 1995. Quantitative monitoring of gene expression patterns with a complementary DNA micro-array. *Science* 270, 467-470.

Schena, M. and Davis, R.W. 1999. "Genes, genomes, and chips". In: *DNA* micro-arrays*: a practical approach*. New York, Oxford University Press, 205:1-16.

Shizuya, H., Birren, B., Kim, U-J., Mancino, V., Slepak, T., Tachiiri, Y. and Simon, M. 1992. Cloning and stable maintenance of 300 kilobase pair fragments of human DNA in *Escherichia coli* using a F-factor based vector. *Proc. Natl. Acad. Sci. USA* 89, 8794-8797.

Smirnoff, N. 1998. Plant resistance to environmental stress. *Curr. Opin. Biotech.* 9, 214-219.

Sterky, F., Regan, S., Karlsson, J., Hertzberg, M., Rohde, A., Holmberg, A., Amini, B., Bhalerao, R., Larsson, M., Villarroel, R., van Montagu, M., Sandberg, G., Olsson, O., Teeri, T.T., Boerjan, W., Gustafsson, P., Uhlen, M., Sundberg, B. and Lundeberg, J. 1998. Gene discovery in the wood-forming tissues of poplar-analysis of 5,692 expressed sequence tags. *Proc. Natl. Acad. Sci. USA* 95, 13330-13335.

Tanksley, S.D., Ganal, M.W. and Martin, G.B. 1995. Chromosome landing - a paradigm for map-based gene cloning in plants with large genomes. *Trends Genet.* 11, 63-68.

Terryn, N., Heijnen, L., De Keyser, A., van Asseldonck, M., De Clercq, R., Verbakel, H., Gielen, J., Zabeau, M., Villarroel, R., Jesse, T., Neyt, P., Hogers, R., van den Daele, H., Ardiles, W., Schueller, C., Mayer, K., Dehais, P., Rombauts, S., Van Montagu, M., Rouze, P. and Vos, P. 1999. Evidence for an ancient chromosomal duplication in *Arabidopsis thaliana* by sequencing and analyzing a 400-kb contig at the APETALA2 locus on chromosome 4. *FEBS Lett.* 445, 237-245.

Thomashow, M.F. 1999. Plant cold acclimation: freezing tolerance genes and regulatory mechanisms. *Annu. Rev. Plant Physiol. Plant Mol. Biol.* 50, 571-599.

The *C. elegans* sequencing consortium 1998. Genome sequence of the nematode *C. elegans*: a platform for investigating biology. *Science* 282, 2012-2018.

Vos, P., Hogers, R., Bleeker, M., Reijans, M., Vandelee, T., Hornes, M., Frijters, A., Pot, J., Peleman, J., Kuiper, M. and Zabeau, M. 1995. AFLP - a new technique for DNA fingerprinting. *Nucleic Acids Res.* 23, 4407-4414.

Vos, P., Simons, G., Jesse, T., Wijbrandi, J., Heinen, L,, Hogers, R., Frijters, A., Groenendijk, J., Diergaarde, P., Reijans, M., Fierensonstenk, J., Deboth, M., Peleman, J., Liharska, T., Hontelez, J. and Zabeau, M. 1998. The tomato MI-1 gene confers resistance to both root-knot nematodes and potato aphids. *Nature Biotech.* 16, 1365-1369.

Wodicka, L., Dong, H.L., Mittmann, M., Ho, M.H. and Lockhart, D.J. 1997. Genome-wide expression monitoring in *Saccharomyces cerevisiae. Nature Biotech.* 15, 1359-1367.

Walbot, V. 1999. Genes, genomes, genomics. What can plant biologists expect from the 1998 National Science Foundation Plant Genome Research Program? *Plant Physiol.* 119, 1151-1155.

Yamamoto, K. and Sasaki, T. 1997. Large-scale EST sequencing in rice. *Plant Mol. Biol.* 35, 135-144.

Zhao, J.M., Williams, C.C. and Last, R.L. 1998. Induction of *Arabidopsis* tryptophan pathway enzymes and camalexin by amino acid starvation, oxidative stress, and an abiotic elicitor. *Plant Cell* 10, 359-370.

Zhu, Y.L., Pilon-Smits, E.A.H., Jouanin, L. and Terry, N. 1999a. Overexpression of glutathione synthetase in Indian mustard enhances cadmium accumulation and tolerance. *Plant Physiol.* 119, 73-79.

Zhu, Y.L., Pilon-Smits, E.A.H., Tarun, A.S., Weber, S.U., Jouanin, L. and Terry, N. 1999b. Cadmium tolerance and accumulation in Indian mustard is enhanced by overexpressing gamma-glutamylcysteine synthetase. *Plant Physiol.* 121, 1169-1177.

Chapter 5

QUANTITATIVE TRAIT LOCI (QTLs) FOR ANALYSIS OF PHYSIOLOGICAL AND BIOCHEMICAL RESPONSES TO ABIOTIC STRESS

Jean-Louis Prioul and Claudine Thévenot

Institut de Biotechnologie des Plantes, Bat. 630, Université de Paris-Sud, 91405 ORSAY Cedex, France.
Jean-Louis.Prioul@ibp.u-psud.fr
Claudine.Thevenot@ibp.u-psud.fr

INTRODUCTION

Quantitative geneticists and plant physiologists have to deal with plant variability however approaches diverse since the geneticists are interested in explaining and in using the genetic variance when the physiologists usually tend to minimise the genetic component in order to concentrate on the environmental variability. Heritability, which estimates the genetic part of the total variance, is used by the geneticist to decide if a trait is valuable for further study (i.e. genetically variable enough) and especially to map loci. When dealing with complex or quantitative traits, several loci are expected for one trait. The availability of molecular markers has allowed mapping of these quantitative trait loci, i.e. QTLs (Paterson *et al.*, 1988). It has enabled the dissection of the main genetic components of a trait into a small number of loci and the evaluation of the contribution of each locus in the trait variability. Early attempts to detect QTL for agronomic traits showed that QTL position was not generally stable from one experiment or one location to the other, implying large environmental effect. Several experiments have tested QTL stability in different environments. In tomato, for example, 350 F2-descendants were grown in three locations (2 in USA and one in Israel);

M.J. Hawkesford and P. Buchner (eds.),
Molecular Analysis of Plant Adaptation to the Environment, 81–101.

29 QTLs were detected, 15 were specific to one location and only 5 were in common to all three locations (Paterson *et al.,* 1991). However lower environmental effect was reported in maize grown in six locations (Stuber *et al.,* 1992) where the proportion of common QTLs was higher.

Thus, it is slightly paradoxical to search for QTLs related to environmental responses, since the so-called environmental variance is rather a residual variance left after elimination of the genetic variance. It means that environmental variance includes not only the proper climatic variance due to various stresses but also a large part of the experimental error encountered in experiment. A careful experimental design with appropriate replication is thus necessary in searching for QTLs related to environmental stresses. Only a few experiments have been reported, in this respect.

Important outputs expected from the QTL analysis are the presence or the absence of co-location between QTL traits related to the same physiological process (e.g. grain filling, stress response) and the co-location of QTLs with candidate genes. A co-location means co-segregation, which may be used by the plant breeder but also by the biotechnologist. Co-locations of QTLs for several trait may reveal co-regulation of expression, if, furthermore, they co-locate with a known function gene, this gene may become a candidate gene for explaining the phenotypic correlation's between traits (review, Prioul *et al.,* 1997 and Prioul *et al.,* 1999).

This approach is presently illustrated by field experiments examining agronomic and grains composition traits and by glasshouse experiments aimed at a more accurate analysis of the effect of various stress, mainly on young plants. Different kinds of responses were examined: yield, root and shoot growth, photosynthetic gas exchange, xylem abscisic acid (ABA) content, carbohydrate content and activity of carbohydrate degradation enzymes (invertases, sucrose synthase and ADP-glucose pyrophosphorylase).

SOME ELEMENTS OF METHODOLOGY

The principle and methodology for QTL detection has been described thoroughly in Prioul *et al.,* (1997). Briefly, the main prerequisites for QTL detection are firstly a segregant population of lines derived from two homozygous parents. The minimum number of lines to be analysed is 100, in order to obtain enough precision. The segregant lines may be F2, pooled F3, double-haploids or recombinant inbred lines. The recombinant inbred lines are obtained by at least 4 selfings from F2, by single seed descent; they have the advantage of being nearly homozygous. The second requirement is a dense mapping of the population in order to get a saturated map i.e. where the number of linkage groups is equal to the number of chromosomes. Under these conditions it is possible, for each marker, to know the genotype (either parent 1 or parent 2) in each of the segregant line.

The principle of QTL detection is to measure the traits under study in the totality of the segregant population and then to examine locus by locus if the mean value of the trait is dependent upon the genotype at this locus. If the result of the statistics is significant, then a QTL explaining the trait variability is located in the vicinity of the marker. This point method based on ANOVA, tends to be replaced by interval mapping which allows location between marker (like MAPMAKER-QTL). A more sophisticated method named composite interval mapping (CIM) combined interval mapping and the introduction of co-factors which increases the sensitivity of detection but also increased the risk of false-positive QTLs (Séne *et al.,* 2000). In addition to the location, a most important parameter issued from the calculation is the R^2 value representing the percentage of the phenotypical variance explained by the QTL.

QTLS AND ABIOTIC STRESS

High temperature (thermotolerance)

Although most data on QTL and stress deals with water stress, one of the earlier experiment aimed at QTL detection for traits related to climate adaptation dealt with pollen thermotolerance (Frova et Sari-Gorla 1994), using a rather small 45 recombinant inbred line population (RIL). Five and six QTLs were detected for pollen germinability and pollen tube growth, respectively.

Drought conditions at vegetative stages

Root system, morphology and shoot traits in rice

Root morphology is considered to be important for drought avoidance/tolerance and, Champoux *et al.,* (1995) have undertaken mapping of five root-related QTLs. The experiments were performed with on rice grown in the glasshouse (203 RILs in triplicate) and in the field. Twelve of the fourteen chromosomal regions having QTLs associated with field drought avoidance/tolerance also contained QTLs associated with root morphology. Most of the identified QTLs were associated with root thickness, root/shoot ratio, and root dry weight per tiller. Root thickness and root dry weight per tiller were the characters found to be the least influenced by environmental differences under greenhouse conditions. Correlations of greenhouse experiments with field drought avoidance/tolerance were significant but not highly predictable. These first results encouraged the authors to select alleles at marker loci associated with the root phenotypes to obtain drought-resistant cultivars. Further studies on the same rice population, but on a much-reduced sample (52 RILs), examined the osmotic adjustment and dehydration tolerance traits more closely. One major locus, associated with osmotic adjustment, appeared to be homeologous to a similar trait in wheat. In the population, osmotic adjustment and dehydration tolerance were negatively correlated to root morphological traits associated to dehydration tolerance, but high osmotic adjustment and dehydration tolerance were associated with one parental allele. This raises the hope for breaking the unfavorable linkage between osmotic adjustment and some root traits (Lilley *et al.,* 1996).

A more recent study based on root morphology, but not considering water stress, utilised 105 rice lines from a double-haploid population. Yadav *et al.,* (1997) noticed that most of the QTLs were concentrated in fairly compact regions on chromosomes 1, 2, 3, 6, 7, 8 and 9. Individual QTLs accounted for 4 to 22% of trait variation. Interaction of QTLs between markers was detected, mostly on different chromosomes, showing antagonistic effects. A comparison with another population showed that one to three QTLs per trait were recovered.

QTLs for shoot traits related to drought avoidance were recently considered by Courtois *et al.,* (2000), in upland rice. Leaf rolling, leaf drying, relative water content and relative growth rate were measured in two sites on 85-105 doubled haploid lines, under different drought intensities. Some of the QTLs were common across traits. Among the eleven possible QTLs for leaf rolling, three QTLs (on chromosomes 1, 5 and 9) were common across the three trials, and four additional QTLs (on chromosomes 3, 4 and 9) were common across two trials. One QTL on chromosome 4 for leaf drying and one QTL on chromosome 1 for relative water content were common across two trials, while no common QTL was identified for relative growth rate under stress. Some of the QTLs detected for leaf rolling, leaf drying and relative water content mapped in the same places as QTLs controlling root morphology, which were identified in a previous study involving the same population. Some QTLs identified in this study were also located similarly to other QTLs for leaf rolling as reported from other populations.

Osmotic adjustment in rice

As already mentioned for rice (Lilley *et al.,* 1996) and as reviewed by Zhang *et al.,* (2000), osmotic adjustment (OA) may play a role in drought tolerance, by providing the possibility of maintaining cell turgor when water potential decreases. Teulat *et al.,* (1998) addressed this problem in barley using a QTL approach. A set of barley recombinant inbred lines, derived from a cross between Tadmor (drought tolerant) and Er/Apm (susceptible), were grown in a growth chamber, at an early growth stage, under two soil moistures (14% and 100% of field capacity, respectively). A linkage map was constructed with 167 RILs genotyped with 78 RFLP, 32 RAPD and three morphological markers. Despite strong environmental effects acting on the traits (heritability ranged from 0.04 to 0.44), interval-mapping and single-marker ANOVA allowed the detection of three QTLs for relative water content (RWC), four QTLs for osmotic potential, two QTLs for osmotic potential at full turgor and one QTL for osmotic adjustment at a 14% soil moisture. For the irrigated treatment, only two QTLs were detected: one for RWC and one for osmotic potential at full turgor. Two chromosomal regions were involved in several OA-related trait variations and could be considered as regions controlling OA; they were located on chromosome 1 (7H) and chromosome 6 (6H), whereas other regions were specific for one trait. No major QTL was found. However, the genomic region involved in OA-related traits on chromosome 1 (7H) in barley seemed to be conserved for OA variation among cereals. Epistatic effects, with or without additive effects, were observed on the traits.

Leaf and ABA in maize

Most studies on drought effects have dealt with vegetative organs, especially leaves. Water deprivation classically produces two classes of reactions, depending on the tissue age: 1) rapid cessation of growth, in young growing organs, 2) decline of photosynthetic rate and alteration of sucrose metabolism, in mature leaves. ABA emitted by root tips and circulating in the xylem sap plays an important triggering role in these responses. Faced with numerous interrelated responses, it is difficult to establish causal links, which explain the whole plant response from a few, supposedly, key elements. This question is especially important for crop plants, since the identification of these key elements would provide a tool for improving adaptation or tolerance to stress. From this view, the QTL approach is the most appropriate since it enables the dissection of correlation between traits. QTL co-location could also provide information on the linkage between traits at different levels of integration (enzyme activity to gas exchange or leaf growth).

A first attempt to detect ABA QTLs in relation to water stress, was undertaken by Quarrie *et al.*, (1994a and b), in wheat and maize F2 populations. Although marker density was rather low (32 in maize), they demonstrated the feasibility of the method. A more thorough analysis by Lebreton *et al.*, (1995) concentrated on 81 maize F2 plants and 84 markers, and the problem was re-examined considering other traits, in addition to

ABA: stomatal conductance, water potential, turgor, root number, root pulling force, chlorophyll fluorescence and anthesis date. Most of the ABA QTLs were confirmed with the increased marker number e.g. on chromosomes 1, 3, 6, 7 and 8, but a new QTL appeared on chromosome 2. When considering all of the traits, a major result was that QTLs for different traits tended to co-localise, thus forming clusters. For example, in the middle of the short arm of chromosome 1, near the umc11 marker, 4 QTLs were mapped: root number, root pulling force, xylem ABA and stomatal conductance. Similarly, QTL for at least 4 traits were co-localised on chromosomes 3 and 7.

In order to try to sort out the causal relationship, Lebreton *et al.*, (1995) proposed a sophisticated method based upon the comparison of the two possible regression lines, obtained when plotting the allele effect for two traits at the 12 markers yielding the largest difference. If trait1/trait2 regression was more significant than trait2/trait1 regression it was assumed that trait2 is controlling trait1. For example $r = -0.67^{*}$ for conductance/ABA and $r = -0.43^{NS}$ for the reciprocal regression could mean that ABA is more likely to control conductance than the opposite. Although that result is consistent with physiological knowledge, the method is not unequivocal, since on theoretical grounds it does not take into account the possible difference in variability of each trait.

Another way to dissect the correlation between traits was proposed by Quarrie *et al.*, (1997) from interspecific comparisons. The principle is that linked genes in one species are likely to be also linked in others. In order to check if the association between gene and QTL, or between genes is not artefactual, the map positions of the homologous genes located at the position under study have to be compared. Quarrie *et al.*, (1997) aimed to re-analyse previous data showing that, in detached rice leaves a negative correlation takes place between ABA accumulation and leaf size (smaller leaves made more ABA). From QTL co-location and the fact that only one ABA QTL and one leaf size QTL was coincident with the expected opposite allele effect, they concluded that association between ABA and leaf size was more likely due to genetic linkage rather than to a direct effect of leaf size on ABA accumulation or vice versa. Special interest was drawn on *Vrn1*, the major verbalisation responsive gene in wheat. Additionally this gene was associated with major effect QTLs on leaf size, tiller number and ABA accumulation. However the homologous gene location in rice did not show any coincidence with the rice ABA and leaf morphology QTLs.

ABA QTLs have also been sought in relation to water stress response in field experiments. Tuberosa *et al.*, (1998) measured L-ABA concentration in leaf samples at two stages (stem elongation and anthesis) in two years. Although neither the number of lines (80) nor the number of map markers (106) were very high they detected sixteen different QTLs, although only four of them were consistent between samples. These four QTLs accounted for 66% of the phenotypic variance. The alleles, which increased L-ABA were contributed by the high ABA parent (Os420). The two most important QTLs were mapped on chromosome 2 near csu133 and csu109a. The effects associated with the QTL near csu133 were more pronounced near anthesis. The support intervals of the four primary QTLs for L-ABA did not overlap the presumed map position of mutants impaired in ABA biosynthesis. This study was also measured stomatal conductance, leaf temperature, leaf relative water content, anthesis-silking interval and grain yield (Sanguineti *et al.,* 1999). The analysis of the effects of each QTL region on the investigated traits indicated that L-ABA mainly represented an indicator of the level of drought stress experienced by the plant at the time of sampling, because an increase in L-ABA was most commonly associated with a decrease in both stomatal conductance and grain yield as well as an increase in leaf temperature. However an opposite result was observed at one QTL region on chromosome 7 near the RFLP locus asg8, thus suggesting a pleiotropic effect.

Cell-membrane stability in rice

Cell-membrane stability (CMS), which is assumed to be of great importance in preventing ion leakage under stress conditions, was considered by Tripathy *et al.,* (2000), who presented the first report using CMS for QTL mapping. They used this trait for analysing drought tolerance in rice. A 104 rice (*Oryza sativa* L.) doubled haploid (DH) lines, derived from a cross between CT9993-5-10-1-M and IR62266-42-6-2, were studied in a greenhouse; on 50-day-old plants a progressive stress was applied by withholding water. The leaf samples were collected from both control (well watered) and stressed plants (at 60-65% of RWC), and the standard test for CMS was carried out. Although no significant difference was observed in RWC between the two parental lines, as well as among the 104 lines under stress conditions, the CMS in the same lines was significantly different. It indicated that the variation in CMS was genotypic in nature, the heritability being 34%. A linkage map of this population comprising of 145 RFLPs, 153 AFLPs and 17 microsatellite markers was used for QTL analysis. Composite interval mapping identified nine putative QTLs for CMS located on chromosomes 1, 3, 7, 8, 9, 11 and 12. The amount of phenotypic variation that was explained by individual QTLs ranged from 13.4% to 42.1%. Four significant ($P<0.05$) pairs of digenic interactions between the detected QTLs for CMS were observed. The results open the possibility of using CMS as a selection trait in breeding for the improvement of drought tolerance in rice.

Drought conditions at maturity stage and QTLs for yield components

Maize

Maize yield is much reduced when a water stress period occurs at flowering time. Thus it is a major challenge in many growing areas to improve yield under drought conditions. However, it is a difficult issue, since drought tolerance traits are obviously multigenic and partly correlated. In order to tackle the problem, Ribaut *et al.*, (1996) have undertaken an extensive QTL approach, with the final aim of developing marker-assisted selection (MAS) strategies. Three growth conditions were established: well watered, intermediate stress and severe stress, over two years. The measured agronomic traits were anthesis-silking interval (ASI), male and female flowering, grain yield (GY), ear number, kernel number and 100-kernel weight. Two tropical inbred lines, presenting contrasting response to drought, were used as parental lines to produce 234 F2 plants which were selfed to obtain the F3 plants used for most measurements. The 234 F2 lines were genotyped at 142 loci.

The first consequence of water stress, was an increase in the genetic variance of ASI and of broad sense heritability of all flowering dates. The numerous QTLs detected for those traits accounted for 47-48% of the phenotypic variance. Under water-stressed conditions, four QTLs were common for the male and female flowering dates, one for ASI and male flowering and four for the ASI and female flowering. The number of co-locations was related to the level of correlation between traits. ASI was a transgressive trait, which means that the extreme values in the lines are lower or higher than the parental values. This observation, which is very frequently made for numerous traits, is useful when seeking improved performance. The drought-susceptible parent contributed alleles reducing ASI (4 days) at two QTL positions. Alleles contributed by the resistant line at the other four QTLs were responsible for a 7-day reduction of ASI. These four QTLs represented around 9% of the linkage map, and were stable over years and stress levels. It is argued by Ribaut *et al.*, (1996), that MAS based on ASI QTLs is a powerful tool for improving drought tolerance of tropical maize inbred lines.

In the second part of their study, Ribaut *et al.*, (1997) analysed yield components: grains yield (GY), ear number, kernel number and 100-kernel weight. Drought resulted in a 60% decrease of GY under severe stress. The families that performed best under control conditions were found to be proportionately more affected by stress and the yield reductions due to severe stress were inversely proportional to the performance under drought. This means that selection for a yield improvement under control conditions would not be very effective for yield improvement under drought. No QTL for yield component accounted for more than 13% of phenotypic variance. Another important observation was that QTL positions were not well conserved across the water regime. This was qualified as "inconsistency" by Ribaut *et al.*, (1997) but it may be more likely an indication that stress specific genes are encountered (see below).

The use of complex interval mapping (CIM) allowed the evaluation of QTL-by-environment interactions (Q x E) and could thus identify 'stable' QTLs across drought environments. Two such QTLs for GY, on chromosomes 1 and 10, coincided with two stable QTLs for kernel number. Moreover, four genomic regions were identified for the expression of both GY and the anthesis-silking interval (ASI). In three of these, the allelic contributions were for short ASI and GY increase, while the allelic contribution for short ASI on chromosome 10 corresponded to a yield reduction. From these results, Ribaut *et al.*, (1997) concluded that to improve yield under drought, through marker-assisted selection (MAS) one should combine the "best" QTLs involved in yield components with the "best" involved in ASI. These QTLs should be stable across target environments and represent the largest possible percentage of the phenotypic variance.

A similar analysis to that of Ribaut *et al.*, (1996, 1997) was made in maize genotypes adapted to temperate areas by Frova *et al.*, (1999) and Sari-Gorla *et al.*, (1999). The measurements of yield components were ear length, ear weight, kernel weight, kernel number, 50-kernel weight and traits related to development and flowering (ASI, flowering dates, plant height). A 142 recombinant inbred population of 142 families originating from a B73xH99 crossing, genotyped at 173 loci (RFLPs, microsatellites and AFLPs) was evaluated in well-watered and water-stressed conditions. A drought tolerance index was calculated as the ratio between the mean value of the traits in the two environments. For the yield component traits, a highly positive correlation was found over the two water regimes, and more than 50% of the quantitative trait loci (QTLs) were in common, the direction of the allelic contribution was always consistent, the allele increasing the trait value being mostly from line B73. Several QTLs were common to two or more traits. For the tolerance index, however, most of the QTLs were specific for a single component and different from these controlling the basic traits; in addition, a large proportion of the alleles increasing tolerance were provided by line H99.

In their analysis of developmental data, Sari-Gorla *et al.*, (1999) not only mapped the corresponding QTLs, but also compared the chromosomal regions where loci for drought tolerance related to plant development and flowering are located. Linkage analysis revealed, for male flowering time and plant height, that most of the QTLs detected were the same under control and stress conditions. In contrast, with respect to female flowering time and ASI, diverse QTLs appeared to be expressed either under control conditions or under stress. All of the QTLs conferring tolerance to drought were located in a different chromosome position as compared to the map position of the factors controlling the trait *per se*. This led Sari-Gorla *et al.*, (1999) to suggest that plant tolerance, in its different components, is not attributable to the presence of favourable allelic combinations controlling the trait, but is based on physiological characteristics not directly associated with the control of the character. This concept must be considered carefully, since it has large implications for defining the strategy of marker assisted selection. A possible logical assumption for explaining the observed disconnection between the so-called factors conferring drought selection and the QTLs may be that these factors are polygenically controlled and that the controlling locus differs depending on the conditions, as also observed by Pelleschi *et al.*, (1999) in the case of invertase.

Sorghum

Sorghum is one of the most drought tolerant grain crops; thus it may be considered as a model for evaluating the resistance mechanisms. Tuinstra *et al.*, (1997) analysed 98 recombinant inbred sorghum lines developed from a cross between two contrasting lines, TX7078 (pre-flowering-tolerant, post-flowering susceptible) and B35 (pre-flowering susceptible, post-flowering-tolerant). The population was characterised under drought and non-drought conditions for traits associated with post-flowering drought tolerance and with components of grain development. QTLs were mapped in 13 genome regions associated with one or more traits of post-flowering drought tolerance. Two QTLs were identified with major effects on yield and 'staygreen' under post-flowering drought. These loci were also associated with yield under fully irrigated conditions suggesting that these tolerance loci have pleiotropic effects on yield under non-drought conditions. QTL analysis indicated many loci that were associated with both rate and duration of grain development. High rate and short duration of grain development were generally associated with larger seed size, but only two of these loci were associated with differences in stability of performance under drought. These results differ notably from those observed in maize (Ribaut *et al.*, 1996, 1997) where the QTLs for yield components were not located at the same position in control non-drought or drought conditions.

The importance of the staygreen character as a post flowering resistance trait was confirmed, in *sorghum* grain, by Crasta *et al.*, (1999). QTLs controlling premature senescence and maturity were measured on recombinant inbred lines, scored with 142 RFLP markers, in order to investigate their possible association under post-flowering drought stress. The RILs and their parental lines were evaluated in four environments.

Simple interval mapping identified seven staygreen QTLs and two maturity QTLs. Three major staygreen QTLs contributed to 42% of the phenotypic variability (LOD 9.0) and four minor QTLs significantly contributed to an additional 25% of the phenotypic variability in staygreen ratings. Two maturity QTLs contributed to 40% and 17% (LOD 10.0 and 4.9) of the phenotypic variability respectively. Composite interval mapping confirmed the above results with an additional analysis of the QTL x environment interaction. Although staygreen ratings were significantly correlated ($r = 0.22$, $P < 0.05$) with maturity, six of the seven staygreen QTLs were independent of the QTLs influencing maturity. Similarly, one maturity QTL was independent of the staygreen QTLs. One staygreen QTL, however, mapped in the vicinity of a maturity QTL, and all markers in the vicinity of the independent maturity QTL were significantly ($P < 0.1$) correlated with staygreen ratings. The molecular genetic analysis of the QTLs influencing staygreen and maturity, together with the association between these two inversely related traits, provides a basis for further study of the underlying physiological mechanisms and for improving drought resistance in plants by pyramiding the favourable QTLs.

The identification of genetic factors involved in staygreen phenotype in relation to plant response to drought in sorghum was recently re-investigated by Xu *et al.*, (2000), with the objective of reducing plant senescence during post flowering stress. QTLs that control the staygreen and chlorophyll content were detected by using a RFLP map, developed from a recombinant inbred line population. Four staygreen QTLs were identified and located on three linkage groups (2 on group A and one on groups Q and J, respectively). The 3 QTLs for chlorophyll content explained 25-30% of the phenotypic variability while the 2 staygreen QTLs on group A explained 13-20% and 20-30% respectively, in all trials at different locations in two years. The three chlorophyll QTLs coincided with three staygreen QTL regions (group A and Q), accounting for 46% of the phenotypic variation. The staygreen QTLs on group A were also located in regions containing the genes for key photosynthetic enzymes, heat shock proteins and ABA-responsive gene. Such spatial arrangement shows that linkage group A is important for drought- and heat-stress tolerance, and yield production in sorghum. This study suggests that high-resolution mapping and cloning of the consistent staygreen QTLs may help to develop drought-resistant hybrids.

Salinity

Tomato plants are very sensitive to salinity, both at germination and at vegetative stages, but wild related species show salt tolerance. Foolad *et al.*, (1998, 1999) aimed to introduce tolerance QTLs in inbred backcrossed lines resulting from an interspecific cross between a salt-sensitive *Lycopersicum esculentum* and a salt tolerant *Lycopersicum pimpinellifolium.* One hundred and ninety families were analysed from the original backcross population, mapped with 151 RFLP markers. Salt stress was applied by a 150 mM NaCl+15 mM $CaCl_2$ solution (- 0.85 MPa). Seed germination and vegetative growth traits were measured. The results indicated the presence of a small but significant correlation ($r = -0.22$, $p < 0.05$) between the rate of seed germination and the percentage of plant survival under salt stress. Seven and five QTLs were identified for salt tolerance during seed germination and vegetative growth, respectively. While in most cases the location of QTLs for germination was different from that for vegetative growth, there were some coincidences in QTL locations; this was consistent with the small phenotypic correlation observed between the two traits. This means, in practical terms, that simultaneous improvement of salt tolerance for germination and vegetative growth should be possible. Seven chromosomal locations with significant effects on salt tolerance were identified. The *L. pimpinellifolium* accession had favourable QTL alleles at six locations. The percentage of phenotypic variation that explained by individual QTLs ranged from 6.5 to 15.6% and the cumulative action of all significant QTLs accounted for 44.5 % of the total phenotypic variance. In addition, a total of 12 pairwise epistatic interactions were identified, including four between QTL-linked and QTL-unlinked regions and eight between QTL-unlinked regions. Transgressive phenotypes were observed in the direction of salt sensitivity. Foolad *et al.,* (1998) concluded that the high correspondence between the phenotypes of the extreme families and their QTL genotypes indicated that tomato salt tolerance could be improved by marker-assisted selection using interspecific variation.

Citrus is another salt sensitive species. As a first step in identifying QTLs having an impact on stress tolerance, Tozlu *et al.,* (1999) considered the effects of 40 mM NaCl on the parental species *Poncirus trifoliata, Citrus grandis* and their F-1 and back-cross progenies. Na^+ and Cl^- analyses in the different tissues provided 38 traits in both salinised and non-salinised conditions. A total of 73 so-called potential QTLs were mapped using a fairly stringent LOD score (> 3.0). Seventeen regions of interest were detected, 8 of them with large effect QTLs. Many QTLs for Na^+ accumulation and Cl^-/Na^+ ratios formed clusters on separate linkage groups.

From the example of tolerance to salinity Flowers *et al.,* (2000) addressed a more general question of the applicability, to any population, of a marker found to be associated to a QTL in a given population. In the absence of adequate candidate genes for salt tolerance, a quantitative trait locus/marker-assisted selection approach has been used. Putative markers for ion transport and selectivity, identified from analysis of AFLPs, had been discovered within a custom-made mapping population of rice. However none of these markers showed any association with similar traits in a closely related population of recombinant inbred lines or in selections of a cultivar. This result cautions against any expectation of a general applicability of markers for physiological traits. Flowers *et al.*, (2000) concluded that a direct knowledge of the genes involved is needed. A way to identify and validate those genes of interest would be to find conditions were they are differentially affected by the experimental treatments.

CANDIDATE GENE APPROACH

Introduction to the candidate gene approach

As mentioned previously by several authors, the use of a QTL as selection trait may be hazardous if the marker is not very close to the actual gene contributing to the trait variation. In fact, the larger the distance between the marker and the pertinent locus, the higher is the chance of breaking the linkage upon recombination. Thus, the best situation would be if the marker was the gene itself. The search for such gene, explaining the nature of the variability of a trait frequently follows a "candidate gene approach". This approach generally consists in two phases. The suggestion of a 'candidate gene' from the examination of the genes mapping around the QTL position in order to choose the gene which is most likely to be related to the trait. Second the validation of the candidate by showing that its molecular polymorphism has an effect on the trait genetic variability. As described, this is known as the positional candidate gene approach. Another non-exclusive method, named "functional candidate gene approach" is based on the *a priori* choice of gene(s), which may be functionally related to the trait. This approach which may bypass the QTL step, is largely used in humans. A correlation between the trait value and the allelic polymorphism of the candidate provides a strong argument in favour of the value of the candidate (see de Vienne *et al.*, 1999 for discussion of methods, and Prioul *et al.*, 1999 for an example). In plants an example of the functional approach is given by the role of the polymorphism in p1, a gene encoding a transcription factor, acting on the maysin content of maize silks (Byrne *et al.*, 1996).

The positional approach may benefit from the use of near-isogenic lines (NILs) which only differ by the chromosome zone of interest. The necessary backcrossing to obtain these lines reduces the size of the zone, through recombination and may allow direct identification of the gene. In the most favourable case (e.g. Fridman *et al.*, 2000), it was demonstrated that the favourable allele effect, in a QTL for glucose and fructose content in tomato, originated from a single nucleotide change in a 484 bp region spanning an exon and intron of a fruit-specific apoplastic invertase. The 'landing precision' was much favoured by the fact that this region was a high recombination hot spot. An important point in relation to this work, is that QTLs were not obtained using RILs but from substitution lines derived from systematic introgression of chromosome fragments of the wild species (*Lycopersicon pennellii*) into the cultivated species through repeated back-crosses and selection of the lines bearing different chromosome fragments. Each line is near-isogenic to the cultivated parent line with exception of the introgressed fragment. The position of this fragment is mapped, and the QTL localisation is simply obtained by the identification of those lines presenting a genotype effect for the trait.

Biochemical QTLs and candidate genes

QTLs for enzyme activities

The actual genes controlling agronomic traits are difficult to identify, whereas the candidate genes involved in biochemical quantitative traits as enzyme activities, substrates or products of reaction, are *a priori* easier for suggesting QTLs. For example, when QTLs for an enzyme activity are detected, one obvious question is whether one of these QTLs co-located with the locus of the structural gene for that enzyme? This approach was first validated, in a preliminary study with a rather small population (65 RILs) by Causse *et al.*, (1995), who measured the activity of key enzymes of carbohydrate metabolism and the concentration of their substrates and products (sucrose, glucose, fructose and starch) in source and sink maize leaves. For each of the measured enzymes, i.e. ADP-glucose pyro-phosphorylase, sucrose synthase (sink), sucrose-phosphate-synthase and two invertases (sink), the corresponding cDNAs were available, and the corresponding loci were mapped in the recombinant inbred lines used for QTL detection. Measurements made at the 3- or 4-leaf stage displayed several QTLs for each trait. Apparent co-locations occurred between QTLs for activities, substrates or products and one of the corresponding structural gene: sucrose-phosphate-synthase activity near *Sps* locus on chromosome 8, and sucrose synthase activity near *Sh1* locus on chromosome 9.

A test on the parental lines further showed that, upon water deprivation, one of the earlier response is an upsurge of the vacuolar invertase activity which is synchronous with an increase of glucose and fructose, whereas photosynthetic rate decreased slightly later (Pelleschi *et al.*, 1997).

This response was closely related to the mRNA level for only one of the invertase genes (*Ivr2*), encoding a vacuolar isoform. As the vacuolar invertase response was different between the two parents, this opened the possibility of searching for QTLs related to the water stress effect of these biochemical traits. On the same plants photosynthetic traits, leaf water status and leaf shoot and root growth traits were measured. Up to 120 RILs were grown in a glasshouse up to adult fourth leaf stage; then watering was withheld for 9 days, and the measurements were made. The cultures were repeated four times. The mean values were used for all calculations, thus the glasshouse effect was minimised.

Although QTLs were detected over the 10 chromosomes, they were not evenly distributed. Apart from chromosomes 2, 3, 8 and 10, QTL clusters were noted on the other 6 chromosomes. A cluster was defined as a region where more than 3 QTLs were within 10 cM. Following this definition, it is possible to define more than 12 of these clusters, frequently consisting of QTLs from different phenotypic classes (carbohydrate metabolism, photosynthesis, water status, growth...). For example two QTL clusters, comprising 6 (bin1.10) and 12 (bin7.01) QTLs from four classes, were present on chromosomes 1 and 7, and six clusters consisting of three classes were located on chromosomes 1, 4, 7, 9. Among all the clusters, five of them may be assigned to stress responses on chromosome 1 (bin1.06 and 1.10), 4 (bin 4.08), 7 (bin 7.04) and 9 (bin 9.08). Conversely, QTLs for a lot of morphological traits (leaf length and width, leaf number, leaf area, plant height, relative water content and transpiration) mapped at the same location under both conditions. These QTLs may be called generalist and may be assumed as house keeping loci. A possible explanation for the clustering could be that the same genes, or closely located loci, are controlling the traits. This is predictable for traits belonging to the same phenotypical group, for example leaf length, width and area, but it is less obvious when traits originate from different groups. In the later case one could assume that some master genes control traits at different organisation levels. Another general observation deals with cluster composition in relation to trait groups. A greater proportion of the growth traits were associated to carbohydrate metabolism traits than to photosynthetic ones (Pelleschi, 1997). This is consistent with the fact that selection for photosynthetic rate or Rubisco does not generally give much improvement in crop yield (Crosbie *et al.,* 1981), whereas alterations of source-sink related processes are more efficient.

Plant responses to water stress may be analysed by comparing QTL distribution under stressed and non-stressed conditions for similar traits. A first observation is that a much larger number of QTLs is detected in stressed plants than in control: 52% of QTLs were water-stress specific and only 16% were in common to both treatments. Stress specific clusters were noted on chromosomes 1 (bin 1.06, 1.09, 1.10), 4 (bin 4.08), 5 (bin 5.04), 7 (bin 7.04) and 9 (bin 9.08). Some consist of physiological traits only (1.06 and 4.08), others of a majority of growth traits (bin 1.09, 5.04) or a mixture of both (bin 1.10, 7.04, 9.08). This stress dependent distribution means that under water restriction a different set of genes is likely to be more important and could allow the plant to adapt its metabolism and growth to the new conditions. The clustering of QTLs from different classes, observed again, may suggest that the genes controlling these QTLs are unlikely to encode single and simple functions, since several functions are involved. The role of regulatory genes may be suggested. An apparent exception to this clustering of genes involved in the same response arises from the comparison of carbon dioxide uptake parameters and ABA content under stress conditions. Both traits should be correlated, since the rise of xylem ABA is assumed to produce stomatal closure. In fact, QTL co-locations were very rare as also noted by Lebreton *et al.,* (1995). A possible explanation in Pelleschi's experiment (1997) could originate from the sampling protocol. Each plant was sampled 9 days after water interruption by which time the ABA concentration effective for stomatal closure probably had been reached already.

Taking advantage of a RFLP map, based on markers largely consisting of known functions, especially in carbohydrate metabolism (Causse *et al.,* 1996), a comparison of the locations for enzyme activity QTLs and for the corresponding structural genes showed some interesting co-locations, suggesting these genes are candidate genes. Three examples may be given of such candidate genes. On chromosome 1, a QTL for the leaf ADP-glucose pyrophosphorylase activity mapped at the same position (bin 1.07) as the locus of the leaf-specific gene for the small subunit of the enzyme (Prioul *et al.,* 1994). On chromosome 8, the co-location mentioned by Causse *et al.,* (1995) between SPS activity and the *Sps* locus was confirmed at bin 8.06. The third main co-location was related to invertase and water stress. As mentioned earlier, one of the first responses to water stress in mature leaves is a large increase in vacuolar invertase and simultaneously in hexose content. This response is genotype dependent and largely variable in the RILs (transgression effect). As a consequence, several QTLs were obtained in both conditions (3 QTLs in control and 7 QTLs for water stress), on chromosomes 1, 2, and 5 to 10. The most interesting location is on chromosome 5 (bin 5.03), since one QTL for control and one for water stress explained 17% and 5% of variability, respectively, and both were located very close to the *Ivr2* gene locus which encodes a vacuolar invertase. Further analysis of the expression of this gene in the parent lines demonstrated it is the only invertase gene among the 6 known genes, either for cell-wall or for vacuolar forms to be inducible by water stress in leaves, to have the same induction pattern as the vacuolar invertase activity, to present the same genotype-dependent expression as the enzyme activity as well as induced in the same ratio. These reasons all supported *Ivr2* as a candidate gene for the invertase QTLs at bin. 5.03 (Peleschi et al 1999). Some other QTLs for invertase activity were found close to carbohydrate QTLs and some of them formed "stress clusters".

QTLs from 2-D protein analysis

Two-dimensional (2-D)-electrophoresis associated with computer-based analysis allows quantification of individual protein spot intensities. As the spots are commonly genetically variable, several hundreds spots are potentially available as quantitative traits. Moreover the dramatic progress in the sequencing of proteins excised from gel blots now allows the identification of more and more unknown proteins. Thus, it becomes possible to determine in parallel, the QTLs for the amount of a protein spot (PQLs, Protein Quantity Loci) and the QTLs for its activity, if it is an enzyme or for a related phenotypical trait. Water stress has a large effect on 2D-protein patterns (Riccardi *et al.,* 1998). Seventy-eight proteins out of a total of 413 showed a significant quantitative variation (increase or decrease), with 38 of them exhibiting a different expression in the two parent genotypes. Eleven proteins that increased by a factor of 1.3 to 5 in stressed plants, and 8 proteins detected only in stressed plants, were selected for internal amino acid microsequencing. By similarity search, 16 were found to be closely related to previously reported proteins. In addition to proteins already known to be involved in the response to water stress (e.g. RAB17 [Responsive to ABA]), several enzymes involved in basic metabolism, such as glycolysis and the Krebs cycle (e.g. enolase and triose phosphate isomerase), were identified, as well as several others, including caffeate O-methyltransferase, the induction of which could be related to lignification. The PQL methodology was illustrated by de Vienne *et al.,* (1999) on phosphoglycerate mutase and some proteins overexpressed under drought stress.

CONCLUSION

Considering different species and the different stress conditions, rather general observations are that QTLs are not evenly distributed over the genome and they frequently form clusters. This clustering not only appeared for correlated traits at a level of organisation, which could be trivial, but also between traits from different organisation levels. This observation confirmed of Khavkin and Coe (1997), who analysed the map distribution of over 800 QTLs for architecture, growth and development, grain yield and ABA content. For interpreting the significance of clustering, Khavkin and Coe (1997) recalled theoretical considerations provided more by 20 years ago by Demarly (1979). He introduced the "linkat concept" as "a set of loci which aggregated in a same chromosomal sector during species differentiation.

These linkats (= clusters) show strong epistasy and generally represent co-adapted functions". Each linkat would be protected against dissociation by a lower recombination rate and its content would be rich in gene duplication, thus maintaining proper function through generation. As our knowledge of genome organisation progresses, Demarly's long sight analysis fits remarkably well to what is uncovered.

Under stress conditions, new QTL clusters frequently showed up, indicating that new genes or groups of genes become more important in explaining the trait variability. This is the case for some regions of chromosomes 1 and 5, for example in maize submitted to water stress. This means that in selecting for improved stress resistance, the favourable allele will not be found at the same loci as in non-stress conditions. Although presently available data are not sufficient, it would be worth checking, in the same species submitted to various stresses (water stress, salinity...), if some stress QTLs would be in common. For example, some biochemical processes, like antioxidative pathways are involved in numerous stress conditions. In any case, a very important element for progress is the identification of the genes corresponding to the QTL. The candidate gene approach provides a means to this aim. For stress, this approach has been mainly limited to structural genes of the carbohydrate metabolism. Although a co-location per trait was observed for three enzymes most of the other QTLs were located elsewhere. These loci are likely involved in stress dependent regulation and thus it is of great importance to identify them.

REFERENCES

Byrne, P.F., Mcmullen, M.D., Snook, M.E., Musket, T.A., Theuri, J.M., Widstrom, N.W., Wiseman, B.R. and Coe, E.H. 1996. Quantitative trait loci and metabolic pathways: Genetic control of the concentration of maysin, a corn earworm resistance factor, in maize silks. *Proc. Natl. Acad. Sci. USA* 93, 8820-8825.

Causse M., Rocher, J.P., Henry, A.M., Charcosset, A. Prioul, J.L. and de Vienne, D. 1995. Genetic dissection of the relationship between carbon metabolism and early growth in maize, with emphasis on key-enzyme loci. *Mol. Breed.* 1, 259-272.

Causse M., Santoni, S., Damerval, C., Maurice, A., Charcosset, A., Deatrick, J. and de Vienne, D. 1996. A composite map of expressed sequences in maize. *Genome* 39, 418-432.

Courtois B., McLaren, G., Sinha, P.K., Prasad, K., Yadav, R. and Shen, L. 2000. Mapping QTLs associated with drought avoidance in upland rice. *Mol. Breed.* 6, 55-66.

Champoux M.C., Wang, G., Sarkarung, S., Mackill, D.J., Otoole, J.C., Huang, N. and McCouch, S.R. 1995. Locating genes associated with root morphology and drought avoidance in rice via linkage to molecular markers. *Theor. Appl. Genet.* 90, 969-981.

Crasta O.R., Xu, W.W., Rosenow, D.T., Mullet, J. and Nguyen, H.T. 1999. Mapping of post-flowering drought resistance traits in grain sorghum: association between QTLs influencing premature senescence and maturity. *Mol. Gen. Genet.* 262, 579-588.

Crosbie T.M., Pearce, R.B. and Mock, J.J. 1981. Selection for high CO_2 exchange rate among inbred lines of maize. *Crop Sci.* 21, 629-631.

Demarly Y. 1979. "The concept of linkat". In: *Proceedings of the Conference on Broadening Genetic Bases of Crops*, eds. A.C. Zeven and A.M. van Harten, pp 257-265. PUDOC, Wageningen.

de Vienne D., Leonardi, A., Damerval, C. and Zivy, M. 1999. Genetics of proteome variation for QTL characterization: application to drought-stress responses in maize. *J. Exp. Bot.* 50, 303-309.

Flowers T.J., Koyama, M. L., Flowers, S.A., Sudhakar, C., Singh K.P. and Yeo, A.R. 2000. QTL: their place in engineering tolerance of rice to salinity, *J. Exp. Bot.* 51, 99-106.

Foolad M. R., Chen F.Q. and Lin, G.Y. 1998. RFLP mapping of QTLs conferring salt tolerance during germination in an interspecific cross of tomato, *Theor. Appl. Genet.* 97, 1133-1144.

Foolad M.R. 1999. Comparison of salt tolerance during seed germination and vegetative growth in tomato by QTL mapping. *Genome* 42, 727-734.

Fridman, E., Pleban, T. and Zamir, D. 2000. A recombination hotspot delimits a wild-species quantitative trait locus for tomato sugar content to 484 bp within an invertase gene. *Proc. Natl. Acad. Sci. USA* 97, 4718-4723.

Frova C. and Sari-Gorla, M. 1994. Quantitative trait loci (QTLs) for pollen thermotolerance detected in maize. *Mol. Gen. Genet.* 245, 424-430.

Frova C., Krajewski, P., diFonzo, N., Villa, M. and Sari-Gorla, M. 1999. Genetic analysis of drought tolerance in maize by molecular markers I. Yield components. *Theor. Appl. Genet.* 99, 280-288.

Khavkin E. and Coe, E. 1997 Mapped genomic locations for developmental functions and QTLs reflect concerted groups in maize (*Zea mays*). *Theor. Appl. Genet.* 95, 343-352.

Lebreton C., Lazicjancic, V., Steed, A., Pekic, S. and Quarrie, S.A. 1995. Identification of QTL for drought responses in maize and their use in testing causal relationships between traits. *J. Exp. Bot.* 46, 853-865.

Lilley J.M., Ludlow, M.M., McCouch, S.R. and Otoole, J.C. 1996. Locating QTL for osmotic adjustment and dehydration tolerance in rice. *J. Exp. Bot.* 47, 1427-1436.

Paterson A.H., Lander, E.S., Hewitt, J.D., Paterson, S., Lincoln, S.E. and Tanksley, S.D. 1988. Resolution of quantitative traits into mendelian factors by using a complete linkage map of restriction fragment length polymorphisms. *Nature* 33, 721-726.

Paterson A.H., Damon, S., Hewitt, J.D., Zamir, D., Rabiowitch, H.D., Lincoln, S.E. Lander, E.S. and Tanksley, S.D. 1991. Mendelian factors underlying quantitative traits in tomato : comparison across species,generations, and environments. *Genetics* 127, 181-197.

Pelleschi S. 1997. Recherche de locus à effet quantitatif, liés au métabolisme glucidique au cours d'une contrainte hydrique, chez le maïs (*Zea mays* L.). *Thèse, Université de Paris-Sud* pp 1-163.

Pelleschi S., Rocher, J.P. and Prioul, J.L. 1997. Effect of water restriction on carbohydrate metabolism and photosynthesis in mature maize leaves. *Plant Cell Envir.* 20, 493-503.

Pelleschi S., Guy, S., Kim, J.Y., Pointe, C., Mahe, A., Barthes, L., Leonardi, A. and Prioul, J.L. 1999. Ivr2, a candidate gene for a QTL of vacuolar invertase activity in maize leaves. Gene-specific expression under water stress. *Plant Mol. Biol.* 39, 373-380.

Prioul J.L., Jeannette, E., Reyss, A., Gregory, N., Giroux, M., Hannah, L.C. and Causse, M. 1994. Expression of adp-glucose pyrophosphorylase in maize (*Zea mays* L.) grain and source leaf during grain filling. *Plant Physiol.* 104, 179-187.

Prioul J.L., Quarrie, S., Causse, M. and de Vienne, D. 1997. Dissecting complex physiological functions through the use of molecular quantitative genetics. *J. Exp. Bot.* 48, 1151-1163.

Prioul J.L., Pelleschi, S., Séne, M., Thévenot, C., Causse, M., de Vienne, D. and Leonardi, A. 1999. From QTLs for enzyme activity to candidate genes in maize. *J. Exp. Bot.* 50, 1281-1288.

Quarrie S., Steed, A., Lebreton, C., Guili, M., Calestani, C. and Marmirolli, N. 1994a. Location of a gene regulating drought-induced abscisic acid production in wheat and maize and associated physiological traits. *Russian J. Plant Physiol.* 41, 565-571.

Quarrie S.A., Gulli, M., Calestani, C., Steed, A. and Marmiroli, N. 1994b. Location of a gene regulating drought-induced abscisic acid production on the long arm of chromosome 5A of wheat. *Theor. Appl.Genet.* 89, 794-800.

Quarrie S.A., Laurie, D.A., Zhu, J.H., Lebreton, C., Semikhodskii, A., Steed, A., Witsenboer, H. and Calestani, C. 1997. QTL analysis to study the association between leaf size and abscisic acid accumulation in droughted rice leaves and comparisons across cereals. *Plant Mol. Biol.* 35, 155-165.

Ribaut J.M., Hoisington, D.A., Deutsch, J.A., Jiang, C. and Gonzalezdeleon, D. 1996. Identification of quantitative trait loci under drought conditions in tropical maize. 1. Flowering parameters and the anthesis-silking interval. *Theor. Appl. Genet.* 92, 905-914.

Ribaut J.M., Jiang, C., Gonzalesdeleon, D., Edmeades, G.O. and Hoisington, D.A. 1997. Identification of quantitative trait loci under drought conditions in tropical maize. 2.Yield components and marker-assisted selection strategies. *Theoret. Appl. Genet.* 94, 887-896.

Riccardi F., Gazeau, P., de Vienne, D. and Zivy, M. 1998. Protein changes in response to progressive water deficit in maize - quantitative variation and polypeptide identification. *Plant Physiol.* 117, 1253-1263.

Sanguineti M. C., Tuberosa, R., Landi, P., Salvi, S., Maccaferri, M., Casarini E. and Conti, S. 1999 QTL analysis of drought related traits and grain yield in relation to genetic variation for leaf abscisic acid concentration in field-grown maize, *J. Exp. Bot.* 50, 1289-1297.

Sari-Gorla M., Krajewski, P., DiFonzo, N., Villa, M. and Frova, C. 1999. Genetic analysis of drought tolerance in maize by molecular markers. II. Plant height and flowering. *Theoret. Appl. Genet.* 99, 289-295.

Séne M., Causse, M., Damerval, C., Thévenot, C. and Prioul, J.L. 2000. Quantitative trait loci affecting amylose, amylopectin and starch content in maize recombinant inbred lines. *Plant Physiol. Biochem.* 38, 459-472.

Stuber C.W., Lincoln, S.E., Wolff, D.W., Helentjaris, T. and Lander, E.S. 1992. Identification of genetic factors contributing to heterosis in a hybrid from two elite maize inbred using molecular marker. *Genetics* 132: 823-839.

Teulat B., This, D., Khairallah, M., Borries, C., Ragot, C., Sourdille, P., Leroy, P., Monneveux, P and Charrier, A. 1998. Several QTLs involved in osmotic adjustment trait variation in barley (*Hordeum vulgare* L.). *Theoret. Appl. Genet.* 96, 688-698.

Tozlu I., Guy C.L. and Moore, G.A. 1999. QTL analysis of Na+ and Cl- accumulation related traits in an intergeneric BC1 progeny of *Citrus* and *Poncirus* under saline and nonsaline environments. *Genome* 42, 692-705.

Tuberosa R., Sanguineti, M.C., Landi, P., Salvi, S., Casarini E. and Conti, S. 1998. RFLP mapping of quantitative trait loci controlling abscisic acid concentration in leaves of drought-stressed maize (*Zea mays* L.), *Theoret. Appl. Genet.* 97, 744-755

Tripathy J.N., Zhang, J., Robin, S., Nguyen T.T. and Nguyen, H.T. 2000. QTLs for cell-membrane stability mapped in rice (*Oryza sativa* L.) under drought stress, *Theoret. Appl. Genet.* 100, 1197-1202.

Tuinstra M.R., Grote, E.M., Goldsbrough, P.B. and Ejeta, G. 1997. Genetic analysis of post-flowering drought tolerance and components of grain development in *Sorghum bicolor* (L.) Moench. *Mol. Breed.* 3, 439-448.

Xu W.W., Subudhi, P.K., Crasta, O.R., Rosenow, D.T., Mullet J.E. and Nguyen, H.T. 2000. Molecular mapping of QTLs conferring stay-green in grain sorghum (*Sorghum bicolor* L. Moench), *Genome* 43, 461-469.

Yadav, R., Courtois, B., Huang, N., and Mclaren, G. 1997. Mapping genes controlling root morphology and root distribution in a doubled-haploid population of rice. *Theoret. Appl. Genet.* 94, 619-632.

Zhang, J.X., Klueva, N.Y., Wang, Z., Wu, R., Ho, T.H.D. and Nguyen, H.T. 2000. Genetic engineering for abiotic stress resistance in crop plants. *In Vitro Cell. & Develop. Biol. Plant.* 36, 108-114.

Chapter 6

MOLECULAR STRATEGIES TO OVERCOME SALT STRESS IN AGRICULTURE

Ilga Winicov

Department of Plant Biology, PO Box 871601, Arizona State University, Tempe, AZ 85287, USA.
winicov@asu.edu

INTRODUCTION

Salinity and drought are responsible for much of the yield reduction in agriculture throughout the world. Furthermore, continued salinization of arable land is becoming widespread because of poor local irrigation practices (Tanji, 1990), thus decreasing the yield from previously productive land. Improving plant resistance to salinity and drought stress, both of which lead to cellular osmotic and oxidative problems, is therefore a challenge to be overcome in order to feed the burgeoning world population. Increased salt-tolerance of crop plants would provide sustainable agriculture on marginal lands and could potentially even improve overall crop yield.

Although many of the molecular changes that occur in plants during osmotic stress are known, the understanding on how to utilise this knowledge to engineer plants with improved salt tolerance is still in the developmental stages. Biochemically it may be possible to overexpress the components of a physiological system that becomes limiting under stress conditions (Winicov, 1990) or provide protective molecules that neutralise biochemical effects of salinity stress. The magnitude of changes in gene expression of plants exposed to salt stress has become apparent with time as more and more gene transcripts are identified to be induced by salt/drought growth condition (Hasegawa *et al.*, 2000; Ingram and Bartels, 1996; Meyer *et al.*, 1990; Winicov, 1998). These findings are being confirmed on a large scale by both QTL and micro-array analysis of the genes involved in stress responses. This complex array of changes in gene expression is further

M.J. Hawkesford and P. Buchner (eds.),
Molecular Analysis of Plant Adaptation to the Environment, 103–129.

overlaid by developmental and tissue specific regulation of many of the stress induced genes and compounded by the fact that many of the responsive genes belong to multigene families. Salt stress inducible genes have been grouped relative to their physiologic or metabolic function as predicted from sequence homology with known proteins and are summarised in Table 1. Detailed information about genes belonging to specific classes of proteins has been reviewed (Hasegawa *et al.,* 2000; Ingram and Bartels, 1996; Winicov, 1998).

Table 1. Diversity of functional groups of genes/proteins activated in salt stress.

1.	Carbon metabolism and energy production/photosynthesis
2.	Cell wall/membrane structural components
3.	Osmoprotectants and molecular chaperones
4.	Water channel proteins
5.	Ion transport
6.	Oxidative stress defences
7.	Detoxifying enzymes
8.	Proteinases
9.	Hormone biosynthesis
10.	Proteins involved in signaling
11.	Transcription factors

While the characterisation to date has focused predominantly on genes induced rapidly by salt stress, little information is available on the expression of genes that are still induced after prolonged stress. However, the salt stress response has clearly an important temporal component to the repertoire of gene activation. Some of the genes are activated very rapidly and transiently, while others may show increased levels of expression for days and weeks in presence of continued salt stress. Examples of genes showing different temporal modes of expression are listed in Table 2. Some of the gene products induced rapidly could be involved in cascades of additional gene activation leading to improved survival and tolerance of saline conditions; others may be responding to generalised non-equilibrium cellular environment and metabolic imbalance common to abiotic stress. Many of the genes encoding proteins involved in signaling pathways and induced by salt-stress are also induced by drought, cold and plant hormones, suggesting interconnected signal pathways or shared functions for these gene products.

The following discussion will provide examples of strategies used to date to alleviate salt stress in plants, primarily based on overexpression of genes activated by salt stress that belong to some of the functional groups described in Table 1.

Table 2. Examples of temporal induction of mRNA accumulation by salt/drought stress.

Gene	Plant	Function	Induction[a]	After 24 hr	Reference
DREB1A	*Arabidopsis*	transcription factor	20 min	non detectable	Liu *et al.*, 1998
DREB2A	*Arabidopsis*	transcription factor	10 min	low	Liu *et al.*, 1998
rd29A	*Arabidopsis*	*lea* like protein	10 min	low	Liu *et al.*, 1998
ERDH	*Arabidopsis*	sugar transport	1 hr	non detectable	Kiyosue *et al.*, 1998
AtPiP5K1	*Arabidopsis*	phosphatidyl inositol kinase	1 hr	low	Mikami *et al.*, 1998
Em	rice	embryo protein	3 hr	decreasing	Bostock and Quatrano, 1992
rd22	*Arabidopsis*		20 min	low	Abe *et al.*, 1997
rd22BP	*Arabidopsis*	transcription factor	10 min	low	Abe *et al.*, 1997
proT2	*Arabidopsis*	proline transporter	4 hr	high	Rentsch *et al.*, 1996
P5CS	*Arabidopsis*	proline synthesis	1 hr	low	Yoshiba *et al.*, 1995
AtPTP1	*Arabidopsis*	protein tyrosine phosphatase	24 hr	decreases after 48 hr	Xu *et al.*, 1998
ATHK1	*Arabidopsis*	protein histidine kinase	10 min	low	Urao *et al.*, 1994
ATPLC1	*Arabidopsis*	phospholipase C	1 hr	high	Hirayama *et al.*, 1995
ARSK1	*Arabidopsis*	protein kinase	12 hr	high	Hwang and Goodman, 1995
ATCDPK1	*Arabidopsis*	protein kinase, Ca^{++}dependent	10 min	non detectable	Urao *et al.*, 1994
ATMEKK1	*Arabidopsis*	MAP kinase	1 hr	high	Mizoguchi *et al.*, 1996
ATMPK3	*Arabidopsis*	MAP kinase	1 hr	high	Mizoguchi *et al.*, 1996
ATPK19	*Arabidopsis*	ribosomal S6 protein kinase	1 hr	high	Mizoguchi *et al.*, 1996
pLE4	tomato	dehydrin	1 hr	low	Cohen and Bray, 1990
pLE16	tomato	nonspecific lipid transfer protein	2 hr	high	Cohen and Bray, 1990
pLE25	tomato	*lea*	2 hr	low	Cohen and Bray, 1990

a- Earliest experimentally measured increase in mRNA levels

In addition, recent focus on the potential for improving salinity-tolerance in crop plants has shifted from single gene manipulation in transgenic plants to an approach, which attempts to boost multiple gene expression of integrated stress defence pathways by manipulation of signaling pathways or transcription factors that participate in stress gene regulation.

PHENOTYPE TARGETED BREEDING/SELECTION FOR SALT STRESS RESISTANCE

Progress in conventional breeding efforts to increase salt tolerance in crop plants has been slow and the results mostly disappointing. With the advent of new techniques in germplasm characterisation with DNA markers, identification of RFLP tags for genomic regions associated with desirable traits, it was envisioned that marker assisted selection of quantitative trait loci (QTLs) would speed up the breeding process for crop plants with improved salt tolerance. The results to date have been less useful than initially predicted (Ribaut and Hoisington, 1998). The alternative strategy in phenotypic selection for improved salt-tolerance of crop plants has utilised regeneration of plants after selection for cellular salt-tolerance in culture (Winicov, 1991; Winicov, 1996), but these plants with heritably improved salt tolerance have not yet been tested under field conditions.

Quantitative trait loci (QTL) assisted breeding

Crop yield is a highly polygenic trait and efforts to map salt/drought stress affected genes in the context of yield have uncovered a highly complex picture in which the effects of individual regions is not easily defined, especially since interactions between regions may depend on other environmental factors.

Molecular technology of QTL mapping for salt tolerance or osmotic adjustment in drought has identified different chromosomal regions associated with ability to withstand these stresses in tomato (Foolad and Jones, 1993) rice (Lilley *et al.*, 1996), barley (Teulat *et al.*, 1998), wheat (Galiba *et al.*, 1992; Morgan and Tan, 1996) and maize (Lebreton *et al.*, 1995). In all species assessed so far, salt/drought tolerance appears to be a quantitative trait. Some of the mapped regions overlap between different plants, but different genes may be involved in the tolerance phenotype in different species and different genes have been identified even in the same species by different investigators (Galiba *et al.*, 1992; Morgan and Tan, 1996). The broad chromosomal distribution of dehydration stress genes from a single gene family, the dehydrins, has been demonstrated in wheat

(Werner-Fraczek and Close, 1998), and in this case allelic variation in one of the dehydrin genes was linked with a phenotype (Ismail *et al.,* 1999). Some of the more promising approaches in mapping such traits have developed near-isogenic lines for a biochemical marker, such as those for glycinebetaine accumulation in maize (Yang *et al.,* 1995), but evidence for a general pattern in marker distribution is still lacking. It is also not clear how much the QTLs identified under drought/salinity conditions are associated with yield under field conditions (Zhang *et al.,* 1999).

Marker assisted selection for quantitative traits such as yield has not provided anticipated results to date. In fact, while mapping of QTLs of various crop plants is being actively reported, data in which these markers have been used to monitor breeding progress for the desired phenotype of increased resistance to salt/drought stress remain scarce. The results are likely to depend on additive gene effects, dominance effects and complementary gene interactions. Example: alfalfa selection based on molecular marker diversity could show no consistent improvement on forage yield (Kidwell *et al.,* 1999), as was found with heritability of water-use efficiency traits (Ray 1999). Also disappointing have been the results in a closely related population of inbred recombinant lines of rice, in which putative markers for ion transport and selectivity, identified from amplified fragment length polymorphism (AFLP) analysis was used to analyse association with salinity tolerance (Flowers *et al.,* 2000). Development of new marker-assisted breeding strategies for improved resistance to salt stress will continue to emerge with development of molecular technologies able to accommodate information about the different QTL regions involved in salt/drought stress resistance and the increased number of plants which will have to be screened to accommodate such manipulations.

Selection at the cellular level and regeneration of plants with improved tolerance to salt stress.

An alternate strategy for salt stress resistance improvement in crop plants was initially developed in our laboratory for alfalfa (Winicov, 1991) and subsequently extended to long grained California grown rice L202 (Winicov, 1996). This strategy utilises the ability to select cell lines in callus culture at mutational frequencies, that have acquired the ability to grow at previously lethal concentrations of NaCl (Winicov *et al.,* 1989) and regenerate fertile plants that show improved salt tolerance compared to the parent plants from which the cells were derived. In each case the trait was passed through seed in a semi-dominant manner (Winicov, 1991; Winicov, 1996). These results demonstrated that cellular tolerance in both alfalfa and rice could be utilised at the whole plant level and is a viable method for obtaining plants with improved salt tolerance. When successful, the cellular selection and regeneration approach relies on selection of mutants optimised for continued survival and productive growth under saline conditions by the plant cell itself, but does not provide ready identification of the mutation that enables this phenotypic change. However, since we have shown that the mutation leads to salt inducible changes in regulation of many genes, the mutation is likely to be in a pathway that signals the transcriptional regulation of a group of functionally related genes. We have cloned and identified a number of the up-regulated genes in these mutant plants. The genes remain salt-inducible in the salt-tolerant alfalfa for months and years and are shown in Table 3, indicating the sustainable nature of this change in gene regulation. Our results were obtained from isogenic cell lines with a single step mutational event and provide additional support for the multigenic nature of salt-tolerance acquisition. However, differences in specific mRNA levels between cell cultures and the salt-sensitive parent and the salt-tolerant mutant plants also demonstrated that tissue specific regulation at the plant level can override the salt dependent regulation of gene expression observed in cultured cells (Winicov and Shirzadegan, 1997).

Table 3. Up-regulated and salt-inducible transcripts in salt-tolerant alfalfa.

Photosynthesis Related Transcripts	**Non-Photosynthesis Gene Transcripts**
PSI, nuclear and chloroplast	*Alfin1* - transcription factor
PSII, nuclear and chloroplast	*MsPRP2*-putative cell wall/membrane protein
Rubisco, nuclear and chloroplast	*H3cI* and *H3cII* (histone H3 isoforms)
ATPase, chloroplast	*pA18* (clathrin, small subunit)
Cyt F, chloroplast	

PSI, photosystem I; PSII, photosystem II; Rubisco, genes encoding large and small subunits of ribulose-bisphosphate carboxylase/oxygenase.

SINGLE GENE TARGETED STRATEGIES TO PROVIDE INCREASED RESISTANCE TO SALT STRESS

While salt induced gene activation has been demonstrated for genes belonging to all the functional groups listed in Table 1, only a limited number of genes have been tested in transgenic plants for their ability to provide increased resistance to salt stress. Current information related to the molecular mechanisms of salinity tolerance among bacteria, yeast and plants have been recently reviewed (Bohnert *et al.,* 1999; Hasegawa *et al.,* 2000; Shinozaki and Yamaguchi-Shinozaki, 1999) and provide details about genes belonging to the different functional groups listed in Table 1. For technical reasons, most of the tests for improved resistance have been short-term acute stress treatments, at high NaCl concentrations, measuring plant survival as evidence for improved salt tolerance. Although these tests for resistance to salt stress are instructive in an experimental sense, long-term stress resistance will ultimately have to be measured in the field under the conditions of one or more growing seasons.

Overexpression of genes to increase cellular concentrations of osmoprotectants

Compatible solutes or osmoprotectants occur in all organisms, including plants and serve to raise osmotic pressure in the cell as well as stabilise proteins and membranes under environmental stress conditions. Osmoprotectants can be sugars, polyols, betaines and similar compounds or amino acids, such as proline and have been shown to play an important role in plant adaptation to salt/drought stress. Thus, increasing the cellular concentration of these compounds by metabolic engineering, often using bacterial or yeast enzymes, has been a logical target to increase plant salt/drought stress resistance.

Manipulation of sugars and polyols

Fructan synthesis has been increased in tobacco (Pilon-Smits *et al.*, 1995) and trehalose synthesis has been engineered in tobacco (Holmstrom *et al.*, 1996; Romero *et al.*, 1997) in each case providing improvements in tolerance to osmotic stress. Fructan accumulation had no significant effect on growth rate under unstressed conditions, however, trehalose accumulation was associated with stunted growth and other pleiotropic effects. Increased osmoprotectant synthesis has been manipulated in plants by overexpression of enzymes leading to mannitol synthesis in tobacco, (Karakas *et al.*, 1997; Tarczynski *et al.*, 1993) and *Arabidopsis* (Thomas *et al.*, 1995). These transgenic plants also showed improved resistance to salt stress, but decreased overall growth was reported for some of the mannitol accumulating transgenic plants. However, it has been suggested that mannitol may exert its protective effects through increasing resistance to oxidative stress through radical scavenging mechanisms, since the cellular mannitol concentrations in transgenic plants are too low to provide for significant osmotic adjustment (Shen *et al.*, 1997).

Introduction of *myo*-inositol O-methyltransferase (IMT1) in tobacco produced transgenic plants with accumulation of the polyol D-ononitol and increased salt and drought tolerance (Sheveleva *et al.*, 1997), although it appears that the endogenous levels of *myo*-inositol are insufficient to maintain high levels of D-ononitol production. High levels of sorbitol accumulation were also achieved in transgenic plants but necrotic lesions in the leaves indicated that this metabolite was toxic when overproduced (Sheveleva *et al.,* 1998) and plants showed other abnormalities at high concentrations of sorbitol accumulation.

Enhanced proline accumulation in stress-resistance: relationship to nitrogen metabolism.

Proline accumulation in salt and water stress of plants has been shown to occur by regulation of both synthetic and degradative enzymes (Delauney and Verma, 1993; Kiyosue *et al.,* 1996), but the true function of this accumulation has been recently debated. Salt-tolerance could not be correlated with increases in proline levels for six independently selected salt-tolerant alfalfa cell lines, nor were there any differences in proline levels of shoots of the salt-tolerant plants regenerated from these cell lines. However, the salt-tolerant calli and plant roots showed a more rapid response in proline accumulation within the first 24 hours of application of salt stress (Petrusa and Winicov, 1997) than the isogenic salt-sensitive plants, suggesting at most a temporal role for proline in the tolerance phenotype. The *Arabidopsis sos1* mutant, defective in high-affinity K uptake, in fact showed an overexpression of the Δ^1 pyrroline-5-carboxylate synthase (P5CS) under conditions of salt stress and had increased proline concentrations (Liu and Zhu, 1997b) despite the NaCl hypersensitive phenotype. Yet, overexpression of P5CS in tobacco led to constitutively enhanced proline concentrations and increased salt tolerance of the transgenic plants (Kishor *et al.,* 1995). Also modest improvements in tolerance were obtained in the R1 transgenic rice plants transformed with mothbean P5CS cDNA (Zhu *et al.,* 1998). Consistent with high levels of proline promoting improved salt tolerance is the results from *Arabidopsis,* where antisense suppression of proline degradation improved tolerance to freezing and salinity (Nanjo *et al.,* 1999a)

The true role for proline in stress protection thus remains to be clarified (Verma, 1999), but may also relate to stress dependent adjustments required for the balance of carbon/nitrogen metabolism. This intricate metabolic balance and the resulting acquired hypersensitivity to proline, which is normally regulated as part of this balance, has been recently demonstrated in studies that show a new link between carbon/nitrogen balance and stress responses (Hellmann *et al.,* 2000). While it has been shown that proline is essential for plants under conditions of osmotic stress and for morphological development (Nanjo *et al.,* 1999b), proline catabolites may also become toxic and lead to apparent proline hypersensitivity, which can be alleviated by salt stress or metabolisable carbohydrates (Hellmann *et al.,* 2000).

Molecular and metabolic controls between carbon and nitrogen metabolism are also likely to function in the observed improvement of drought recovery by transgenic tobacco plants overexpressing nitrate reductase (NR) (Ferrario-Mery *et al.,* 1998). Overexpression of NR allows the plants to maintain N assimilation during stress and thus allow the plants to overcome the limiting physiological condition caused by stress increase of NR protein turnover.

Betaine and related osmoprotectant engineering for stress resistance

Betaines are methylated amino acid derivatives, differentially present or absent in higher plants (McNeil *et al.,* 1999). Choline is the precursor for biosynthesis of glycine-betaine and choline-O-sulfate, but others are formed by methylation of the corresponding amino acid. The maize mutation that abolishes Glycin-betaine synthesis has been exploited to develop isogenic maize lines with and without this compound to evaluate for salt tolerance in the field (Saneoka *et al.,* 1995), indicating a positive correlation of Gly-betaine content with growth and survival, however no yield data was obtained in this study.

Choline-oxidizing enzymes from bacteria have been inserted in plants that normally do not accumulate Glycin-betaine with small but significant increases in salt-stress resistance of the transgenic plants (Hayashi *et al.,* 1997; Lilius *et al.,* 1996; Sakamoto *et al.,* 1998). However, the amount of Gly-betaine produced in the transgenic plants is quite low, because of the limited choline supply. This could potentially be overcome by increasing the choline flux as demonstrated by increased Gly-betaine levels in response to an exogenous choline supply (Nuccio *et al.,* 1998). However, additional factors could become limiting in continued recycling of S-adenosyl-homocysteine (SAH), a potent inhibitor of S-adenosylmethionine dependent methylation reactions that are required for choline precursor synthesis. SAH catabolism is dependent on SAH hydrolase and adenosine kinase, both salt responsive enzymes in glycine betaine accumulating plants, but not in non-accumulating plant species (Dr. E. Weretilnyk, personal communication).

Discovery of the osmoprotectant 3-dimethylsulfonipropionate (DMSP) in marine algae and characterization of its biosynthetic pathway in algae (Gage *et al.,* 1997) and chloroplasts of some flowering plants (Trossat *et al.,* 1998) has been intriguing with the potential for osmoprotectant production not requiring nitrogen. However, novel pathway introduction in plants has so far been only modestly fruitful, because of unforeseen limitations in endogenous metabolism for enhanced product production (Huang *et al.,* 2000) and frequent manifestation of these limitations as decreased growth of the transgenic plants under control conditions. Thus successful engineering of these pathways for stress resistance will require a more comprehensive understanding of the endogenous regulation of potential precursor and co-factor pools.

Overexpression of genes to counteract oxidative stress

Oxidative stress injury has been considered to be a common component of abiotic stress damage in plants (Noctor and Foyer, 1998). The reactive oxygen species (ROS) can be scavenged by nonenzymatic antioxidants and cellular enzymes with specific functions in detoxification of ROS. Many of the genes for antioxidant enzymes have been cloned and expressed in transgenic plants to test their role in plant protection against oxidative stress and against oxidative damage as a result from salinity and drought stress (Allen, 1995; Foyer *et al.,* 1994; Foyer *et al.,* 1998). However, specific protection has been somewhat variable between genes and different plant species and the results may be complicated by alterations in the H_2O_2 flux caused by the transgenes leading to additional modifications of cellular signaling processes.

Ion uptake, transfer across membranes and compartmentalisation in the plant appear as an integral metabolic change in salt stress and adaptation (Niu *et al.,* 1995). Genes encoding membrane components of ion pumps could thus be a group of genes which, when activated, would alleviate salt stress. This was first demonstrated in yeast, where increased Na^+ tolerance was shown to be dependent on the mutated transmembrane domain of the high affinity K^+- transporter, HKT1 (Rubio *et al.,* 1996). Increasing amount of data on ion homeostasis in yeast and plants, suggest a similarity in their transport determinants (Serrano *et al.,* 1999). It has been further supported by identification of mutants in a salt-hypersensitivity assay that are deficient potassium uptake (Wu *et al.,* 1996). The salt- hypersensitivity assay also enable Photoinhibition of photosynthesis, accompanies plant salt/drought stress. Molecular mechanisms to prevent photoinhibition, or damage to photosystem II (PSII) have focused on overproduction of enzymes with potential to eliminate reactive oxygen intermediates in transgenic plants with the expectation that these plants would be more resistant to oxidative and other abiotic stresses. Overexpression of superoxide dismutase (SOD) in tobacco was shown to protect plants from oxidative stress as induced by high light and low temperature. These plants were also able to maintain better photosynthetic rates under stress conditions (Gupta *et al.,* 1993). Transgenic alfalfa-overexpressing SOD was reported to show reduced injury from water deficit stress while maintaining the requisite field performance (McKersie *et al.,* 1996) and to date this remains the only study in which field performance of the transgenic crop has been tested. Mn-SOD was shown to reduce cellular damage from oxidative stress in transgenic tobacco (Bowler *et al.,* 1991) and Fe-SOD overproduction in tobacco chloroplasts also provided enhancement of oxidative stress tolerance (Van Camp *et al.,* 1996). Subsequent work with poplar, overexpressing chloroplast-targeted Fe-SOD suggested that increased levels of the enzyme can protect PSII from overreduction at low CO_2 concentrations, but could not protect PSII efficiency under these conditions (Arisi *et al.,* 1998). Nor was photoinhibition of photosystem II prevented in the Fe-SOD overproducing poplar (Tyystjarvi *et al.,* 1999), indicating indirect effects in chloroplast protection by ROS scavenging enzymes.

Increased protection to salt/drought stress has also been reported by overexpression of cytosolic ascorbate peroxidase (APX), but not the chloroplast targeted isoform (Torsethaugen *et al.,* 1997). Glutathione-S-transferase, when expressed constitutively in tobacco, provided protection against both cold and salt stress in the transgenic seedlings (Roxas *et al.,* 1997). Glutathione reductase overproduction in cytosol showed somewhat less photoinhibition in transgenic tobacco, but did not enhance repair of photoinhibitory damage (Tyystjarvi *et al.,* 1999). Interestingly, oxidative stress protection has also been obtained in transgenic plants producing the iron storage protein ferritin (Deak *et al.,* 1999). Presently, cellular compartmentalisation of the different antioxidant enzymes and the complexity of the interacting pathways add to the difficulty of comparing the levels of stress resistance obtained with each construct.

Importance of ion transport and compartmentalisation in the salt-stress response

Ion uptake, transfer across membranes and compartmentalisation in the plant appear as an integral metabolic change in salt stress and adaptation (Niu *et al.,* 1995). Genes encoding membrane components of ion pumps could thus be a group of genes which, when activated, would alleviate salt stress. This was first demonstrated in yeast, where increased Na^+ tolerance was shown to be dependent on the mutated transmembrane domain of the high affinity K^+- transporter, HKT1 (Rubio *et al.,* 1996). Increasing amount of data on ion homeostasis in yeast and plants, suggest a similarity in their transport determinants (Serrano *et al.,* 1999). It has been further supported by identification of mutants in a salt-hypersensitivity assay that are deficient potassium uptake (Wu *et al.,* 1996). The salt-hypersensitivity assay also enabled identification of additional mutants related to ion uptake functions indicating the essential role of the high affinity K^+ uptake system in mediating the harmful effects of saline environments in *Arabidopsis* and demonstrating an important link between Ca^{++} regulation, K^+ deficiency and Na^+ stress in plants (Liu and Zhu, 1997a). It will be interesting to see if the genes identified in the hypersensitivity assay to NaCl are limiting under normal conditions and will be able to increase salinity tolerance in wild type *Arabidopsis.*

Compartmentalisation of salt in the plant vacuole has been suggested as one of the mechanisms by which plant cells may overcome salt stress (Binzel *et al.,* 1988). Recent introduction of a cDNA encoding an

Arabidopsis Na^+/H^+ antiport, normally localised in the prevacuolar compartment, has shown that overexpression of this gene in *Arabidopsis* can increase salt tolerance of the transgenic plants (Apse *et al.,* 1999). Although salt induction of the endogenous gene was undetectable, modest overexpression of this antiport did provide survival to salt stress of the transgenic plants and the extent of its function undoubtedly will be tested in other plant species.

Dehydrins/late embryogenesis abundant (lea) gene products as molecular tools for salt/drought stress improvement.

Dehydrins are an immunologically related family of proteins that were first described as late embryogenesis abundant (LEA) proteins in plants (Close, 1997; Close, 1996; Dure, 1992). They were also found to accumulate in response to the plant hormone abscisic acid (ABA) and environmental stress conditions that involve dehydration, such as drought, salinity and cold. Although their biochemical function in plants still remains unclear, they have been postulated to interact with hydrophobic surfaces through their amphipatic α-helices and thus protect macromolecules under conditions of decreasing water activity and during seed desiccation. The dehydrins constitute an extensive multigene family in most plant species studied to date (Close, 1997).

The protective effect of enhanced levels of LEA proteins was tested in transgenic rice by introducing high level expression of the *HVA1* gene from barley (Xu *et al.,* 1996) regulated by the rice actin 1 gene promoter. High levels of constitutive accumulation of the HVA1 protein were obtained in both leaves and roots and the transgenic rice plants showed significantly increased tolerance to salinity and water deficit. Because constitutive expression of high levels of the HVA1 protein requires metabolic channelling with a potential yield cost and has led to frequent gene silencing in this system (Cheng *et al.,* 1998); subsequent experiments have aimed to engineer promoter elements that would be stress inducible. This approach appears feasible by modifying the rice actin 1 promoter with the ABA response (ABRC1) element from the barley HVA1 gene to make the actin 1 promoter salt/drought stress-regulated for future transgene expression in rice (Su *et al.,* 1998), where it would be only expressed under stress.

Other potential transgenes for modification of the salt/drought stress response

Among the classes of genes (Table 1), are certainly other potential candidates that will be tested for their ability to provide protection against salt/drought stress in transgenic plants. Many of these genes unfortunately belong to sizeable multigene families and it is therefore difficult to select new candidates until more is known about the specificity of their expression and regulation.

Aquaporins are water-selective channel proteins, which are constituents of plant membranes and have been shown to be among the salt/drought inducible genes (Kjellbom *et al.,* 1999). Antisense reduction of *PIP1* aquaporin mRNA levels in *Arabidopsis* have demonstrated the importance of these molecules in plant water transport (Kaldenhoff *et al.,* 1998). Interesting to note though, while reduced expression of the aquaporin decreased the cellular osmotic water permeability coefficient three fold, xylem pressure measurements suggested that the plants were able to compensate for the reduced cellular water permeability by increasing the size of the root system by five fold.

Similarities of many of the salt/drought responses between yeast and plants suggest other areas of transgene exploration for protection against salt and drought. This relationship has been exploited productively in identifying many genes involved in the signaling pathways and will be discussed in a subsequent section. Other proteins involving structural components of the cell, cell wall extensibility (Cosgrove, 1997), chaperones, engineered membrane composition (Allakhverdiev *et al.,* 1999) and a variety of transport proteins are also likely to be tested in the future as more information becomes available about the large variety of genes induced by salt/drought stress.

STRATEGIES THAT TARGET ACTIVATION OF GROUPS OF GENES RELATED BY FUNCTION TO SALT/DROUGHT STRESS RESISTANCE

Native differences in resistance to salt stress that exist between varieties of different crop plants indicate that different levels of utilisation of similar genetic information can have a positive effect on survival and productive growth under stress conditions. The current information explosion on genes induced by salt stress, as well as the molecular systems that allow us to test the physiological relatedness and function of these gene products in emerging pathways, has suggested that it might be possible to approach the problem of improved salt tolerance in crop plants with a physiologically more integrated view. In view of the apparent multigenic characteristic of salt tolerance it is possible to project engineering improvements in the tolerance phenotype by manipulating: 1) the signaling systems that lead to enhanced expression of genes that can counteract the stress response, or 2) the transcription factors that are limiting in expression of genes that could provide tolerance to salt stress. Presently some progress has been made on both types of manipulations and will be discussed below. Moreover, these approaches will continue to evolve, as we are able to integrate additional molecular and physiological data associated with improved salt tolerance in plants.

Manipulation of stress signal components to improve salt/drought stress resistance

The observation of specific and co-ordinate mRNA accumulation in response to salt and water stress in plants has been well-documented (Bray, 1997; Ingram and Bartels, 1996; Shinozaki and Yamaguchi-Shinozaki, 1999; Winicov, 1998). The signaling pathways and molecular mechanisms responsible for this co-ordinate transcript accumulation have been reviewed recently (Hasegawa *et al.,* 2000; Shinozaki *et al.,* 1999). Current models for stress gene activation through these pathways are built largely on analogy to yeast osmosensing and signaling pathways, since many of the gene products can complement mutations in the yeast system. Genes have been cloned from *Arabidopsis* that are similar to the two-component phosphorelay-mediator like osmosensing system from yeast (Urao *et al.,* 1998). Phospholipase D has been implicated in the early events leading to desiccation survival in *Craterostigma plantiagineum* (Frank *et al.,* 2000) and this indicated a role for phospholipids in the signal transduction process. Also, model plants able to withstand salt/drought stress could be instrumental in identifying the significant signaling events necessary for metabolic conversion to tolerance (Taybi and Cushman, 1999) and could suggest means by which targeted groups of salt/drought stress regulated genes could be manipulated in crop plants.

Although an increasing number of kinases and phosphatases have been identified that respond to both salt and drought stress, the pathways themselves have remained unresolved, in part because the large number of genes encoded by these gene families. The pathways involve both calcium dependent protein kinases (CDPK) (Harmon *et al.,* 2000) and the mitogen-activated protein kinase (MAPK) cascade (Mizoguchi *et al.,* 2000). Many of these genes have been shown to be induced by a variety of abiotic stresses and thus are presumed to participate in signaling necessary for gene activation in response to stress. The challenges in this field will be to establish the specificity and selectivity of these pathways for normal development and stress resistance, since it is now clear that the complexity of interactions and intersections of the signaling pathways lie not only between the stresses imposed by salt, drought, cold, light, ABA and other plant hormones, but may even connect to responses caused by pathogen infections (Moller and Chua. 1999).

Modification of salt-stress responses through manipulation of Ca^{2+}/calmodulin-dependent phosphatase and kinase interactions

The role of Ca^{2+} dependent signaling in salt-stress responsive gene activation through calmodulin and calcineurin like activities is being supported by results from a number of laboratories. Calcium dependent protein kinases have been shown to activate a stress-inducible promoter (Sheen, 1996) responding to cold, salt, dark and ABA, indicating a broad range of activities for these proteins in plant stress signal transduction. Other experiments have revealed a more dynamic mode of action and distinct Ca^{2+} signaling pathways regulating calmodulin expression by different stress conditions (van der Luit *et al.*, 1999). The SOS3 mutation rendering *Arabidopsis* hypersensitive to salt also suggests, that interruption of calcium signaling through a calcineurin-like pathway mediates the hypersensitivity response (Liu and Zhu, 1998).

The relevance of these signaling pathways to improving salt-tolerance in plants was tested by introducing a truncated form of the catalytic subunit and the regulatory subunit of yeast calcineurin (CaN) to express a constitutively activated Ca^{2+} and calmodulin-dependent protein phosphatase in tobacco (Pardo *et al.*, 1998). Transgenic expression of CaN in the tobacco plants was associated with improved resistance to salt stress and grafting experiments indicated a significant function of the transgenic roots in the improved stress response. Despite the yeast CaN function in tobacco salt tolerance, the homologous system has not been cloned from plants. Instead, recent identification of the protein kinase (Liu *et al.*, 2000) encoded by the SOS2 gene in *Arabidopsis* that is activated by the calcium binding protein SOS3 (Halfter *et al.*, 2000) suggests that in plants resistance to salt hypersensitivity may be mediated by the SOS3/SOS2 kinase complex. The salt hypersensitivity assay has identified three loci in *Arabidopsis* that are involved in essential signaling processes necessary to maintain K^+ nutrition at levels adequate for average NaCl concentrations. It will be interesting to see if manipulation of this pathway can improve salt tolerance above the average limits observed for individual plant species.

Introduction of improved salt tolerance through mitogen-activated protein kinase cascade and connections to hormone and cell cycle signal transduction

An interesting development in manipulation of mitogen-activated protein kinases that mediate signal transduction of growth control, cell cycle and stress responses has emerged from a comparison of MAPKKKs from *Arabidopsis* (ANP1) and tobacco (NPK1) and their constitutive expression on a variety of stress cascades (Kovtun *et al.*, 2000). Both kinases mediate H_2O_2 signal transduction in oxidative stress. Constitutively expressed NPK1 was shown to function in oxidative stress signaling, had a negative effect on auxin induced gene expression and provided enhanced tolerance to abiotic stress, including salt stress, but did not activate some previously described drought, cold and ABA signaling pathways. The transgenic plants showed no apparent decrease in growth under normal conditions.

These results emphasise the need for better understanding of the connectivity and specific targeting of the many kinases and phosphatases. While it is recognised that hormones, such as ABA, ethylene (Kieber *et al.*, 1993) and auxin can play a significant role in salt/drought stress responses, the limiting steps in metabolic activation (Qin and Zeevaart, 1999) and signal transduction pathways are not yet sufficiently clear to provide predictable options for their manipulation.

Manipulation of tolerance to salt/drought stress by enhancing expression of transcription factors that induce stress gene transcription

Relatively few transcription factors have been identified to date that bind to promoter elements in genes regulated by salt/drought stress and little information is available on transcription factors that function to induce gene expression in plants that counteract the stress responses. The plant bZIP protein EmBP-1 (Guiltinan *et al.*, 1990) that recognises an ABA response element in a number of genes is ABA regulated during embryo maturation. Stress induced expression of EmBP-1 in other tissues initially suggested its involvement in ABA regulated gene expression during salt/drought stress. However, introduction of a truncated dominant negative version of the EmBP-1 protein in transgenic tobacco instead revealed an important developmental function for this protein in vegetative tissues (Eckardt *et al.*, 1998). Two homeodomain-leucine zipper proteins were found to be inducible in *Craterostigma plantagineum* during the early stages of dehydration (Frank *et al.*, 1998), but their function has not been yet tested in transgenic plants. Expression from STZ, the TFIIIA-type zinc-finger factor, increases in *Arabidopsis* after salt treatment. It was found to activate transcription and complement the salt-sensitive phenotype in yeast, but its function in salt tolerance of plants remains to be demonstrated (Lippuner *et al.*, 1996).

Recently two laboratories have shown that overexpression of a single transcription factor associated with salt/drought stress (Kasuga *et al.*, 1999) or salt tolerance (Winicov and Bastola, 1999) can improve salt tolerance of the transgenic plants. Transcription factors DREB1A and DREB2A were initially cloned as DNA binding factors that recognized the ABA independent dehydration-responsive element (DRE) in stress induced genes (Liu *et al.*, 1998). DREB1A and its two homologs responded to cold stress, while DREB2A and its homolog responded to salt stress. Both of these genes encode the EREBP/AP2 DNA binding domain. Overexpression of DREB1A led to strong expression of target genes under unstressed conditions and improved tolerance to freezing and dehydration, while overexpression of DREB2A led to a weaker expression of the target genes. The ubiquitous expression of these transcription factors under the control of the CaMV 35 S promoter however, led to growth retardation and a dwarf phenotype in *Arabidopsis*. The growth inhibition was overcome when the cDNA encoding DREB1A was put under the control of the stress inducible promoter *rd29A*, one of its target genes. The introduction of the stress inducible promoter to control expression of this transcription factor allowed for normal growth under control conditions while providing the transgenic plants with enhanced levels of DREB1A mRNA under stress conditions and improved stress resistance to freezing, drought and salinity (Kasuga *et al.*, 1999).

The other factor, *Alfin1* was initially cloned by differential screen from a cDNA library constructed with mRNA from salt tolerant alfalfa callus (Winicov, 1993), was found to be salt inducible in the tolerant callus and was expressed predominantly in roots. *Alfin1* contained sequence information for adjacent Cys_4 and His/Cys_3 zinc-finger domains analogous to those found in leukaemia-associated-protein (LAP) (Saha *et al.*, 1995) or plant homeodomain (PHD) associated fingers (Aasland *et al.*, 1995). It was subsequently shown to bind DNA in a sequence specific manner to G rich triplets, as well as promoter fragments of the salt inducible *MsPRP2* gene, also predominantly expressed in alfalfa roots (Bastola *et al.*, 1998). *Alfin1* is a low copy gene and is conserved in rice and *Arabidopsis* (Winicov and Bastola, 1997) and was of interest for both root gene expression and its potential role in salt tolerance. Overexpression of *Alfin1* under the control of the CaMV 35S promoter in the transgenic alfalfa produced normal plants, which nevertheless were more tolerant to several weeks of salt stress than untransformed parent plants. These transgenic plants also showed increased mRNA accumulation from the endogenous *MsPRP2* gene in the roots, indicating that Alfin1 was enhancing other gene expression in the transformed plants (Winicov and Bastola, 1999). A broad based gene activation in Alfin1 overexpressing transgenic plants was also indicated, since the transgenic plants show a two to four fold increased root growth under normal and saline conditions when compared to untransformed and vector transformed alfalfa (Winicov, 2000). These results suggest an important function for Alfin1 and root growth for salinity tolerance in alfalfa. *Alfin1* essential function in root growth was further demonstrated by defective root development in *Alfin1* antisense expressing transgenic alfalfa. However, Alfin1 function must depend on a complement of other genes found in the root, since high levels of Alfin1 in the shoot of the transgenic plants overexpressing *Alfin1* did not induce endogenous shoot *MsPRP2* expression at the level observed in the roots.

These results obtained from plants in which a single-transcription factor was able to provide increased tolerance to salt-stress, indicate that it is feasible to manipulate the expression of a number of genes in a transgenic plant and obtain increased resistance salt/drought stress.

FUTURE CHALLENGES

Plant cells manifest an immense and dynamic range of molecular reactions when exposed to salt/drought stress. The scientific literature has chronicled an increasing number of these reactions, some of which have been exploited by biotechnology to investigate their possible use for improving salinity tolerance in crop plants. The experiments to date have helped to clarify our understanding of the multiplicity of biochemical systems involved in the stress responses as well as to alert us of the substantial interactions among these systems, which can significantly influence the outcome of engineering crop plants for improved salt/drought tolerance.

Since the overall goal in manipulation of crop plants is to retain most of their desirable agronomic characteristics and just manipulate those that will add increased resistance to salt/drought stress, we need to consider the varied aspect of the stress response in order to target those areas that will be beneficial and potentially bypass those that may lead us to unanticipated complications. The schematic of Figure 1 summarises in only the broadest terms the responses and interactions that play a molecular role during salt/drought stress and indicate the complexity of directed channelling them toward the desired salt-tolerant phenotype of a crop plant. As any metabolic system, it has products or metabolic steps that are limiting to the overall system and others that feed back or modify connecting pathways. Our ability to identify those products that would provide maximum protection may improve with tissue and development specific micro-array analyses of stress responsive genes. However, we also need to develop new technologies that utilise the plant cell to screen and optimise our selection of useful modifications in gene expression/regulation for improved salt tolerance.

Single gene transfer experiments carried out to date that provided incremental improvements in salt tolerance, indicate that these changes are beneficial, but it would be extremely labor intensive to pyramid a multiplicity of traits in different backgrounds and maintain an acceptable crop yield. The recent examples in manipulation of signaling pathways are encouraging despite the potential complexity of signaling pathway interactions and the high copy number of signaling molecules, with their variation in tissue and cell specific expression. Manipulation of transcription factors targeted to induce genes functionally related to withstand or counteract the stress response seems simplistic and yet desirable at this point. This strategy might bypass the many intersecting pathways shown in Figure 1 that control the early responses to stress.

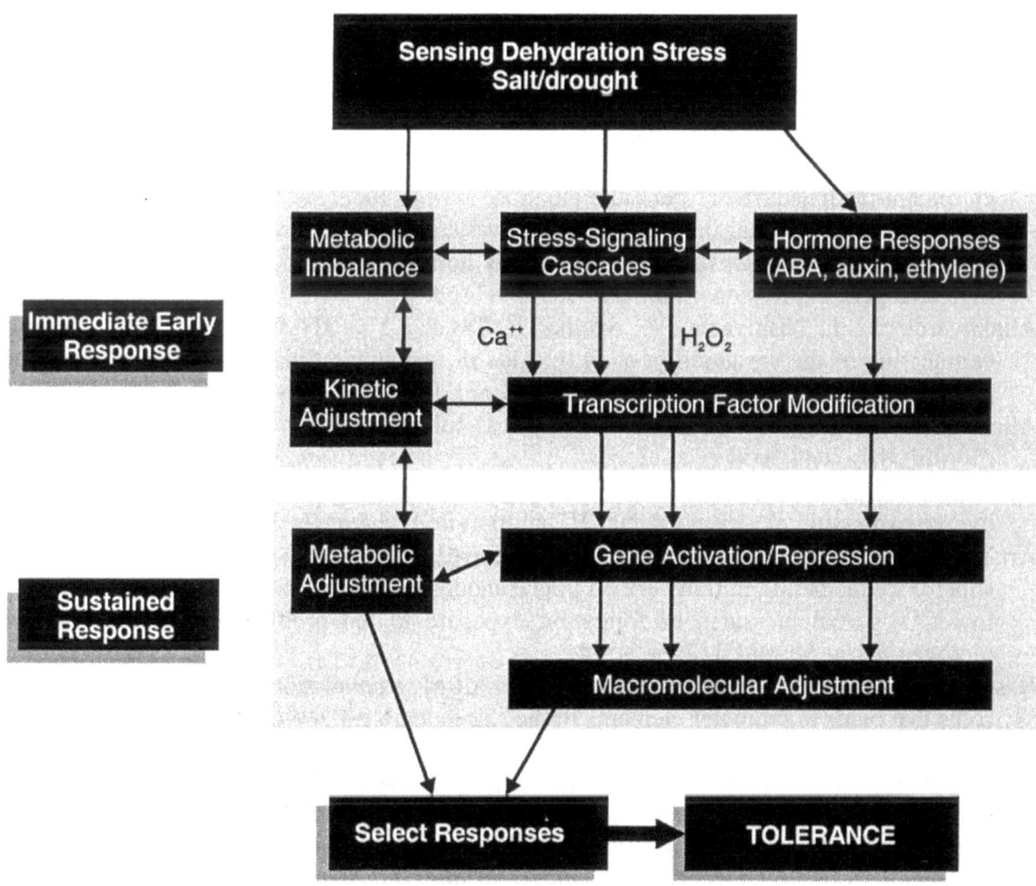

Figure 1. Intersecting pathways in acute and long term adjustment to dehydration stress add to the challenge of selecting only those reactions necessary and sufficient for engineering improved tolerance.

Future experiments will undoubtedly discover additional transcription factors related to the salt/drought response. A major challenge also will be to determine the specificity of functional groups of genes that can be induced for improved tolerance in different plants and the extent to which their expression will need to be regulated in order to maintain acceptable yield for crop plants.

Acknowledgments

Because of space constraints, not all references could be included and my apologies in this regard. I wish to thank colleagues for discussions and making available unpublished data. The work in my laboratory has been supported in the past by US Department of Agriculture (NRI) and by the National Science Foundation EPSCoR (Nevada) Women in Science and Engineering Program.

REFERENCES

Aasland, R., Gibson, T.J. and Stewart, A.F. 1995. The PHD finger: implications for chromatin-mediated transcriptional regulation. *Trends Biochem. Sci.* 20, 56-59.

Abe, H., Yamaguchi-Shinozaki, K., Urao, T., Iwasaki, T., Hosokawa, D. and Shinozaki, K. 1997. Role of *Arabidopsis* MYC and MYB homologs in drought- and abscisic acid-regulated gene expression. *Plant Cell* 9, 1859-1868.

Allakhverdiev, S.I., Nishiyama, Y., Suzuki, I., Tasaka, Y. and Murata, N. 1999. Genetic engineering of the unsaturation of fatty acids in membrane lipids alters the tolerance of *Synechocystis* to salt stress. *Proc. Natl. Acad. Sci. USA.* 96, 5862-5867.

Allen, R.D. 1995. Dissection of oxidative stress tolerance using transgenic plants. *Plant Physiol.* 107, 1049-1054.

Apse, M. P., Aharon, G. S., Snedden, W. A. and Blumwald, E. 1999. Salt tolerance conferred by overexpression of a vacuolar Na^+/H^+ antiport in *Arabidopsis. Science* 285, 1256-1258.

Arisi, A.-C.M., Cornic, G., Jouanin, L. and Foyer, C.H. 1998. Overexpression of iron superoxide dismutase in transformed poplar modiefies the regulation of photosynthesis at low CO_2 partial pressures or following exposure to the prooxidant herbicide methyl viologen. *Plant Physiol.* 117, 565-574.

Bastola, D.R., Pethe, V.V. and Winicov, I. 1998. Alfin1, a novel zinc-finger protein in alfalfa roots that binds to promoter elements in the salt-inducible *MsPRP2* gene. *Plant Mol. Biol.* 38, 1123-1135.

Binzel, M.L., Hess, F.D., Bressan, R.A. and Hasegawa, P.M. 1988. Intracellular compartmentation of ions in salt adapted tobacco cells. *Plant Physiol.* 86, 607-614.

Bohnert, H.J., Su, H. and Shen, B. 1999. "Molecular mechanisms of salinity tolerance". In: *Molecular Responses to Cold, Drought, Heat and Salt Stress in Higher Plants*, ed. K. Shinozaki, K. and Yamaguchi-Shinozak.i R.G. Landes Co. Austin.

Bostock, R.M. and Quatrano, R.S. 1992. Regulation of Em gene expression in rice:interaction between osmotic stress and abscisic acid. *Plant Physiol.* 98, 1356-1363.

Bowler, C., Slooten, L., Vandenbranden, S., Rycke, R.D., Botterman, J., Sybesma, C., Montagu, M.V. and Inze, D. 1991. Manganese superoxide dismutase can reduce cellular damage mediated by oxygen radicals in transgenic plants. *EMBO* 10, 1723-1732.

Bray, E. A. 1997. Plant responses to water deficit. *Trends in Plant Sci.* 2, 48-54.

Cheng, W., Su, J., Zhu, B., Jayaprakash, T.L. and Wu, R. 1998. "Development of transgenic cereal crop plants that are tolerant to high salt, drought and low temperature". In: *Frontiers in Biology: The Challenges of Biodiversity.*, ed. C. H. Chou and K. T. Shao. pp. 115-122. Academia Sinica, Taipei.

Close, T. 1997. Dehydrins: a commonality in the response of plants to dehydration and low temperature. *Physiolog. Plant.* 100, 291-296.

Close, T.J. 1996. Dehydrins: emergence of a biochemical role of a family of plant dehydration proteins. *Physiolog. Plant.* 97, 795-803.

Cohen, A. and Bray, E. A. 1990. Characterization of three mRNAs that accumulate in wilted tomato leaves in response to elevated levels of endogenous abscisic acid. *Planta* 182, 27-33.

Cosgrove, D.J. 1997. Relaxation in a high-stress environment: the molecular bases of extensible cell walls and cell enlargement. *Plant Cell* 9, 1031-1041.

Deak, M., Horvath, G.V., Daveletova, S., Torok, K., Sass, L., Vass, I., Barna, B., Kiraly, Z. and Dudits, D. 1999. Plants ectopically expressing the iron-binding protein ferritin, are tolerant to oxidative damages and pathogens. *Nat. Biotechnol.* 17, 192-196.

Delauney, A.J. and Verma, D. P. S. 1993. Proline biosynthesis and osmoregulation in plants. *Plant J.* 4, 215-223.

Dure, L.I. 1992. "The LEA proteins of higher plants". In: *Control of Plant Gene Expression*, ed. D. P. S. Verma. pp. 325-335. CRC Press, Bocca Raton.

Eckardt, A.N., McHenry, L. and Guiltinan, M. J. 1998. Overexpression of EmBP, a dominant negative inhibitor of G-box-dependent transactivation, alters vegetative development in transgenic tobacco. *Plant Mol. Biol.* 27, 411-418.

Ferrario-Mery, S., Valadier, M.-H. and Foyer, C.H. 1998. Overexpression of nitrate reductase in tobacco delays drought-induced decreases in nitrate reductase activity and mRNA. *Plant Physiol.* 117, 293-302.

Flowers, T.J., Koyama, M.L., Flowers, S.A., Sudhakar, C., Singh, K. P. and Yeo, A. R. 2000. QTL: their place in engineering tolerance of rice to salinity. *J. Exp. Botany* 51, 99-106.

Foolad, M.R. and Jones, R.A. 1993. Mapping salt-tolerance genes in tomato (Lycopersicon esculentum) using trait-based marker analysis. *Theor. Appl. Genet.* 87, 184-192.

Foyer, C.H., Descourvieres, P. and Kunert, K. J. 1994. Protection against oxygen radicals: an important defence mechanism studied in transgenic plants. *Plant Cell Envir.* 17, 507-523.

Foyer, C.H., Kingston-Smith, A.H., Harvinson, J., Arisis, A.-C. M., Jouanin, L. and Noctor, G. 1998. "The use of transformed plants in the assesment of physiological stress responses". In: *Responses of plant metabolism to air polution and global change.*, eds. L. J. De Kok and I. Stulen. pp. 251-261. Backhuys, Leiden.

Frank, W., Munnik, T., Kerkmann, K., Salamini, F. and Bartels, D. 2000. Water deficit triggers phospholipase D activity in the resurection plant *Craterostigma plantagineum*. *Plant Cell* 12, 111-123.

Frank, W., Phillips, J., Salamini, F. and Bartels, D. 1998. Two dehydration-inducible transcripts from the resurrection plant *Craterostigma plantagineum* encode interacting homeodomain-leucine zipper proteins. *Plant J.* 15, 413-421.

Gage, D.A., Rhodes, D., Nolte, K.D., Hicks, W.A., Leustek, T., Cooper, A.J.L. and Hanson, A.D. 1997. A new route for synthesis of dimethyl-sulphoniopropionate in marine algae. *Nature* 387, 891-894.

Galiba, G., Simon-Sarkadi, L., Kocsy, L., Salgo, G. and Sutka, A. 1992. Possible chromosomal location of genes determining the osmoregulation of wheat. *Theor. Appl. Genet.* 85, 415-418.

Guiltinan, M.J., Marcotte, W.R. and Quatrano, R.S. 1990. A plant leucine zipper protein recognizes an abscisic acid response element. *Science* 250, 267-270.

Gupta, A.S., Webb, R.P., Holaday, A.S. and Allen, R. D. 1993. Overexpression of superoxide dismutase protects plants form oxidative stress. *Plant Physiol.* 103, 1067-1073.

Halfter, U., Ishitani, M. and Zhu, J.-K. 2000. The *Arabidopsis* SOS2 protein kinase phsically interacts with and is activated by the calcium-binding protein SOS3. *Proc. Natl. Acad. Sci. USA.* 97, 3735-3740.

Harmon, A.C., Gribskov, M. and Harper, J.F. 2000. CDPKs-a kinase for every Ca^{+2} signal. *Trends Plant Sci.* 5, 154-159.

Hasegawa, P.M., Bressan, R.A., Zhu, J.-K. and Bohnert, H.J. 2000. Plant cellular and molecular responses to high salinity. *Annu. Rev. Plant Physiol. Plant Mol. Biol.* 51, 436-499.

Hayashi, H., Mustardy, L., Deshnium, P., Ida, M. and Murata, N. 1997. Transformation of *Arabidopsis thaliana* with the *codA* gene for choline oxidase accumulation of glycinebetaine and enhanced tolerance to salt and cold stress. *Plant J.* 12, 133-142.

Hellmann, H., Funck, D., Rentsch, D. and Frommer, W. B. 2000. Hypersensitivity of an *Arabidopsis* sugar signaling mutant toward exogenous proline application. *Plant Physiol.* 122, 357-367.

Hirayama, T., Ohto, C., Mizoguchi, T. and Shinozaki, K. 1995. A gene encoding a phosphatidylinositol-specific phospholipase C is induced by dehydration and salt stress in *Arabidopsis thaliana. Proc. Natl. Acad. Sci. USA.* 92, 3903-3907.

Holmstrom, K.-O., Mantyla, E., Welin, B., Mandal, A., Paiva, E.T., Tunnela, O.E. and Longdesborough, J. 1996. Drought tolerance in tobacco. *Nature* 379, 683-684.

Huang, J., Hiriji, R., Adam, L., Rozwadowski, K.L., Hammerlindl, J.K., Keller, W.A. and Selvaraj, G. 2000. Genetic engineering of glycinebetaine production toward enhancing stress tolerance in plants: metabolic limitations. *Plant Physiol.* 122, 747-756.

Hwang, I. and Goodman, H.M. 1995. An *Arabidopsis thaliana* root-specific kinase homolog is induced by dehydration, ABA, and NaCl. *Plant J.* 8, 37-43.

Ingram, J. and Bartels, D. 1996. The molecular basis of dehydration tolerance in plants. *Annu. Rev. Plant Physiol. Plant Mol. Biol.* 47, 377-403.

Ismail, A.M., Hall, A.E. and Close, T.J. 1999. Allelic variation of a dehydrin gene cosegregates with chilling tolerance during seedling emergence. *Proc. Natl. Acad. Sci. USA.* 96, 13566-13570.

Kaldenhoff, R., Grote, K., Zhu, J.-J. and Zimmermann, U. 1998. Significance of plasmalemma aquaporins for water transport in *Arabidopsis thaliana. Plant J.* 14, 121-128.

Karakas, B., Ozias-Akins, P., Stushnoff, C., Suefferheld, M. and Rieger, M. 1997. Salinity and drought tolerance of mannitol-accumulating transgenic tobacco. *Plant Cell Envir.* 20, 609-616.

Kasuga, M., Liu, Q., Miura, S., Yamaguchi-Shinozaki, K. and Shinozaki, K. 1999. Improving plant drought, salt, and freezing tolerance by gene transfer of a single stress-inducible transcription factor. *Nature Biotechnol.* 17, 287-291.

Kidwell, K.K., Hartweck, L.M., Yandell, B.S., Crump, P.M., Brummer, J.E., Moutray, J. and Osborn, T.C. 1999. Forage yields of alfalfa populations derived from parents selected on the basis of molecular marker diversity. *Crop Sci.* 39, 223-227.

Kieber, J.J., Rothenberg, M., Roman, G., Feldmann, K.A. and Ecker, J.R. 1993. CTR1: a negative regulator of the ethylene response pathway in *Arabidopsis*, encodes a member of the RAf family of protein kinases. *Cell* 72, 427-441.

Kishor, P.B.K., Hong, Z., Miao, G.-H., Hu, C.-A.A. and Verma, D.P.S. 1995. Overexpression of Δ^1-pyrroline-5-carboxylate synthetase increases proline production and confers osmotolerance in transgenic plants. *Plant Physiol.* 108, 1387-1394.

Kiyosue, T., Abe, H., Yamaguchi-Shinozaki, K. and Shinozaki, K. 1998. ERD6, a cDNA clone for an early dehydration-induced gene of *Arabidopsis*, encodes a putative sugar transporter. *Biochim. Biophys. Acta* 1370, 187-191.

Kiyosue, T., Yoshiba, Y., Yamaguchi-Shinozaki, K. and Shinozaki, K. 1996. A nuclear gene encoding mitochondrial proline dehydrogenase, an enzyme involved in proline metabolism, is upregulated by proline but downregulated by dehydration in *Arabidopsis*. *Plant Cell* 8, 1323-1335.

Kjellbom, P., Larsson, C., Johansson, I., Karlsson, M. and Johanson, U. 1999. Aquaporins and water homeostasis in plants. *Trends Plant Sci.* 4, 308-314.

Kovtun, Y., Chiu, W.-L., Tena, G. and Sheen, J. 2000. Functional analysis of oxidative stress-activated mitogen-activated protein kinase cascade in plants. *Proc. Natl. Acad. Sci, USA.* 97, 2940-2945.

Lebreton, C., Jazic-Jancic, V., Steed, A., Pekic, S. and Quarrie, S.A. 1995. Identification of QTL's for drought responses in maize and their use in testing causal relationships between traits. *J. Exp. Bot.* 46, 853-865.

Lilius, G., Homberg, N. and Bulow, L. 1996. Enhanced NaCl stress tolerance in transgenic tobacco expressing bacterial choline dehydrogenase. *Bio-Technology* 14, 177-180.

Lilley, J.M., Ludlow, M.M., McCouch, S.R. and O'Toole, J.C. 1996. Locating QTL for osmotic adjustment and dehydration tolerance in rice. *J. Exp. Bot.* 47, 1427-1436.

Lippuner, V., Cyert, M.S. and Gasser, C.S. 1996. Two classes of plant cDNA clones differentially complement yeast calcineurin mutants and increase salt tolerance in wild-type yeast. *J. Biol. Chem.* 271, 12859-12866.

Liu, J., Ishitani, M., Halfter, U., Kim, C.-S. and Zhu, J.-K. 2000. The *Arabidopsis thaliana* SOS2 gene encodes a protein kinase that is required for salt tolerance. *Proc. Natl. Acad. Sci. USA.* 97, 3730-3734.

Liu, J. and Zhu, J.-K. 1997a. An *Arabidopsis* mutant that requires increased calcium for potassium nutrition and salt tolerance. *Proc. Natl. Acad. Sci. USA.* 94, 14960-14964.

Liu, J. and Zhu, J.-K. 1997b. Proline accumulation and salt-stress-induced gene expression in a salt-hypersensitive mutant of *Arabidopsis. Plant Physiol.* 114, 591-596.

Liu, J. and Zhu, J.-K. 1998. A calcium sensor homolog required for plant salt tolerance. *Science* 280, 1943-1945.

Liu, Q., Kasuga, M., Sakuma, Y., Abe, H., Miura, S., Yamaguchi-Shinozaki, K. and Shinozaki, K. 1998. Two transcription factors, DREB1 and DREB2, with and EREBP/AP2 DNA binding domain separate two cellular signal transduction pahtways in drought- and low- temperature-responsive gene expression, respectively, in *Arabidopsis*. *Plant Cell* 10, 1391-1406.

McKersie, B.D., Bowley, S.R., Harjanto, E. and Leprince, O. 1996. Water-deficit tolerance and field performance of transgenic alfalfa overexpressing superoxide dismutase. *Plant Physiol.* 111, 1177-1181.

McNeil, S.D., Nuccio, M.L. and Hanson, A.D. 1999. Betaines and related osmoprotectants. Targets for metabolic engineering of stress resistance. *Plant Physiol.* 120, 945-949.

Meyer, G., Schmitt, J.M. and Bohnert, H.J. 1990. Direct screening of a small genome: estimation of the magnitude of plant gene expression changes during adaptation to high salt. *Molec. Gen. Genet.* 224, 347-356.

Mikami, K., Katagiri, T., Iuchi, S., Yamaguchi-Shinozaki, K. and Shinozaki, K. 1998. A gene encoding phosphatidylinositol-4-phosphate 5-kinase is induced by water stress and abscisic acid in *Arabidopsis thaliana. Plant J.* 15, 563-568.

Mizoguchi, T., Ichimura, K., Yoshida, R. and Shinozaki, K. 2000. "MAP kinase cascades in *Arabidopsis*: their roles in stress and hormone responses". In: *MAP Kinases in Plant Signal Transduction.*, ed. H. Hirt. pp. 29-38. Results and Problems in Cell Differentiation. Vol. 27. Berlin, Springer-Verlag, Heidelberg.

Mizoguchi, T., Irie, K., Hirayama, T., Hayashida, N., Yamaguchi-Shinozaki, K., Matsumoto, K. and Shinozaki, K. 1996. A gene encoding a mitogen-activated protein kinase kinase kinase is induced simultaneously with genes for a mitogen-activated protein kinase and an S6 ribosomal protein kinase by touch, cold, and water stress in *Arabidopsis thaliana*. *Proc. Natl. Acad. Sci. USA.* 93, 765-769.

Moller, S.G. and Chua, N-H. 1999. Interactions and intersections of plant signaling pathways. *J. Mol. Biol.* 283, 219-234.

Morgan, J.M. and Tan, M.K. 1996. Chromosomal location of a wheat osmoregulation gene using RFLP analysis. *Aust. J. Plant Physiol.* 23, 803-806.

Nanjo, T., Kobayashi, M., Yoshiba, Y., Kakubari, Y., Yamaguchi-Shinozaki, K. and Shinozaki, K. 1999a. Antisense suppression of proline degradation improves tolerance to freezing and salinity in *Arabidopsis thaliana. FEBS Lett.* 461, 205-210.

Nanjo, T., Kobayashi, M., Yoshiba, Y., Sanada, Y., Wada, K., Tsukaya, H., Kakubari, Y., Yamaguchi-Shinozaki, K. and Shinozaki, K. 1999b. Biological functions of proline in morphogenesis and osmotolerance revealed in antisense transgenic *Arabidopsis thaliana. Plant J.* 18, 185-193.

Niu, X., Bressan, R.A., Hasegawa, P.M. and Prado, J.M. 1995. Ion homeostasis in NaCl stress environments. *Plant Physiol.* 109, 735-742.

Noctor, G. and Foyer, C. H. 1998. Ascorbate and glutathione: Keeping active oxygen under control. *Annu. Rev. Plant Physiol. Plant Mol. Biol.* 49, 249-279.

Nuccio, M.L., Russell, B.L., Nolte, K.D., Rathinasabapathi, B., Gage, D.A. and Hanson, A.D. 1998. The endogenous choline supply limits glycine betaine synthesis in transgenic tobacco expressing choline monooxygenase. *Plant J.* 16, 487-498.

Pardo, J.M., Reddy, M.P., Yang, S., Maggio, A., Huh, G.-H., Matsumoto, T., Coca, M.A., Paino-D'Urzo, M., Koiwa, H., Yun, D.-J., Watad, A.A., Bressan, R.A. and Hasegawa, P.M. 1998. Stress signaling through Ca^{2+}/calmodulin-dependent protein phosphatase calcineurin mediates salt adaptation in plants. *Proc. Natl. Acad. Sci. USA.* 95, 9681-9686.

Petrusa, L.M. and Winicov, I. 1997. Proline status in salt-tolerant and salt-sensitive alfalfa cell lines and plants in response to NaCl. *Plant Physiol. Biochem.* 35, 303-310.

Pilon-Smits, E.A.H., Ebskamp, M.J.M., Paul, M.J., Jeuken, M.J.W., Weisbeek, P.F. and Smeekens, S.C.M. 1995. Improved performance of transgenic fructan-accumulating tobacco under drought stress. *Plant Physiol.* 107, 125-130.

Qin, X. and Zeevaart, J.A.D. 1999. The 9-cis-epoxycarotenoid cleavage reaction is the key regulatory step of abscisic acid biosynthesis in water-stressed bean. *Proc. Natl. Acad. Sci. USA.* 96, 15354-15361.

Ray, I. M., Townsend, M.S., Muncy, C.M. and Henning, J.A.. 1999. Heritabilities of water-use efficiency traits and correlations with agronomic traits in water-stressed alfalfa. *Crop Sci.* 39, 494-498.

Rentsch, D., Hirner, B., Schmelzer, E. and Frommer, W.B. 1996. Salt stress-induced proline transporters and salt stress-repressed broad specificity amino acid permeases identified by suppression of a yeast amino acid permease-targeting mutant. *Plant Cell* 8, 1437-1446.

Ribaut, J.-M. and Hoisington, D. 1998. Marker-assisted selection: new tools and strategies. *Trends Plant Sci.* 3, 236-239.

Romero, C., Belles, J.M., Vaya, J.L., Serrano, R. and Culianez-Macia, F. 1997. Expression of the yeast *trehalose-6-phosphate synthase* gene in transgenic tobacco plants: pleiotropic phenotypes include drought tolerance. *Planta* 201, 293-297.

Roxas, V.P., Smith, R.K., Allen, E.R. and Allen, R.D. 1997. Overexpression of glutathione S-transferase/glutathione peroxidase enhances the growth of transgenic tobacco during stress. *Naure. Biotechnol.* 15, 988-991.

Rubio, F., Gassmann, W. and Schroeder, J. I. 1996. Sodium-driven potassium uptake by the plant potassium transporter HKT1 and mutations conferring salt tolerance. *Science* 270, 1660-1663.

Saha, V., Chaplin, T., Gregorini, A., Ayton, P. and Yound, B.D. 1995. The leukemia-associated-protein (LAP) domain, a cysteine-rich motif, is present in a wide range of proteins, including MLL, AF10, and MLLT6 proteins. *Proc. Natl. Acad. Sci. USA.* 92, 9737-9741.

Sakamoto, A., Alia and Murata, N. 1998. Metabolic engineering of rice leading to bio-synthesis of glycinebetaine and tolerance to salt and cold. *Plant Mol. Biol.* 38, 1011-1019.

Saneoka, H., Nagasaka, C., Hahn, D.T., Yang, W.-J., Premachandra, G.S., Joly, R.J. and Rhodes, D. 1995. Salt tolerance of glycinebetaine-deficient and containing maize lines. *Plant Physiol.* 107, 631-638.

Serrano, R., Culianz-Macia, A. and Moreno, V. 1999. Genetic engineering of salt and drought tolerance with yeast regulatory genes. *Scientia Hort.* 78, 261-269.

Sheen, J. 1996. Ca^{2+} -dependent protein kinases and stress signal transduction in plants. *Science* 274, 1900-1902.

Shen, B., Jensen, R.G. and Bohnert, H.J. 1997. Increased resistance to oxidative stress in transgenic plants by targeting mannitol biosynthesis to chloroplasts. *Plant Physiol.* 113, 1177-1183.

Sheveleva, E., Chmara, W., Bohnert, H.J. and Jensen, R.G. 1997. Increased salt and drought tolerance by D-ononitol production in transgenic *Nicotiana tabacum* L. *Plant Physiol.* 115, 1211-1219.

Sheveleva, E.V., Marquez, S., Chmara, W., Zegeer, A., Jensen, R.G. and Bohnert, H.J. 1998. Sorbitol-6-phosphate dehydrogenase expression in transgenic tobacco. *Plant Physiol.* 117, 831-839.

Shinozaki, K. and Yamaguchi-Shinozaki, K. 1999. "Molecular responses to drought stress". In: *Cold, drought, heat and salt stress in higher plants*, eds. K. Shinozaki, K and Yamaguchi-Shinozaki. R.G. Landes Co., Austin.

Shinozaki, K., Yamaguchi-Shinozaki, K., Liu, Q., Kasuga, M., Ichimura, K., Mizoguchi, T., Urao, T., Miyata, S., Nakashima, K., Shinwari, Z.K., Sakuma, Y., Ito, T. and Seki, M. 1999. "Molecular responses to drought stress in plants: regulation of gene expression and signal transduction". In: *Plant Responses to Environmental Stress.*, eds. M. F. Smallwood, C. M. Calvert and D. J. Bowles. pp. 133-143. BIOS Scientific Publishers. Oxford.

Su, J., Shen, Q., Ho, T.-H.D. and Wu, R. 1998. Dehydration-stress-regulated transgene expression in stably transformed rice plants. *Plant Physiol.* 117, 913-922.

Tanji, K.K. 1990. *Agricultural salinity and management* NY, USA: Irrigation and Drainage Division, American Society of Civil Engineers.

Tarczynski, M.C., Jensen, R.G. and Bohnert, H.J. 1993. Stress protection of transgenic tobacco by production of the osmolyte mannitol. *Science* 259, 508-510.

Taybi, T. and Cushman, J.C. 1999. Signaling events leading to crassulacean acid metabolism induction in the common ice plant. *Plant Physiol.* 121, 545-555.

Teulat, B., This, D., Khairallah, M., Borries, C., Ragot, C., Sourdille, P., Leroy, P., Monneveux, P. and Charrier, A. 1998. Several QTLs involved in osmotic-adjustment trait variation in barley (Hordeum vulgare L.). *Theor. Appl. Gen.* 96, 688-698.

Thomas, J.C., Sepahi, M., Arendall, B. and Bohnert, H.J. 1995. Enhancement of seed germination in high salinity by engineering mannitol expression in *Arabidopsis thaliana*. *Plant Cell Envir.* 18, 801-806.

Torsethaugen, G., Pitcher, L.H., Zilinskas, B.A. and Pell, E.J. 1997. Overproduction of ascorbate peroxidase in the tobacco chloroplast does not provide protection against ozone. *Plant Physiol.* 114, 529-537.

Trossat, C., Rathinasabapathi, B., Weretilnyk, E.A., Shen, T.-L., Huang, Z.-H., Gage, D.A. and Hanson, A.D. 1998. Salinity promotes accumulation of 3-dimethylsulfonio-proprionate and its precursor S-methylmethionine in chloroplasts. *Plant Physiol.* 116, 165-171.

Tyystjarvi, E., Riikonen, M., Arisi, A.-C.M., Kettunen, R., Jouanin, L. and Foyer, C. H. 1999. Photoinhibition of photosystem II in tobacco plants overexpressing glutathione reductase and poplars overexpressing susperoxide dismutase. *Physiol. Plant.* 105, 409-416.

Urao, T., Katagiri, T., Mizoguchi, T., Yamaguchi-Shinozaki, K., Hayashida, N. and Shinozaki, K. 1994. Two genes that encode Ca^{2+} -dependent protein kinases are induced by drought and high-salt stresses in *Arabidopsis* thaliana. *Mol. Gen. Genet.* 244, 331-340.

Urao, T., Yakubov, B., Yamaguchi-Shinozaki, K. and Shinozaki, K. 1998. Stress-responsive expression of genes for two-component response regulator-like proteins in *Arabidopsis thaliana*. *FEBS Lett.* 427, 175-178.

Van Camp, W., Capiau, K., Van Montagu, M., Inze, D. and Slooten, L. 1996. Enhancement of oxidative stress tolerance in transgenic tobacco plants overproducing Fe-superoxide dismutase in chloroplasts. *Plant Physiol.* 112, 1703-1714.

van der Luit, A.H., Olivari, C., Haley, A., Knight, M.R. and Trewawas, A.J. 1999. Distinct calcium signaling pathways regulate clamodulin gene expression in tobacco. *Plant Physiol.* 121, 705-714.

Verma, D.P.S. 1999. "Osmotic stress tolerance in plants: role of proline and sulfur metabolisms". In: *Molecular Responses to Cold, Drought, Heat and Salt Stress in Higher Plants*, eds. K. Shinozaki, K. and Yamaguchi-Shinozaki. pp. 153-168. R.G. Landes Co., Austin.

Werner-Fraczek, J.E. and Close, T.J. 1998. Genetic studies of *Triticeae* dehydrins: assignment of seed proteins and a regulatory factor to map positions. *Theor. Appl. Genet.* 97, 220-226.

Winicov, I. 1990. Gene expression in salt tolerant alfalfa cell cultures and the salt tolerant plants regenerated from these cultures. In: *Progress in Plant Cellular and Molecular Biology*. pp. 301-310. Kluwer Academic Press, Dordrecht.

Winicov, I. 1991. Characterization of salt tolerant alfalfa (Medicago sativa L.) plants regenerated from salt tolerant cell lines. *Plant Cell Rep.* 10, 561-564.

Winicov, I. 1993. cDNA encoding putative zinc finger motifs from salt-tolerant alfalfa (*Medicago sativa* L.) cells. *Plant Physiol.* 102, 681-682.

Winicov, I. 1996. Characterization of rice (*Oryza sativa* L.) plants regenerated from salt-tolerant cell lines. *Plant Sci.* 113, 105-111.

Winicov, I. 1998. New molecular approaches to improving salt tolerance in crop plants. *Ann. Bot.* 82, 703-710.

Winicov, I. 2000. Alfin1 transcription factor overexpression enhances plant root growth under normal and saline conditions and improves salt tolerance in alfalfa. *Planta* 210, 416-422.

Winicov, I. and Bastola, D.R. 1997. Salt tolerance in crop plants: new approaches through tissue culture and gene regulation. *Acta Physiol. Plant.* 19, 435-449.

Winicov, I. and Bastola, D.R. 1999. Transgenic overexpression of the transcription factor Alfin1 enhances expression of the endogenous MsPRP2 gene in alfalfa and improves salinity tolerance of the plants. *Plant Physiol* 120, 473-80.

Winicov, I. and Shirzadegan, M. 1997. Tissue specific modulation of salt inducible gene expression: callus versus whole plant response in salt tolerant alfalfa. *Physiol. Plant.* 100, 314-319.

Winicov, I., Waterborg, J.H., Harrington, R.E. and McCoy, T.J. 1989. Messenger RNA induction in cellular salt tolerance of alfalfa (Medicago sativa). *Plant Cell Rep.* 8, 6-11.

Wu, S.-J., Ding, L. and Zhu, J.-K. 1996. SOS1, a genetic locus essential for salt tolerance and potassium acquisition. *Plant Cell* 8, 617-627.

Xu, D., Duan, X., Wang, B., Hong, B., Ho, T.-H.D. and Wu, R. 1996. Expression of a late embryogenesis abundant protein gene, HVA1, from barley confers tolerance to water deficit and salt stress in transgenic rice. *Plant Physiol.* 110, 249-257.

Xu, Q., Fu, H.-H., Gupta, R. and Luan, S. 1998. Molecular characterization of a tyrosine-specific protein phosphatase encoded by a stress-responsive gene in *Arabidopsis*. *Plant Cell* 10, 849-857.

Yang, W.J., Nadolska-Orczyk, A., Wood, K.V., Hahn, D.T., Rich, P.J., Wood, A.J., Saneoka, H., Premachandra, G.S., Bonham, C.C., Rhodes, J.C., Joly, R.J., Samaras, Y., Godsbrough, P.B. and Rhodes, D. 1995. Near-isogenic lines of maize differing for glycinebetaine. *Plant Physiol.* 107, 621-630.

Yoshiba, Y., Kiyosue, T., Katagiri, T., Ueda, H., Mizoguchi, T., Yamaguchi-Shinozaki, K., Wada, K., Harada, Y. and Shinozaki, K. 1995. Correlation between the induction of a

gene for Δ^1-pyrroline-5-carboyxylate synthetase and accumulation of proline in *Arabidopsis thaliana* under osmotic stress. *Plant J.* 7, 751-760.
Zhang, J., Nguyen, H.T. and Blum, A. 1999. Genetic analysis of osmotic adjustment in crop plants. *J. Exp. Bot.* 50, 291-302.
Zhu, B., Su, J., Chang, M., Verma, D.P.S., Fan, Y.-L. and Wu, R. 1998. Overexpression of a Δ^1 -pyrroline-5-carboxylate synthetase gene and analysis of tolerance to water- and salt-stress in transgenic rice. *Plant Sci.* 139, 41-48.

Chapter 7

THE RESPONSES OF PLANTS TO PATHOGENS

David B. Collinge[1] Jonas Borch[1], Kenneth Madriz-Ordeñana[2] and Mari-Anne Newman[1]

[1]*Department of Plant Biology, Royal Veterinary and Agricultural University 1871-Frederiksberg C, Denmark.*
dbc@kvl.dk
[2]*Centro de Investigación en Biología Celular y Molecular (CIBCM) Universidad de Costa Rica San José, Costa Rica.*
kmadriz@cibcm.ucr.ac.cr

INTRODUCTION

Some of the most serious and universal challenges faced by plants come from pathogenic microorganisms. These represent highly diverse types of organisms ranging from viruses, bacteria, Oomycetes, protozoa and fungi *sensu stricto* (ascomycetes, basidiomycetes), see Agrios (1997), for an overview. In addition, aphids and nematodes often induce similar responses in the plant as microorganisms. Plants have responded to this onslaught by evolving a plethora of defence mechanisms. These represent visible physical attributes and inducible alterations in the structure of exposed organs and tissues as well as the more cryptic production of chemicals and proteins which can damage or inhibit the development of the pathogen (see table 1). The defence mechanisms can be induced following the perception of the pathogen, and/or constitutively present in the host (figure 1). Recent advances in molecular techniques have led to an increased understanding of the regulation of the defence mechanisms and the role of different signal transduction pathways in their regulation. The knowledge gained has also led to the demonstration that manipulation (addition, alteration in regulation or removal by antisense technology) of a single defence factor in a plant can alter the outcome of the interaction, for example, resulting in reduced levels of infection by a pathogen (see Cornelissen and Schram, 2000).

M.J. Hawkesford and P. Buchner (eds.), Molecular Analysis of Plant Adaptation to the Environment, 131–158.

Table 1. Examples of induced defences in plants

Defence mechanism	Examples/description	Examples of induction	Activity
PR proteins	PR-1 β-1,3-glucanase Chitinase	Salicylic acid, constitutively present in some tissues	Antimicrobial proteins with specific activities against e.g. Oomycetes or fungi
Defensins	Antimicrobial proteins	Jasmonic acid	Antimicrobial proteins
Phytoalexins	Pisatin Rishitin	Microbial polysaccharides	Antimicrobial secondary metabolites
Lignin	Polyphenolic compounds	Pathogens, elicitors	Physical barrier e.g. cell wall apposition
Callose	β-1,3-glucan	Pathogens, wounding	Physical barrier e.g. cell wall apposition
Cell wall proteins	Hydroxyproline rich glycoprotein, Glycine-rich protein	Pathogens, elicitors	Physical barrier e.g. cell wall apposition
Hypersensitive response	Programmed localised cell death	Avr gene products, active oxygen species	Restricts pathogen growth (particularly biotrophs). Releases secondary signals
SAR	Systemic acquired resistance	pathogens, HR, SA	Broad-spectrum induced resistance
ISR	Induced systemic resistance	non-pathogenic rhizobacteria	Broad-spectrum induced resistance

The study of defence mechanisms started with microscopic observations, which demonstrated that infection with pathogenic microbes is not always successful, but can be arrested by visible changes in the host cells. These morphological changes include induced programmed cell death and structural alterations of the cell wall of the challenged cell. The cell wall appositions or papillae can be surrounded by a halo, as in the cereal-powdery mildew interactions (see Thordal-Christensen *et al.*, 2000).

Interestingly, the defence response induced in a particular host is essentially qualitatively the same irrespective of the type of the infecting pathogen though the timing and degree of the responses differ. Each individual defence mechanisms are effective against only a few specific organisms. For example, chitinase is most effective against ascomycete fungi but not against viruses. The significance of these observations lies in the means by which pathogen attack is perceived. Pathogens inadvertently produce many chemically highly diverse molecules during the infection process. Plants have developed an ability to perceive some of these molecules and use the perception event to activate their defence responses. The molecules detected in this way are termed elicitors. The same elicitor can be produced in an interaction with taxonomically distinct pathogens that require radically different plant defence mechanisms to prevent their attack. In different situations, the elicitor may be a molecule produced and secreted by the pathogen, a molecule released from the pathogen following the action of the host defences or a molecule released from the host cell following the action of enzymes produced by the host. Furthermore, different types of elicitor can be present in the same interaction (see Ebel and Cosio, 1994).

The interaction between pathogens and plants fall into two general categories: compatible, leading to sustained spread of the pathogen and symptom development in the host (a susceptible host infected by a virulent pathogen), or incompatible, resulting in very limited pathogen growth, absence of observable disease symptoms and often a localised necrotic lesion. This very rapid resistance reaction is termed the hypersensitive response (HR) (Klement, 1982; Klement, 1963), and is believed to result in the limitation of pathogen spread and growth. In this case, the interaction is incompatible, the pathogen is termed avirulent and the host resistant. Recognition between plants and the potential pathogen is believed to determine the ultimate result of the interaction. Two levels of specificity leading either to resistance or susceptibility are recognised. The first is referred to as "non-host resistance". In this case all the cultivars of a given plant species are resistant to infection by a particular pathogen (Heath, 2000b). The second is "race-cultivar" specificity, which refers to specificity shown by different cultivars of the same plant species to different races of the same pathogenic species. Race specificity is based on the concepts of the "gene-for-gene hypothesis" (Flor, 1971). A pathogenicity factor or virulence determinant is a factor such as a toxin, hydrolytic enzyme or inhibitor which is necessary for the pathogen to attack the host. Note that the term virulence is also used in relation to race-specificity (above) and is also used to describe the aggressiveness of the pathogen.

Biological studies very early demonstrated that many plants can develop resistance ("immunity") to bacteria, viruses or fungi following an initial inoculation with a necrotising pathogen (Chester, 1933). Several distinct phenomena are implicated; these are under the regulation of distinct signal transduction pathways. The resistance is initially localised to the primary infected leaf, in 3-4 days it extends to plant tissue distant from the initial infection, and can persist for several weeks to months providing protection against a broad spectrum of plant pathogens. This localised acquired resistance (LAR) and systemic acquired resistance (SAR) was extensively characterised by Ross (1961a,b). He demonstrated that inoculation of a tobacco leaf with tobacco mosaic virus (TMV) which caused a local lesion, reduced the severity of subsequent infections with TMV in the distal, untreated portion of the plant. SAR has since been demonstrated in many plant species in response to bacterial, fungal and viral pathogens, and can also be induced following treatment with various chemicals. LAR has been shown as an effective form of protection in leaves of many cereals to subsequent infection with powdery mildew fungi (Thordal-Christensen *et al.*, 2000). A second induced systemic resistance was first described in 1991. It was demonstrated that rhizosphere-colonising *Pseudomonas* spp. have the potential to enhance pathogen resistance of the host plant (Van Peer *et al.*, 1991; Wei *et al.*, 1991). This type of induced resistance, termed rhizobacteria-mediated induced systemic resistance (ISR) (Pieterse *et al.*, 1996), has been demonstrated in different plant species under conditions in which the rhizobacteria remain spatially separated from the challenging pathogen (van Loon *et al.*, 1998). ISR resembles SAR in that both types of induced resistance render uninfected plant parts more resistant towards a broad spectrum of plant pathogens. They differ in their signal transduction pathways: in contrast to SAR, the rhizobacteria-mediated ISR is independent of salicylic acid (SA) and pathogenesis related (PR) gene activation, instead ISR requires responsiveness to jasmonic acid and ethylene (Pieterse *et al.*, 1998). Collectively, the induced resistance phenomena demonstrate that the reaction of the host to the pathogen includes components that have a role in a successful defence. The development of molecular techniques allowing the study of the genes induced in defence of plants has led to the cataloguing of many genes upregulated as part of the defence. A recent review (Rushton and Somssich, 1998) lists over 100 defence-related genes from a wide range of plant species, many of which are present in all species examined to date. The challenge is to demonstrate the importance of individual genes in a specific plant microbe interaction.

A number of terms have been coined to describe the gene transcripts that accumulate following attack by pathogens (Dixon and Harrison, 1990; Gregersen *et al.*, 1997; Rushton and Somssich, 1999; Scott *et al.*, 1990; Gregersen *et al.*, 1997). Throughout this review we refer to the plant response to pathogens as the defence response. This choice of term reflects the fact that the same defence-related gene transcripts are expressed in both compatible and incompatible interactions. There is usually, but not always (Gregersen *et al.*, 1997), a difference in timing so the defences are induced later in compatible than in incompatible interactions; in the diseased state the host plant cannot therefore be regarded as a passive partner. Other forms of abiotic stress also induce many defence-related genes.

Two approaches have been taken for identifying the components of defence responses and a third is now emerging at the dawn of the post genomic era. The targeted approach is used when a gene encoding a component of a known mechanism is needed, for example, (1) to obtain a known defence-related gene in a new host species, (2) to obtain the gene for an enzyme in a biosynthetic pathway, (3) to follow a lead given by histological data, or (4) to fill in the gaps left by the shotgun approach (see below), where these can be identified. While this approach is invaluable for increasing knowledge of individual systems, molecular biologists and geneticists favour the second approach: the shotgun approach. The plant is inoculated with a pathogen, then transcripts (and proteins) which are only present in the inoculated tissue but not in the uninoculated tissue are isolated and identified (see table 2 for examples). Previously this approach has lead to the identification of novel genes in defence, and the demonstration that genes that were not known to have a defence-related role were actually upregulated. The disadvantage of this approach is that it depends on the quality of the sequence databases. Clearly this is a problem which is decreasing as the databases increase in extent.

Even though the completion of sequenced genomes has led to the presence of still many unidentified genes, the post-genomic approach is a logical development of the shotgun approach and is inherently less prone to stochastic problems caused by sample size since all genes are (in theory) investigated simultaneously. The post-genomic approach includes sequence analyses of whole genomes and the use of DNA chip technology to study plant gene expression, often termed functional genomics or the transcriptome (see Richmond and Somerville, 2000; Zhu and Wang, 2000). This approach allows a greater understanding of plant defence responses. The use of micro-array analysis demonstrated that up to 705 genes of 2375 from *Arabidopsis* were either up- or down-regulated after inoculation with the non-host fungal pathogen *Alternaria brassicicola* or after treatment with the defence signal compounds salicylic acid, methyl jasmonate or ethylene (Schenk *et al.,* 2000). This kind of analysis is also being extended to determine the effects of mutation on the regulation of the defence response (Petersen *et al.,* 2000; Schenk *et al.,* 2000). The completion of the *Arabidopsis* sequence (The Arabidopsis Genome Initiative, 2000) means that it is now relatively trivial to determine the frequency of specific gene families in the genome. New hypotheses are emerging through *in silico* analyses of the data bases (Michelmore, 2000).

Table 2. Examples of application of shotgun cloning techniques to plant-pathogen interactions

Technique	**System**	**Transcript species identified**	**Reference**
Differential hybridisation of cDNA libraries	Barley/powdery mildew	Many species including PR-proteins	Collinge *et al.*, 2000
Subtractive hybridisation of cDNA libraries	barley/powdery mildew	4 spp. including 14-3-3 proteins, peroxidase & GRP94	Collinge *et al.*, 2000
Differential display	*Helianthus annuus* - *Plasmopara halstedii*	Full-length, pathogen, wounding, auxin, 2,4-D, and SA-induced	Mazeyrat *et al.*, 1999
Differential display	Soybean suspension cultures - *Pseudomonas syringae* pv glycinea	6 induced by HR	Seehaus and Tenhaken, 1998
cDNA-AFLP	Tobacco suspension cultures - *Cladosporium fulvum*	290 of 30,000	Durrant *et al.*, 2000
DNA chip technology	*Arabidopsis thaliana* - *Alternaria brassicicola*	705 of 2375 genes either up- or down-regulated	Petersen *et al.*, 2000; Schenk *et al.*, 2000

THE DEFENCES OF PLANTS

The defences of plants, though numerous, can be grouped in classes on the basis of their regulation or mode of action. In this section we describe them by mode of action and describe their regulation in the next section.

The hypersensitive response

It was observed over 100 years ago that some resistant hosts exhibit localised necrosis of infected cells, which is often restricted to a single cell (Stakman, 1915). This phenomenon is now known as the hypersensitive response (HR) (Klement, 1963). HR usually restricts the growth of the pathogen and the pathogen often dies in the necrotic tissue, especially when this is a biotroph. HR can kill the host cell in a few hours and it is triggered by pathogens, not saprophytes. This is probably related to pathogenicity factors: pathogens need to interact with the host to grow and thus induce host defence responses whereas saprophytes do not. The HR is associated both with race-specific resistance and with non-host resistance (see introduction). HR has been described as programmed cell death. The term apoptosis, which is used for this kind of programmed cell death in animal systems (Wyllie, 1981), has been used to describe HR in plant-pathogen systems. Animal apoptosis and plant HR clearly share many features, such as DNA laddering and the accumulation of active oxygen species (Gilchrist, 1998). The regulation of its induction is not entirely understood though several genes have been identified which contribute (Dangl *et al.,* 1996; Greenberg, 2001; Heath, 2000a). In some plant/pathogen systems, active oxygen species are involved in its regulation (see below). There is also a considerable body of evidence for a role for ion fluxes in the regulation of defence, especially in relation to the HR. For example, K^+ efflux and H^+ influxes occur early in the HR, and these seem dependent on H^+-ATPase activity, which is thought to result in the opening of a Ca^{2+} channel (Atkinson and Baker, 1987). The blockage of calcium channels with inhibitors resulted in reduced HR, while forced opening using an ionophore resulted in the stimulation of HR (Levine *et al.,* 1996). There is also evidence for an involvement of lipids in the signaling process regulating HR (Croft *et al.,* 1990; Herbers *et al.,* 1996).

A receptor-mediated signal transduction event is responsible for triggering HR and other associated events, how this works is still not understood in its entirety. In tomato, HR mediated by the specific resistance gene *Pto*, the protein kinase *Pti1* is implicated in the defence pathway which leads to HR (Zhou *et al.,* 1995). It is not only HR *per se* which is necessarily responsible for resistance. For example, a rapid and localised, massive accumulation of phytoalexins (see below) can be important in restricting pathogen growth.

Finally, resistance is not always associated with HR, for instance the *mlo* mutant of barley operates at the stage of penetration of the powdery mildew fungus (reviewed by (Jørgensen, 1994; Thordal-Christensen *et al.,* 2000). Whereas *mlo* is race non-specific, resistance governed by the *DND1* locus of *Arabidopsis* controls the phenotype of race-specific resistance without HR (Yu *et al.,* 1998). *Arabidopsis* mutants which show increased levels of resistance but lack the manifestation of localised cell death have been described; these *cim* (constitutive immunity) mutants show no obvious HR phenotype but exhibit high levels of gene expression involved in SAR and resistance to virulent fungal and bacterial pathogens (Ryals *et al.,* 1996). These and other mutational studies suggest that more than one pathway may lead to programmed cell death in plants (Heath, 2000a).

There is increasing evidence that some host-specific toxins produced by certain necrotrophic pathogens act by causing an HR- or apoptotic-like necrosis (Gilchrist, 1998). The best-documented examples of this mode of action for a pathogen comes from the interaction between oats and *Cochliobolus victoriae* (Navarre and Wolpert, 1999) and tomato with *Alternaria alternata* (Wang *et al.*, 1996). Similarly, the broad host range necrotrophic fungi *Botrytis cinerea* and *Sclerotinia sclerotiorum* appear to induce an HR-like necrosis in order to live off the dead tissue (Govrin and Levine, 2000). Though the mechanism by which they achieve this is not understood, it would appear to involve active oxygen species (AOS), and, in *Arabidopsis*, to be dependent on the DND1 gene.

Modifications of the cell wall

Resistance of plants to pathogens depends on both preformed and induced mechanisms. Examples of preformed barriers include the presence of the plant cuticle. Papillae are induced fortifications of the cell wall (cell wall appositions), which form subjacent to a penetration attempt or infection site by a pathogen in many plant species (Brown *et al.*, 1993; Thordal-Christensen *et al.*, 2000). They, depending on the plant species, comprise a broad spectrum of substances ranging from elemental *sulfur*, silicon, callose (β-1,3-glucan), structural proteins, and phenolic compounds such as lignin. The components of papillae are often oxidatively cross-linked to each other. This process is believed to use hydrogen peroxide and the enzyme peroxidase. Hydrogen peroxide can be detected in papillae as they form (Thordal-Christensen *et al.*, 1997). In barley responding to the powdery mildew fungus, many of the same components accumulate both in the walls of cells undergoing a hypersensitive response and in papillae, namely phenolic compounds such as lignin, guanidine-containing compounds, proteins and H_2O_2 (Aist and Bushnell, 1991; Thordal-Christensen *et al.*, 2000).

Papillae can arrest the growth of fungi apparently by preventing penetration. It has been suggested that papillae arrest the growth of fungi both through the oxidative cross-linking which restrains penetration, and by allowing the mobilisation of chemical defences. Some of the components of the cell wall (*e.g.* lignin monomers, hordatines) possess a direct antimicrobial activity (von Röpenack *et al.*, 1998), others can inhibit the action of microbial hydrolytic enzymes involved in the breakdown of the cell wall (Ishihara *et al.*, 2000). The larger papillae formed in *mlo* barley interacting with the powdery mildew fungus correlates with enhanced resistance (Skou, 1982). However, there is no experimental evidence for their exact role in defence. Mutants of *Arabidopsis* exhibiting reduced penetration resistance to *Blumeria graminis* exist but the nature of the mutations is not understood (H. Thordal-Christensen *pers. comm.*).

Antimicrobial and PR proteins

Approximately twenty classes of antimicrobial proteins are now known which differ substantially in structure and mode of action. Despite many years of effort, the biochemical activities and mode of action has not been resolved for many antimicrobial proteins. These classes can include both induced and preformed species (Broekaert *et al.,* 2000). The term pathogenesis-related (or PR-) protein is used to describe proteins which accumulate specifically in response to pathogens (van Loon *et al.,* 1994). Fourteen families of PR-protein are currently recognised (van Loon and van Strien, 1999). The name originates from the high levels of these proteins observed in virus-infected tissues (van Loon, 1970). Thus they were originally suggested to be induced by the virus to facilitate infection (Antoniw *et al.,* 1980). Whereas many of these possess *in vitro* antimicrobial activities, antimicrobial activities have not been demonstrated for all PR-proteins. To add to their complexity several PR-protein families have both preformed and induced members. In addition, the same protein can be under developmental regulation, being constitutively present in specific tissues, such as seed, flower parts or roots, and also accumulate in response to pathogens in other tissues, such as leaves. Finally, the accumulation of several PR-proteins is induced by the hormone ethylene. The antimicrobial proteins (including the preformed species) have been reviewed in detail recently (Broekaert *et al.,* 2000). The "classic" PR-proteins, PR-1 to PR-5, have received the greatest attention over the years and are now described in many plants.

Phytoanticipins and Phytoalexins

Phytoanticipins and Phytoalexins are chemically highly diverse, low molecular weight antimicrobial compounds (*i.e.* not proteins) (see Hammerschmidt, 1999; Mansfield, 2000 for review). Phytoanticipins are preformed antimicrobial compounds and include substances as diverse as saponins, phenylpropanoids such as stilbenes, alkaloids, cyanogenic glucosides and glucosinolates (VanEtten *et al.,* 1994). Phytoalexins, in contrast, accumulate in the plant in response to pathogen attack, or by some other forms of stress which imitate pathogen attack. Phytoalexins encompass compounds as diverse as phenylpropanoids, alkaloids, terpenes and elemental *sulfur* (see *e.g.* Agrios, 1997; Goodwin and Mercer, 1983; Lucas, 1998; Mansfield, 2000 for structures), there is a considerable overlap between the two classes: a phytoalexin in one species may be a phytoanticipin in another. Evolution to radically different new structures is arguably easier for phytoanticipins and phytoalexins than for proteins since a new structure can be achieved by an addition of a radical following the duplication and subsequent mutation of an existing gene encoding an enzyme. It has previously been suggested that stilbenes represent a good example of this as they appear to have evolved as phytoalexins/ phytoanticipins independently several times: phylogenetic analysis suggested that the enzyme stilbene synthase had evolved from chalcone synthase (Tropf *et al.*, 1994). However, a recent analysis contradicts this view (Goodwin *et al.*, 2000).

There is considerable evidence for the importance of both phytoanticipins and phytoalexins in defence. In particular, it is likely that they play a role in limiting the host range of pathogens. In some pathosystems phytoalexin accumulation is associated with the cell collapse occurring during the development of HR (see below). Their importance in defence has been challenged for the simple reason that they accumulate in both compatible and incompatible interactions. Closer study of the accumulation patterns generally demonstrates that phytoalexins accumulate more rapidly, and locally to higher concentrations in an incompatible interaction than in a compatible interaction (Tsuji *et al.*, 1992). In some cases it has been demonstrated that the local concentration of a phytoalexin is high enough to inhibit the pathogen (Keen, 1971). In a very few cases it is proved possible to inhibit the production of phytoalexins by mutation, for example, the camalexin-deficient *pad* mutants of *Arabidopsis* (Glazebrook and Ausubel, 1994; Glazebrook *et al.*, 1997a). However, when the inhibition of specific enzymes early in the pathways (especially phenylalanine ammonia-lyase, PAL) is achieved either pharmacologically of by antisense gene silencing technology, it is difficult to ascertain whether the effect is due to the inability of the host to produce phytoalexin, lignin or even the signaling molecule, salicylic acid (Blount *et al.*, 2000; Mauch-Mani and Slusarenko, 1996).

The best-documented examples of pathogen adaptation come from the study of phytoalexins and the preformed phytoanticipins (VanEtten *et al.*, 1994). It has been shown that many of the pathogens of pea (*Pisum sativum*) produce an enzyme, pisatin demethylase, which can disarm the phytoalexin pisatin to a form which is much less toxic (VanEtten *et al.*, 1995). Transfer of pisatin demethylase from the pea pathogen *Nectria haematococca* to *Cochliobolus heterostrophus* allowed this maize pathogen to cause an, albeit limited, infection on pea (Schaefer *et al.*, 1989). This indirectly suggests that phytoalexins have a role in response to fungal pathogens.

Gaeumannomyces graminis is a cereal pathogen causing the disease Take-all. Isolates (*G.g.* var. *avenae*) adapted to oat (*Avena sativa*) can infect both oat and wheat (*Triticum aestivum*), but wheat isolates (*G.g.* var. *tritici*) can only attack wheat. Oats produce the saponin phytoanticipins known as avenacins. *G.g.* var. *avenae* possesses an enzyme, avenacinase, which hydrolyses the avenacins to less toxic forms (Mansfield, 2000). Mutant *G.g.* var. *avenae* in which the gene encoding the avenacinase enzyme has been disrupted are no longer capable of infecting oats (Bowyer *et al.*, 1995). Thus in both these examples, it seems that the inability to detoxify a phytoalexin or phytoanticipin results in a restriction of the host range of a particular pathogen.

The extreme diversity of phytoalexins makes them an attractive option for improving resistance to pathogens using a transgenic approach. The main problem with this approach is the complexity of the biosynthesis of most phytoalexins; these are compounds produced by multi-step biosynthetic pathways, which means that their production requires the presence and co-regulation of a number of genes encoding the enzymes for these pathways. Nevertheless, one phytoalexin, the stilbene resveratrol, is produced by a single enzyme, stilbene synthase, from precursors that commonly accumulate in response to pathogens. It has therefore proved straightforward to engineer plants which accumulate resveratrol and demonstrate that these exhibit enhanced tolerance or resistance to certain pathogens (Hain *et al.*, 1993; Leckband and Lorz, 1998; Stark-Lorenzen *et al.*, 1997; Thomzik *et al.*, 1997).

THE REGULATION OF SIGNAL TRANSDUCTION

The nature of plant-pathogen interactions

The regulation of plant defences uses essentially the same kinds of signal transduction pathways as encountered in plant development and in other physiological processes. However, the regulation and nature of plant defences against pathogens (and pests) is more complex than against abiotic stresses. Firstly, there are essentially two different lifestyles used by pathogens. The necrotrophs obtain nutrition by killing and then consuming host tissue, using toxins and tissue-destroying enzymes in large quantities. Examples include soft rot bacteria (*Erwinia* spp.) and many major fungal pathogens. In contrast, the biotrophs invade living tissue and acquire nutrients from the host without causing excessive damage. Some biotrophs can actually stimulate their hosts to provide extra nutrients, and, at the same time, they need to avoid being detected by the host and damaged by its defences. Examples include viruses, rust fungi, powdery and downy mildews. The biotrophs, *sensu stricto*, are obligate pathogens whilst the necrotrophs are facultative pathogens that can be cultured *in vitro*. There are many intermediate and combined life styles (organisms which start as a biotrophs and complete their life cycle as a necrotrophs – e.g. *Septoria tritici* and *Cladosporium fulvum*). These extremes illustrate that the necrotrophs, through causing a lot of damage, cause the release of large amounts of, for example, cell wall components that are not released in an interaction with a biotrophic pathogen. Indeed the host can exploit this to detect pathogen attack to activate its defences. The biotrophs need to be able to suppress the activation of the host defences through manipulation of the perception and/or signal transduction pathways of the host, as well as being able to withstand the antimicrobial defences produced by their chosen host. There is increasing molecular genetic evidence for the existence of pathogenicity factors (also called virulence determinants) which suppress host defences. The best studied being those dependent on the type III secretory pathway of bacteria encoded by the *hrp/hrc* regulon (Alfano and Collmer, 1996; Kjemtrup *et al.*, 2000), which injects pathogenicity factors into the host cell. Although this is a difficult area to study at the physiological and biochemical level, it has been demonstrated that the pathogenicity factors injected are targeted to different host compartments, and, at least in one case, the product of the *avrRpt2* avirulence gene, can function to suppress host defences (Kjemtrup *et al.*, 2000). Some of these are recognised as specific elicitors by the host (*i.e.* represent avirulence gene products).

Both the necrotrophic and biotrophic lifestyles require the host to recognise the presence of the pathogen and mount a defence against it. The existence of host defences places selective pressure on the pathogen to develop mechanisms which both disguise its presence and suppress host responses. This co-evolution makes the determination of the regulation of plant defence responses more complicated than those encountered for abiotic stresses. Secondly, pathogens represent very different kinds of organisms, viruses, bacteria, Oomycetes (algae), protozoa and fungi (ascomycetes and basidiomycetes). The biotrophic and necrotrophic lifestyles are found in taxonomically diverse pathogens. This means that whereas the same defences need to be overcome, and some of the same strategies may be employed to do so, the details differ; *i.e.* the molecules exhibited on the surface of the pathogen, or injected into the host to suppress defences can be radically different. There is therefore a need for diversity in the detection systems used against different pathogens. Indeed, plants detect foreign molecules in the apoplast, inside their cytoplasm and the products of pathogenic activity, such as the oligosaccharides released by the action of hydrolytic enzymes from the cell wall (see Alfano and Collmer, 1996). This is reflected in the large numbers of individual putative receptors that have been detected in plants (see figure 1). For instance, with the completion of sequencing of the *Arabidopsis* genome, approximately 120 examples of the NBS-LRR (Nucleotide Binding Sites-Leucine Rich Repeat) super-family of disease resistance genes are now known (The Arabidopsis Genome Initiative, 2000) and it is estimated that there are perhaps as many as 750 to 1500 in rice (Meyers *et al.*, 2001). *Arabidopsis* also possesses about 30 homologues of the "Clavata/Cf" classes of genes; these are known to function as resistance genes in tomato, 174 receptor kinases and 860 serine-threonine protein kinases. Furthermore, plants do not recognize that a particular molecule is always produced by a specific taxon (*e.g.* an Oomycete) and do not rely on producing a defence response that is only effective against that particular type of pathogen (PR-1 in the case of Oomycetes). The entire arsenal of defences is therefore essentially always induced.

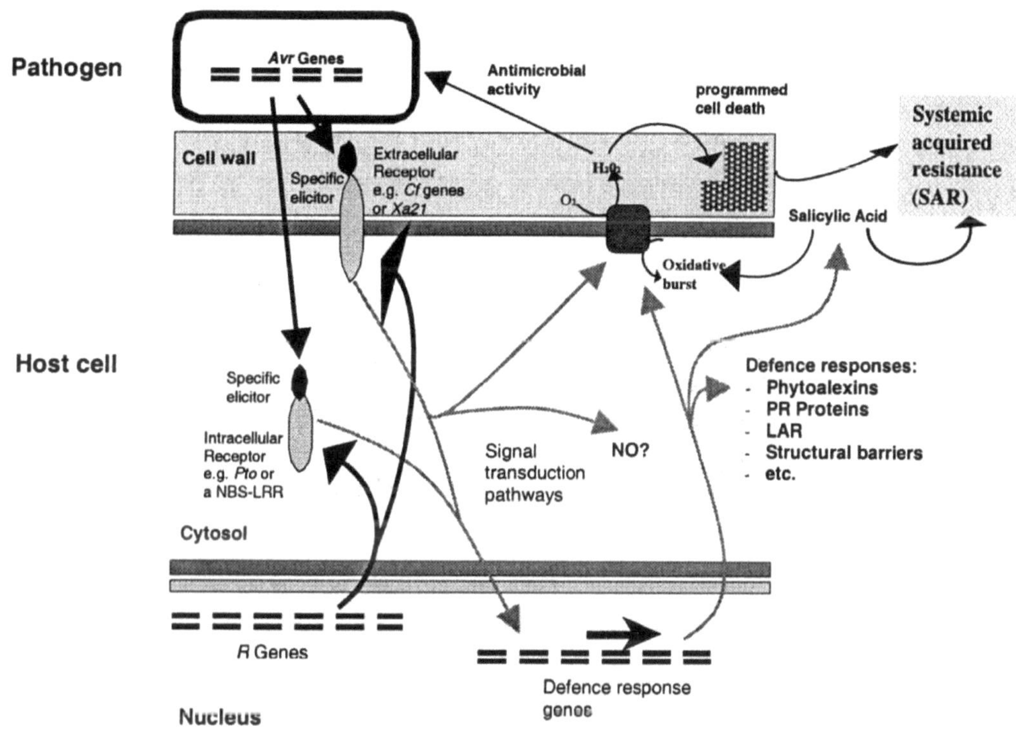

Figure 1. Simplified model for the activation of defence responses in host cells. The model is based on the current knowledge of the race-specific disease resistance and the induction of systemic and local acquired resistance.

Genes in the regulation of defence

Several signal transduction pathways have been identified which regulate the defence responses. Race specific resistance systems represent the most intensively studied forms. However mutational studies especially using *Arabidopsis* as a model plant (Feys and Parker, 2000; Glazebrook *et al.,* 1997b; Parker *et al.,* 2000), and to a lesser extent barley (Piffanelli *et al.,* 1999), have recently demonstrated that the genetic regulation of resistance is much more complicated than hitherto suspected. Interestingly, a number of different mutant screens have independently yielded several regulatory genes (*e.g. Npr1, eds1* and *pad4*) (Cao *et al.,* 1998), which are central in the regulation of both race-specific resistance, SAR and/or ISR. Furthermore, some of these mutations, namely *eds1* and *pad4,* result in enhanced susceptibility to a virulent pathogen that is associated with a compromised defence response (Aarts *et al.,* 1998; Glazebrook *et al.,* 1996). The contribution of the mutational approaches can be limited by the occurrence of lethal phenotypes and redundancy within the signaling pathways: if two genes encode similar products, then it can be extremely difficult to identify a mutant affecting only one of them (unless an ameliorated phenotype can be observed), and it is unlikely that the double mutant can be identified. Alternative molecular genetic approaches can overcome this problem. These include the use of the yeast two hybrid system (Brent and Finley, 1997) and modifications of this or biochemical screening methods based on the general principal of affinity chromatography. These approaches can, in principal, lead to the identification of components of the signal transduction pathways, which regulate or are regulated by a gene identified in a mutant screen. However, the biochemical approach is hampered by the relative scarcity of components of signal transduction in the cell. The final problem is to demonstrate that a component in a signal transduction pathway, which interacts with an important gene, is itself important: there is a risk that this component would not emerge in a mutant screen where there is redundancy in the pathway or it encodes an essential component. In this situation, the use of double-stranded RNA to silence a gene in a transient expression system can be a useful tool (Schweizer *et al.,* 2000).

Race specificity

One of the major advances in our understanding of plant-pathogen interactions is the cloning of a number of disease resistance genes. Plant resistance genes have been cloned from disparate species. For recent reviews on cloned plant resistance genes see Ellis *et al.,* (2000a), Parker *et al.,* (2000), Takken and Joosten, (2000). These genes are responsible for the recognition of some strains of a particular pathogen through the recognition of a specific elicitor produced by the pathogen, when the pathogen carries the corresponding avirulence gene (*avr* gene). The model of a ligand-receptor mechanism of recognition based on the concepts of the gene-for-gene hypothesis (Flor, 1971) has been proposed previously (Gabriel and Rolfe, 1990; Lamb, 1999). In this model, the resistance gene-encoded receptor was believed to span the plasma membrane. This still appears to be true for some cloned resistance genes such as the *Cf* family from tomato which confers resistance to several races of the biotrophic fungus *Cladosporium fulvum* (De Wit and Joosten, 1999). It now transpires that in a majority of cases of cloned resistance genes, the recognition event appears to take place inside the host cell. This could be dependent on a process analogous to the type III secretory system possessed by pathogenic bacteria. This surprising observation of recognition inside the host cell, which was initially made with bacterial *avr* genes (Scofield *et al.,* 1996) implies that a number of mainly biotrophic fungal pathogens, though not all: *Magnaporthe grisea* is an obvious exception (Jia *et al.*, 2000), also inject pathogenicity factors into the host cell (Alfano and Collmer, 1996; Kjemtrup *et al.*, 2000). According to the structural features predicted, the majority of plant resistance genes possess domains of leucine-rich repeats (LRR) of around 20-30 amino acids (Pan *et al.*, 2000) which apparently mediate protein-protein interactions (Kajava, 1998; Kobe and Deisenhofer, 1995; Leister and Katagiri, 2000). These leucine-rich domains are repeated a variable number of times: between 14 and 38 in alleles of the L locus of flax conferring resistance to rust (Ellis *et al.*, 1999). The LRR regions show a high degree of variation which can account for *avr* recognition specificity in several (Ellis *et al.*, 2000b), though not all cases (Luck *et al.*, 2000). Some of the resistance proteins contain nucleotide-binding sites (NBS), which, are predicted to be involved in signaling through the activation of kinase cascades (Ellis *et al.*, 2000b; Hammond-Kosack and Jones, 1997).

Beyond the identification of mutants necessary for the function of NBS-LRR disease resistance genes (and in some cases cloning the corresponding gene), little is known about the mechanisms by which resistance genes function (Parker *et al.*, 2000; Piffanelli *et al.*, 1999). Progress has been greatest with understanding the mode of action of the *Pto* gene of tomato which confers resistance against the bacterial pathogen *Pseudomonas syringae* pv. *tomato* (reviewed by Martin, 1999). *Pto,* the first resistance gene to be cloned, encodes a protein kinase. A gene, *Prf,* was identified in a mutant screen, which was necessary for the action of the *Pto* gene. *Prf* is a NBS-LRR gene like the major class of resistance genes. However, despite considerable effort, it has not proved possible to determine the function of the PRF protein in the signal transduction cascade. The approach, which provided the breakthrough in this system, was the use of the yeast two-hybrid system to isolate cDNAs encoding genes interacting with PTO or the AVRPTO-PTO complex. It has been demonstrated that PTO phosphorylates two of these, a protein kinase, PTI1 and a transcription factor, PTI4 *in vitro*. However, it has not proved possible to obtain evidence for their importance as both represent small gene families. PTI1 is apparently responsible for inducing HR in this system, and PTI4 (along with PTI5 and PTI6) are apparently involved in PR-gene transcription. Nevertheless, this system is significant since it seems to represent the only intact signal transduction pathway from the pathogen to the defence response identified to date.

Other components

The shotgun cloning approach has led to the identification of several other proteins as being upregulated following pathogen attack; these are therefore implicated in the regulation of the defence response independently of race specificity. Little is known of their significance.

Protein kinases

A number of protein kinases have been identified in the defence response and are implicated in different signal transduction pathways. These have been identified among the resistance genes, in mutant screens and as transcripts accumulating as part of the defence response. Thus two race – specific resistance genes, *Pto* and *Xa21* are protein kinases of different classes (see *e.g.* Baker *et al.*, 1997). Others are invoked in nitric oxide (NO) and salicylic acid (SA) signaling (Kumar and Klessig, 2000; Liu *et al.*, 2000). These and other protein kinases appear to be post-translationally activated, however, others have been identified for which the transcripts accumulate in response to pathogen attack. Thus (Lange *et al.*, 1999) identified a receptor-like protein kinase in bean roots which is induced by *Fusarium* attack. A similar receptor kinase has been identified in barley following attack by *Blumeria graminis* (Gregersen and Jensen, *pers. comm.*). Several protein kinases of the MAPK family have been implicated in the regulation of defence responses (Romeis *et al.*, 1999; Zhang and Klessig, 1998a; Zhang and Klessig, 1998b). However, it is clear that our understanding of the role of individual protein kinases in the defence response is still limited.

14-3-3 proteins

The 14-3-3 proteins are a family of eukaryotic regulatory proteins which regulate many diverse physiological processes by binding target proteins after phosphorylation (reviewed by Finnie *et al.*, 1999). One of the first demonstrations of a 14-3-3 protein in a plant species came from barley inoculated with the powdery mildew fungus (Brandt *et al.*, 1992). The transcripts for several barley 14-3-3 proteins accumulate in the infected epidermis and not in the uninfected mesophyll tissues of barley leaves, where they are constitutively present (Gregersen *et al.*, 1997). The 14-3-3 proteins of tomatoes undergo differential regulation in a defence response against *Cladosporium fulvum* and the fungal toxin fusicoccin (Roberts and Bowles, 1999). It is suggested that the interaction of the 14-3-3 proteins with the H^+-ATPase is the key event in both interactions, though other relevant targets exist (Finnie *et al.*, 1999).

Chemical inducers of resistance

A number of substances have been identified which can induce resistance against pathogens. These include active oxygen species (AOS), salicylic acid (SA, and various analogues), jasmonates (JA) and ethylene. AOS act locally and are involved in the induction of the hypersensitive response (see Scandalios this volume and Jabs (1999)); salicylates and jasmonates are involved in the systemic induction of resistance. The role of ethylene is less clear. Furthermore, SA has been suggested to work by amplifying other defence responses that are induced when an agonist like a pathogen or phosphatase inhibitor induces these (Shirasu *et al.*, 1997). With the development of *Arabidopsis* as a model for studying plant microbe interactions it has been possible to approach an understanding of the genetic regulation of these overlapping phenomena. It is now clear that, at least in *Arabidopsis*, these substances act through partly independent, partly overlapping signal transduction pathways which in some cases also involve components of the signal transduction pathways activated by race-specific resistance. These topics have been reviewed recently by (Durner *et al.*, 1997; Glazebrook *et al.*, 1997b; van Loon and van Strien, 1999).

The salicylate pathway and SAR

Systemic acquired resistance (SAR) is an inducible plant defence state, the induction of which often requires the accumulation of salicylic acid (SA). SA was first suggested to be involved in SAR signaling based on the observation that exogenously applied SA induced resistance associated with the accumulation of certain PR proteins (Uknes *et al.*, 1992; Ward *et al.*, 1991; White, 1979). SA accumulation was observed in both tobacco and cucumber tissue reacting hypersensitively to a viral infection as well as accumulating to a lesser extend in distal parts of these plants concomitant with SAR development (Malamy *et al.*, 1990; Métraux *et al.*, 1990). Further evidence for a key role of SA in SAR came from an analysis using transgenic plants expressing the *NahG* gene of *Pseudomonas putida*. This gene encodes a salicylate hydroxylase which converts salicylate to catechol (Friedrich *et al.*, 1995). Transgenic tobacco and *Arabidopsis* constitutively expressing *NahG* are blocked in their ability to accumulate significant amounts of SA and also blocked in the expression of pathogen-induced SAR, indicating that endogenous SA accumulation is an essential prerequisite for SAR development (Gaffney *et al.*, 1993; Lawton *et al.*, 1995). SA has been considered a candidate for the systemically transported SAR signal. (Shulaev *et al.*, 1995) showed that SA molecules synthesised in the primary infected leaf were subsequently found systemically throughout the plant. In contrast to these findings, grafting experiments with tobacco suggested that SA is not the systemically transported signal. Vernooij *et al.*, (1994) showed that a non-transformed scion grafted onto a TMV-infected SA-deficient NahG rootstock expressed SAR, whereas a NahG scion grafted onto a TMV-infected non-transformed rootstock failed to develop SAR. Similar results were obtained using grafting experiments between non-transformed tobacco and transgenic tobacco exhibiting suppression of the *PAL* gene, in which the biosynthesis of SA is blocked (Pallas *et al.*, 1996). Upon primary infection of a single cucumber leaf, accumulation of SA in the phloem fluids was preceded by a transient increase in PAL activity in the stems and petioles, these results suggest that SA is synthesised *de novo* in response to an early mobile signal from the inoculated leaf (Smith-Becker *et al.,* 1998). Taken together, this data suggest that even though SA is transported within the plant, it is not the systemically transported SAR signal. The subject of SA and its involvement in SAR has recently been extensively reviewed by Delaney (2000) Dempsey *et al.,* (1999) and Durner *et al.,* (1997).

Both low and high affinity SA binding proteins have been identified and cloned in tobacco. One low affinity receptor from tobacco is a catalase, which is inhibited by SA. It has been suggested that as catalase activity reduces levels of hydrogen peroxide (H_2O_2), the reduction in catalase activity would result in increased H_2O_2 levels. H_2O_2 also induces accumulation of PR proteins associated with SAR. However, as inhibition of catalase required SA levels far in excess of those present in induced tissues, it seems therefore unlikely that this catalase plays a central role in SAR signaling. A high affinity SA-binding protein (SABP) has been identified (Durner *et al.,* 1997). SABP has an even higher affinity for the SA functional analogue benzothiadiazole (BTH, commercialised as Bion), which is more effective at inducing PR gene expression than SA. Furthermore there is some evidence for the involvement of protein phosphorylation events in the SA signal transduction pathway (Conrath *et al.,* 1997; Kumar and Klessig, 2000).

The jasmonate pathway and ISR

An additional mechanism of systemic resistance is the induced systemic resistance (ISR). ISR has been described in several plants and appears to be a phenomenon similar to SAR, *i.e.* acts locally as well as systemically and induced plants show a broad-spectrum disease resistance. However, ISR operates via the jasmonic acid and ethylene pathways and in most cases is independent of SA and PR protein accumulation (Bowling *et al.,* 1997; Pieterse and van Loon, 1999). Non-pathogenic microorganisms from the rhizosphere, particularly the plant growth-promoting rhizobacteria (PGPR), are able to activate ISR, which, at least in *Arabidopsis*, is effective against fungi, bacteria and viruses (van Loon *et al.,* 1998). Using *P. fluororescens* strain WCS417r as the ISR-inducing strain and *P. syringae* pv. *tomato* DC3000 as the challenging pathogen, ISR was fully maintained in NahG *Arabidopsis* plants, and not associated with transcriptional activation of genes encoding SA-inducible PR proteins (Pieterse *et al.,* 1996). Furthermore, treatment of roots with WCS417r failed to induce ISR in both jasmonate- and ethylene-insensitive *Arabidopsis* mutant plants. Together these results indicate a SA-independent pathway, while both the jasmonate and ethylene response pathways are essential for establishment of ISR (Pieterse *et al.*, 1998).

Active oxygen and nitric oxide species

A very early event following stimulation of plants with an elicitor, in both non-host and cultivar-specific HR, is the generation of active oxygen species (AOS). AOS are toxic intermediates that result from successive one-electron steps in the reduction of molecular O_2. The free radical species detected in plant-pathogen interactions are superoxide anion (O_2^-), hydrogen peroxide (H_2O_2) the most stable of the active oxygen species and hydroxyl radical ($OH^•$) (Mehdy, 1994). The release of active oxygen species has been termed "the oxidative burst" and is correlated with HR in several plant-pathogen interactions.

The oxidative burst is thought to be initiated by receptor-mediated recognition of a pathogen (Jabs, 1999). In incompatible interactions, AOS production consists of two phases; phase I is non-specific and occurs immediately after addition of most pathogens. The phase II active oxygen response occurs just after the very early event of H^+, K^+, and Ca^{2+} ion fluxes (Jabs *et al.*, 1997). These ion-fluxes are required for the subsequent oxidative burst. Other proteins such as G-proteins and phospholipase A have also been suggested to act in signaling upstream of the oxidative burst (Chandra *et al.*, 1996; Vera-Estrella *et al.*, 1994).

Whether the signaling compound is H_2O_2, superoxide or a different AOS is not fully understood, it is most likely that several of the AOS are active. H_2O_2 is produced from superoxide by superoxide dismutase as well as non-enzymatically. The enzyme(s) that is responsible for the formation of AOS is not clear (Bolwell, 1999; Grant and Loake, 2000). NAD(P)H oxidase has often been suggested to be involved in the oxidative burst. However, many of these studies rely on the use of the NAD(P)H oxidase inhibitor DPI, which, unfortunately, also inhibits nitric oxide synthase (NOS) (Bolwell, 1999). All the active oxygen species are predicted to yield a change in the cellular redox state. It may be this change, and not a single enzyme activity *per se* that is responsible for the cellular events. A change in the cellular redox state (towards oxidation) triggers the antioxidative defence. Repression of the antioxidative defence could be another mechanism for accumulation of AOS. In the interaction between tobacco and TMV, the expression of a cytosolic form of the antioxidative enzyme ascorbate peroxidase is repressed post-transcriptionally (Mittler *et al.*, 1999). Furthermore, the amount of antioxidative compounds and enzymes change in the apoplast of barley leaves inoculated with powdery mildew (Vanacker *et al.*, 1998).

Interestingly, the antioxidative enzyme activities increased more in susceptible than in resistant lines, suggesting that the antioxidative status influence the balance between biotrophy and resistance. In support of this study, it has been demonstrated that pathogen defences are initiated in transgenic antisense catalase tobacco plants by H_2O_2 (Chamnongpol *et al.*, 1996). The molecular mechanisms of recognising the AOS signal are poorly understood in plants (see Scandalios in this book). Recently, a few genes have been described which are regulated by H_2O_2, however, the exact role of these genes in the induction of HR is not fully understood (Desikan *et al.*, 1999; Durrant *et al.*, 2000).

In animals, nitric oxide (NO) is an important redox-active signaling molecule. NO has recently been demonstrated to have an effect as a signaling molecule in plants complementary and synergistic to H_2O_2. NO is necessary and sufficient for induction of hypersensitive cell-death and necessary for race specific resistance of soybean and *Arabidopsis*. Furthermore it induces and is required for high PAL-gene expression. PAL is the first enzyme in the biosynthetic pathway for SA, an increase in SA production is observed after treatment with NO donor. Two secondary messengers known to serve in the NO pathway in mammals, cyclic ADP-ribose and cyclic GMP was demonstrated to accumulate after NO administration and induce PAL and PR-1 gene expression and SA accumulation. Though these results suggest that plants possess a NO signaling pathway similar to mammals, and though NO synthase (NOS) activity is induced in leaves attacked by pathogens, no homologues of the mammalian NOS are found in the *Arabidopsis* genome database. Other enzymes may have taken over the NO production in plants (Bolwell, 1999).

CONCLUDING REMARKS

There has been an enormous increase in the amount of detailed knowledge of plant responses to pathogen attack over the last decade. The number of genes which have been identified as pathogen-induced, has increased enormously, but more interestingly, so has the knowledge of their regulation by different defence pathways. The development of techniques for studying the expression of individual genes in transgenic plants has led to the realisation that specific defence related products have functions in resistance against specific pathogen species (or rather taxonomic groups) or in limiting the ability of opportunistic pathogens in attacking a specific host. This identification and isolation of novel defence-related genes has also led to the demonstration that transgenic plants have real potential for plant protection (Cornelissen and Schram, 2000). The cloning of some of the genes responsible for their regulation opens the possibility for new strategies. Indeed it has already been demonstrated in the laboratory that enhanced expression of *Prf* and *Npr1* leads to increased resistance against a number of pathogens (Cao *et al.*, 1998; Oldroyd and Staskawicz, 1998). There are, however, no published accounts of field trials with these or other regulatory genes, and none of these transgenic disease resistant plants have been exploited commercially so far. It will be interesting to see whether this will be possible over the next decade.

Acknowledgements

We are grateful to Uwe Conrath, Tine Ebstrup and Karin Olsen for critical comments to the manuscript, and to Hans Thordal-Christensen and Per Gregersen for allowing us to cite unpublished data.

REFERENCES

Aarts, N., Metz, M., Holub, E., Staskawicz, B.J., Daniels, M.J. and Parker. J.E. 1998. Different requirements for EDS1 and NDR1 by disease resistance genes define at least two R gene-mediated signaling pathways in *Arabidopsis*. *Proc. Natl. Acad. Sci. USA* 95, 10306-10311.

Agrios, G. N. 1997. *Plant Pathology*. Academic Press, San Diego, Calif.

Aist, J. R. and Bushnell. W. R. 1991. "Invasion of plants by powdery mildew fungi, and cellular mechanisms of resistance". In: *The Fungal spore and disease initiation in plants and animals*. eds. G.T. Cole and H.C. Hoch, pp. 321-343. Plenum Press, New York.

Alfano, J. R. and Collmer, A. 1996. Bacterial pathogenicity in plants: life up against the wall. *Plant Cell* 8, 1683-1698.

Antoniw, J.F., Ritter, C.E.,. Pierpoint, W.S, and van Loon, L.C. 1980. Comparison of three pathogenesis-related proteins from plants of two cultivars of tobacco infected with TMV. *J. Gen.Virol.* 47, 79-87.

Atkinson, M.M. and Baker, C.J. 1987. Alteration of plasmalemma sucrose transport in *Phaseolus vulgaris* by *Pseudomonas syringae* pv *syringae* and its association with K^+/H^+ exchange. *Phytopathology* 77, 1573-1578.

Baker, B., Zambryski, P., Staskawicz, B. and Dinesh-Kumar, S.P. 1997. Signaling in plant-microbe interactions. *Science* 276, 726-733.

Blount, J.W., Korth, K.L., Masoud, S.A., Rasmussen, S., Lamb, C. and Dixon, R.A. 2000. Altering expression of cinnamic acid 4-hydroxylase in transgenic plants provides evidence for a feedback loop at the entry point into the phenylpropanoid pathway. *Plant Physiol.* 122, 107-116.

Bolwell, G.P. 1999. Role of active oxygen species and NO in plant defence responses. *Curr. Opin. Plant Biol.* 2, 287-294.

Bowling, S.A., Clarke, J.D., Liu, Y.D., Klessig, D.F., and Dong, X.N. 1997. The *cpr5* mutant of *Arabidopsis* expresses both NPR1-dependent and NPR1-independent resistance. *Plant Cell* 9, 1573-1584.

Bowyer, P., Clarke, B.R., Lunness, P., Daniels, M.J. and Osbourn, A.E. 1995. Host-range of a plant-pathogenic fungus determined by a saponin detoxifying enzyme. *Science* 267, 371-374.

Brandt, J., Thordal-Christensen, H., Vad, K., Gregersen, P.L. and Collinge, D.B. 1992. A pathogen-induced gene of barley encodes a protein showing high similarity to a protein kinase regulator. *Plant Journal* 2, 815-820.

Brent, R. and Finley, R.L. 1997. Understanding gene and allele function with two-hybrid methods. *Ann. Rev. Genet.* 31, 663-704.

Broekaert, W.F., Terras, F.R.G. and Cammue, B.P.A. 2000. "Induced and Preformed Antimicrobial Proteins". In: *Mechanisms of Resistance to Plant Diseases.* eds. A.J. Slusarenko, R.S.S. Fraser, and L.C. van Loon, pp. 371-477. Kluwer Academic Publishers, Dordrecht.

Brown, I., Mansfield, J. Irlam, I. Conrads-Strauch, J. and Bonas, U. 1993. Ultrastructure of interactions between *Xanthomonas campestris* pv *vesicatoria* and pepper, including immunocytochemical localization of extracellular polysaccharides and the AvrBs3 protein. *Mol. Plant Microbe Interact.* 6, 376-386.

Cao, H., Li, X. and Dong, X.N. 1998. Generation of broad-spectrum disease resistance by overexpression of an essential regulatory gene in systemic acquired resistance. *Proc. Natl. Acad. Sci. USA* 95, 6531-6536.

Chamnongpol, S., Willekens, H., Langebartels, C., van Montagu, M., Inzé, D., and van Camp, W. 1996. Transgenic tobacco with a reduced catalase activity develops necrotic lesions and induces pathogenesis-related expression under high light. *Plant J.* 10,491-503.

Chandra, S., Heinstein, P.F. and Low, P.S. 1996. Activation of phospholipase A by plant defense elicitors. *Plant Physiol.* 110, 979-986.

Chen, Z.Y., Kloek, A.P. Boch, J. Katagiri, F. and Kunkel, B.N. 2000. The *Pseudomonas syringae avrRpt2* gene product promotes pathogen virulence from inside plant cells. *Mol. Plant Microbe Interact.* 13, 1312-1321.

Chester, K. 1933. The problem of acquired physiological immunity in plants. *Quarterly Review of Biology* 8, 275-324.

Collinge, D.B., Gregersen, P.L. and Thordal-Christensen, H. 2000. "The nature and role of defence response genes in cereals". In: *The Powdery Mildews: A Comprehensive Treatise.* eds. R.R. Belanger and W.R. Bushnell. APS Press, St. Paul, Minnesota, USA.

Conrath, U., Silva, H. and Klessig, D.F. 1997. Protein dephosphorylation mediates salicylic acid-induced expression of PR-1 genes in tobacco. *Plant Journal* 11, 747-757.

Cornelissen, B.J.C. and Schram, A. 2000. "Transgenic approaches to control epidemic spread of diseases". In: *Mechanisms of Resistance to Plant Diseases.* eds. A.J. Slusarenko, R.S.S. Fraser, and L.C. van Loon, pp. 575-599. Kluwer Academic Publishers, Dordrecht.

Croft, K.P.C., Voisey, C.R. and Slusarenko, A.J. 1990. Mechanism of hypersensitive cell collapse - correlation of increased lipoxygenase activity with membrane damage in leaves of *Phaseolus vulgaris* (L) inoculated with an avirulent race of *Pseudomonas syringae* pv *phaseolicola. Physiol. Mol. Plant Path.* 36, 49-62.

Dangl, J.L., Dietrich, R.A. and Richberg, M.H. 1996. Death don't have no mercy: Cell death programs in plant-microbe interactions. *Plant Cell* 8, 1793-1807.

de Wit, P.J. and Joosten, M.H. 1999. Avirulence and resistance genes in the *Cladosporium fulvum*-tomato interaction. *Curr. Opin Microbiol.* 2, 368-373.

Delaney, T.P. 2000. New mutants provide clues into regulation of systemic acquired resistance. *Trends Plant Sci.* 5, 49-51.

Dempsey, D.A., Shah, J. and Klessig, D.F. 1999. Salicylic acid and disease resistance in plants. *Crit. Rev. Plant Sci.* 18, 547-575.

Desikan, R., Clarke, A., Atherfold, P., Hancock, J.T., and Neill, S.J. 1999. Harpin induces mitogen-activated protein kinase activity during defence responses in *Arabidopsis thaliana* suspension cultures. *Planta* 210, 97-103.

Dixon, R.A. and Harrison, M.J. 1990. Activation, structure and organization of genes involved in microbial defence of plants. *Adv. Genet.* 28, 165-234.

Du, L. and Chen, Z. X. 2001. Identification of genes encoding receptor-like protein kinases as possible targets of pathogen- and salicylic acid-induced WRKY DNA-binding proteins in *Arabidopsis*. *Plant J.* 24, 837-847.

Durner, J., Shah, J. and Klessig, D.F. 1997. Salicylic acid and disease resistance in plants. *Trends Plant Sci.* 2, 266-274.

Durrant, W.E., Rowland, O., Piedras, P., Hammond-Kosack, K.E. and Jones, J.D.G. 2000. cDNA-AFLP reveals a striking overlap in race-specific resistance and wound response gene expression profiles. *Plant Cell* 12, 953-977.

Ebel, J. and Cosio, E.G. 1994. Elicitors of plant defense responses. *Int. Rev. Cytol.* 148, 1-36.

Ellis, J., Dodds, P.N. and Pryor, T. 2000a. Structure, function and evolution of plant disease resistance genes. *Curr. Opin. Plant Biol.* 3, 278-284.

Ellis, J., Dodds, P.N. and Pryor, T. 2000b. The generation of plant disease resistance gene specificities. *Trends Plant Sci.* 5, 373-379.

Ellis, J.G., Lawrence, G.J., Luck, J.E., and Dodds, P.N. 1999. Identification of regions in alleles of the flax rust resistance gene L that determine differences in gene-for-gene specificity. *Plant Cell* 11, 495-506.

Feys, B..J. and Parker, J.E. 2000. Interplay of signaling pathways in plant disease resistance. *Trends Genet.* 16, 449-455.

Finnie, C., Borch, J. and Collinge, D.B. 1999. 14-3-3 proteins: eukaryotic regulatory proteins with many functions. *Plant Mol. Biol.* 40, 545-554.

Flor, H.H. 1971. Current status of the gene-for-gene concept. *Ann. Rev. Phytopath.* 9, 275-296.

Friedrich, L., Vernooij, B., Gaffney, T., Morse, A. and Ryals, J. 1995. Characterization of tobacco plants expressing a bacterial salicylate hydroxylase gene. *Plant Mol. Biol.* 29, 959-968.

Gabriel, D.W. and Rolfe, B.G. 1990. Working models of specific recognition in plant-microbe interactions. *Ann. Rev. Phytopath.* 28, 365-391.

Gaffney, T., Friedrich, L., Vernooij, B., Negrotto, D., Nye, G., Uknes, S., Ward, E., Kessmann, H. and Ryals, J. 1993. Requirement of salicylic acid for the induction of systemic acquired resistance. *Science* 261, 754-756.

Gilchrist, D.G. 1998. Programmed cell death in plant disease: The purpose and promise of cellular suicide. *Ann. Rev. Phytopath.* 36, 393-414.

Glazebrook, J. and Ausubel, F.M. 1994. Isolation of phytoalexin-deficient mutants of *Arabidopsis thaliana* and characterization of their interactions with bacterial pathogens. *Proc. Natl. Acad. Sci. USA* 91, 8955-8959.

Glazebrook, J., Rogers, E.E. and Ausubel, F.M. 1996. Isolation of *Arabidopsis* mutants with enhanced disease susceptibility by direct screening. *Genetics* 143, 973-982.

Glazebrook, J., Rogers, E.E. and. Ausubel, F.M 1997b. Use of *Arabidopsis* for genetic dissection of plant defense responses. *Ann. Rev. Genet.* 31, 569.

Glazebrook, J., Zook, M., Mert, F., Kagan, I., Rogers, E.E., Crute, I.R., Holub, E.B., Hammerschmidt, R. and Ausubel, F.M. 1997a. Phytoalexin-deficient mutants of *Arabidopsis* reveal that PAD4 encodes a regulatory factor and that four PAD genes contribute to downy mildew resistance. *Genetics* 146, 381-392.

Goodwin, P.H., Hsiang, T. and Erickson, L. 2000. A comparison of stilbene and chalcone synthases including a new stilbene synthase gene from *Vitis riparia* cv. Gloire de Montpellier. *Plant Sci.* 151, 1-8.

Goodwin, T.W. and Mercer, E.I. 1983. *Introduction to plant biochemistry*. Pergamon, Oxford.

Govrin, E.M. and Levine, A. 2000. The hypersensitive response facilitates plant infection by the necrotrophic pathogen *Botrytis cinerea*. *Curr. Biol.* 10, 751-757.

Grant, J.J. and Loake, G.J. 2000. Role of reactive oxygen intermediates and cognate redox signaling in disease resistance. *Plant Physiol.* 124, 21-30.

Greenberg, J.T. 2001. Programmed cell death in plant pathogen interactions. *Ann. Rev. Plant Physiol. Plant Mol. Biol.* 48, 525-545.

Gregersen, P.L., Thordal-Christensen, H., Forster, H., and Collinge. D.B., 1997. Differential gene transcript accumulation in barley leaf epidermis and mesophyll in response to attack by *Blumeria graminis* f.sp. *hordei* (syn. *Erysiphe graminis* f.sp. *hordei*). Physiol. Mol. Plant Path. 51, 85-97.

Hain, R., Reif, H.J., Krause, E., Langebartels, R., Kindl, H., Vornam, B., Wiese, W., Schmelzer, E., Schreier, P.H., Stocker, R.H., and Stenzel, K. 1993. Disease resistance results from foreign phytoalexin expression in a novel plant. *Nature* 361, 153-156.

Hammerschmidt, R. 1999. Phytoalexins: what have we learned after 60 years? *Ann. Rev. of Phytopath.* 37, 285-306.

Hammond-Kosack, K.E. and Jones, J.D.G. 1997. Plant disease resistance genes. *Ann. Rev. Plant Physiol. Plant Mol. Biol.* 48, 575-607.

Heath, M.C. 2000a. Hypersensitive response-related death. *Plant Mol. Biol.* 44, 321-334.

Heath, M.C. 2000b. Nonhost resistance and nonspecific plant defenses. *Curr. Opin. Plant Biol.* 3, 315-319.

Herbers, K., Meuwly, P., Frommer, W. B., Metraux, J. P. and Sonnewald, U. 1996. Systemic acquired resistance mediated by the ectopic expression of invertase: possible hexose sensing in the secretory pathway. *Plant Cell* 8, 793-803.

Ishihara, M., Hasegawa, M., Taira, T. and Toyama, S. 2000. Isolation and antimicrobial activity of feruloyl oligosaccharide ester from pineapple stem residues. *Journal of the Japanese Society for Food Science and Technology-Nippon Shokuhin Kagaku Kogaku Kaishi* 47, 23-29.

Jabs, T. 1999. Reactive oxygen intermediates as mediators of programmed cell death in plants and animals. *Biochem. Pharm.* 57, 261-245.

Jabs, T., Tschope, M., Colling, C., Hahlbrock, K., and Scheel, D. 1997. Elicitor-stimulated ion fluxes and O_2^- from the oxidative burst are essential components in triggering defense gene activation and phytoalexin synthesis in parsley. *Proc. Natl. Acad. Sci. USA* 94, 4800-4805.

Jia, Y.L., McAdams, S.A., Bryan, G.T., Hershey, H.P. and Valent, B. 2000. Direct interaction of resistance gene and avirulence gene products confers rice blast resistance. *EMBO J.* 19, 404-4014.

Jørgensen, J.H. 1994. Genetics of powdery mildew resistance in barley. *Crit. Rev. Plant Sci.* 13, 97-119.

Kajava, A.V. 1998. Structural Diversity of Leucine-rich Repeat Proteins. *J. Mol. Biol.* 277, 519-527.

Keen, N.T. 1971. Hydroxyphaseollin production by soybeans resistant and susceptible to *Phytophthora megasperma* var. *soyae*. *Physiol. Plant Path.* 1, 265-275.

Kjemtrup, S., Nimchuk, Z. and Dangl, J. 2000. Effector proteins of phytopathogenic bacteria: bifunctional signals in virulence and host recognition. *Curr. Opin. Microbiol.* 3, 73-78.

Klement, Z. 1963. Rapid detection of the pathogenicity of phytopathogenic *Pseudomonas*. *Nature* 199, 300.

Klement, Z. 1982. "Hypersensitivity". In: *Phytopathogenic Procaryotes.* eds. M.S. Mount and G.S. Lacy, pp. 150-178. Academic Press, New York.

Kobe, B. and Deisenhofer, J. 1995. The leucine rich repeat: a versatile binding motif. *Trends Biochem. Sci.* 19, 415-421.

Kumar, D. and Klessig, D.F. 2000. Differential induction of tobacco MAP kinases by the defense signals nitric oxide, salicylic acid, ethylene, and jasmonic acid. *Mol. Plant Microbe Interact.* 13, 347-351.

Lamb, C. 1996. A ligand-receptor mechanism in plant-pathogen recognition. *Science* 274, 2038-2039.

Lange, J., Xie, Z.-P., Broughton, W.J., Vögeli-Lange, R. and Boller, T. 1999. A gene encoding a receptor-like protein kinase in the roots of common bean is differentially regulated in response to pathogens, symbionts and nodulation factors. *Plant Sci.* 142, 133-145.

Lawton, K., Weymann, K. Friedrich, L. Vernooij, B. Uknes, S. and Ryals, J. 1995. Systemic acquired resistance in *Arabidopsis* requires salicylic acid but not ethylene. *Mol. Plant Microbe Interact.* 8, 863-870.

Leckband, G. and Lorz, H. 1998. Transformation and expression of a stilbene synthase gene of *Vitis vinifera* L. in barley and wheat for increased fungal resistance. *Theor. Appl. Genet.* 96, 1004-1012.

Leister, R.T. and Katagiri, F. 2000. A resistance gene product of the nucleotide binding site — leucine rich repeats class can form a complex with bacterial avirulence proteins *in vivo*. *Plant J.* 22, 345-354.

Levine, A., Pennell, R.I., Alvarez, M.E., Palmer, R., and Lamb, C. 1996. Calcium-mediated apoptosis in a plant hypersensitive disease resistance response. *Curr. Biol.* 6, 427-437.

Liu, Y.H., S.Q. Zhang, and D.F. Klessig. 2000. Molecular cloning and characterization of a tobacco MAP kinase kinase that interacts with SIPK. *Mol. Plant Microbe Interact.* 13, 118-124.

Lucas, J.A. 1998. *Plant Pathology and Plant Pathogens*. Blackwell Science, Oxford, UK.

Luck, J.E., Lawrence, G.J., Dodds, P.N., Shepherd, K.W., and Ellis. J.G., 2000. Regions outside of the leucine-rich repeats of flax rust resistance proteins play a role in specificity determination. *Plant Cell* 12, 1367-1377.

Malamy, J., Carr, J.P., Klessig, D.F. and Raskin, I. 1990. Salicylic acid: a likely endogenous signal in the resistance response of tobacco to viral infection. *Science* 250, 1002-1004.

Mansfield, J.W. 2000. "Antimicrobial compounds and resistance. the role of phytoalexins and phytoanticipins". In: *Mechanisms of Resistance to Plant Diseases.* eds. A.J. Slusarenko, R.S.S. Fraser, and L.C. van Loon, pp. 325-370. Kluwer Academic Publishers, Dordrecht.

Martin, G.B. 1999. Functional analysis of plant disease resistance genes and their downstream effectors. *Curr. Opin. Plant Biol.* 2, 273-279.

Mauch-Mani, B. and Slusarenko, A.J. 1996. Production of salicylic acid precursors is a major function of phenylalanine ammonia-lyase in the resistance of *Arabidopsis* to *Peronospora parasitica*. *Plant Cell* 8, 203-212.

Mazeyrat, F., Mouzeyar, S., Courbou, I., Badaoui, S., Roeckel-Drevet, P., de Labrouhe, D.T. and Ledoigt, G. 1999. Accumulation of defense related transcripts in sunflower hypocotyls (*Helianthus annuus* L.) infected with *Plasmopara halstedii*. *Eur. J. Plant Path.* 105, 333-340.

Mehdy, M.C. 1994. Active oxygen species in plant defence against pathogens. *Plant Physiol.* 105, 467-472.

Meyers, B.C., Dickerman, A.W., Michelmore, R.W., Sivaramakrishnan, S., Sobral, B.W. and Young, N.D. 2001. Plant disease resistance genes encode members of an ancient and diverse protein family within the nucleotide-binding superfamily. *Plant J.* 20, 317-332.

Métraux, J.-P., Signer, H., Ryals, J., Ward, E., Wyssbenz, M., Gaudin, J., Raschdorf, K., Schmid, E., Blum, W. and Inverardi, B. 1990. Increase in salicylic-acid at the onset of systemic acquired resistance in cucumber. *Science* 250, 1004-1006.

Michelmore, R.W. 2000. Genomic approaches to plant disease resistance. *Curr. Opin. Plant Biol.* 3, 125-131.

Mittler, R., Lam, E. Shulaev, V. and Cohen, M. 1999. Signals controlling the expression of cytosolic ascorbate peroxidase during pathogen-induced programmed cell death in tobacco. *Plant Mol. Biol.* 39, 1025-1035.

Navarre, D.A. and Wolpert, T.J. 1999. Victorin induction of an apoptotic/senescence-like response in oats. *Plant Cell* 11, 237-249.

Oldroyd, G.E.D. and Staskawicz, B. J. 1998. Genetically engineered broad-spectrum disease resistance in tomato. *Proc. Natl. Acad. Sci. USA.* 95, 10300-10305.

Pallas, J.A., Paiva, N.L, Lamb, C. and Dixon, R.A. 1996. Tobacco plants epigenetically suppressed in phenylalanine ammonia-lyase expression do not develop systemic acquired resistance in response to infection by tobacco mosaic virus. *Plant J.* 10, 281-293.

Pan, Q., Wendel, J. and Fluhr, R. 2000. divergent evolution of plant nbs-lrr resistance gene homologues in dicot and cereal genomes. *J. Mol. Evol.* 50, 203-213.

Parker, J.E., Feys, B.J., van der Biezen, E.A., Noël, L., Aarts, N., Austin, M.J., Botella, M.A., Frost, L.N., Daniels, M.J. and Jones, J.D.G. 2000. Unravelling R gene-mediated disease resistance pathways in *Arabidopsis. Mol. Plant Path.* 1, 17-24.

Petersen, M., Brodersen, P., Naested, H., Andreasson, E., Lindhart, U., Johansen, B., Nielsen, H.B., Lacy, M., Austin, M.J., Parker, J.E., Sharma, S.B., Klessig, D.F., Martienssen R., Mattsson, O., Jensen, A.B. and Mundy, J. 2000. *Arabidopsis* MAP kinase 4 negatively regulates systemic acquired resistance. *Cell* 103, 1111-1120.

Pieterse, C.M.J. and van Loon, L.C. 1999. Salicylic acid-independent plant defence pathways. *Trends Plant Sci.* 4, 52-58.

Pieterse, C.M.J., van Wees, S.C.M. Hoffland, E. van Pelt, J.A. and van Loon, L.C. 1996. Systemic resistance in *Arabidopsis* induced by biocontrol bacteria is independent of salicylic acid accumulation and pathogenesis-related gene expression. *Plant Cell* 8, 1225-1237.

Pieterse, C.M.J., van Wees, S.C.M., van Pelt, J.A., Knoester, M., Laan, R., Gerrits, N., Weisbeek, P.J. and van Loon, L.C. 1998. A novel signaling pathway controlling induced systemic resistance in *Arabidopsis. Plant Cell* 10, 1571-1580.

Piffanelli, P., Devoto, A. and Schulze-Lefert, P. 1999. Defence signaling pathways in cereals. *Curr. Opin. Plant Biol.* 2, 295-300.

Richmond, T. and Somerville, S. 2000. Chasing the dream: plant EST micro-arrays. *Curr. Opin. Plant Biol.* 3, 108-116.

Roberts, M.R. and Bowles, D.J. 1999. Fusicoccin, 14-3-3 proteins, and defense responses in tomato plants. *Plant Physiol.* 119, 1243-1250.

Romeis, T., Piedras, P., Zhang, S.Q., Klessig, D.F., Hirt, H. and Jones, J.D.G. 1999. Rapid *Avr9*- and *Cf-9*-dependent activation of MAP kinases in tobacco cell cultures and leaves: convergence of resistance gene, elicitor, wound, and salicylate responses. *Plant Cell* 11, 273-287.

Ross, A.F. 1961a. Localized acquired resistance to plant virus infections in hypersensitive hosts. *Virology* 14, 329-339.

Ross, A.F. 1961b. Systemic acquired resistance induced by localized virus infections in plants. *Virology* 14, 340-358.

Rushton, P.J. and Somssich, I.E. 1998. Transcriptional control of plant genes responsive to pathogens. *Curr. Opin. Plant Biol.* 1, 311-315.

Rushton, P.J. and Somssich, I.E. 1999. "Transcriptional regulation of plant genes responsive to pathogens and elicitors". In: *Plant-Microbe Interactions.* eds. G. Stacey and N.T. Keen, pp. 251-274. APS Press, St. Paul, Minnesota.

Ryals, J.A., Neuenschwander, U.H., Willits, M.G., Molina, A., Steiner, H.Y. and Hunt, M.D. 1996. Systemic acquired resistance. *Plant Cell* 8, 1809-1819.

Schenk, P.M., Kazan, K., Wilson, I., Anderson, J.P., Richmond, T., Somerville, S., and Manners, J.M. 2000. Coordinated plant defense responses in *Arabidopsis* revealed by micro-array analysis. *Proc. Natl. Acad. Sci. USA* 97, 11655-11660.

Schweizer, P., Pokorny, J. Schulze-Lefert, P. and Dudler, R. 2000. Double-stranded RNA interferes with gene function at the single-cell level in cereals. *Plant Journal* 24, 895-903.

Scofield, S.R., Tobias, C.M., Rathjen, J.P., Chang, J.H., Lavelle, D.T., Michelmore, R.W. and Staskawicz, B.J. 1996. Molecular basis of gene-for-gene specificity in bacterial speck disease of tomato. *Science* 274, 2063-2065.

Scott, K.J., Davidson, A.D., Jutidamrongphan, W., Mackinnon G. and Manners, J.M. 1990. The activation of genes of wheat and barley by fungal phytopathogens. *Aust. J. Plant Physiol.* 17, 229-238.

Seehaus, K. and Tenhaken, R. 1998. Cloning of genes by mRNA differential display induced during the hypersensitive reaction of soybean after inoculation with *Pseudomonas syringae* pv. *glycinea. Plant Mol. Biol.* 38, 1225-1234.

Shirasu, K., Nakajima, H., Rajasekhar, V.K., Dixon, R.A. and Lamb, C. 1997. Salicylic acid potentiates an agonist-dependent gain control that amplifies pathogen signals in the activation of defense mechanisms. *Plant Cell* 9, 261-270.

Shulaev, V., León, J. and Raskin, I. 1995. Is salicylic acid a translocated signal of systemic acquired- resistance in tobacco. *Plant Cell* 7, 1691-1701.

Skou, J.P. 1982. Callose formation responsible for the powdery mildew resistance in barley with genes in the ml-o locus. *Phytopathologische Zeitschrift* 104, 90-95.

Smith-Becker, J., Marois, E., Huguet, E.J., Midland, S.L., Sims, J.J. and Keen, N.T. 1998. Accumulation of salicylic acid and 4-hydroxybenzoic acid in phloem fluids of cucumber during systemic acquired resistance is preceded by a transient increase in phenylalanine ammonia- lyase activity in petioles and stems. *Plant Physiol.* 116, 231-238.

Stakman, E.C. 1915. Relation between *Puccinia graminis* and plants highly resistant to its attack. *J. Agricult. Res.* 4, 193-200.

Stark-Lorenzen, P., B. Nelke, G. Hanssler, H.P. Muhlbach, and J.E. Thomzik. 1997. Transfer of a grapevine stilbene synthase gene to rice (*Oryza sativa* L). *Plant Cell Rep.* 16, 668-673.

Takken, F.L.W. and Joosten, M.H. 2000. Plant resistance genes: their structure, function and evolution. *Eur. J. Plant Path.* 106, 699-713.

Tang, X.Y., Frederick, R.D., Zhou, J.M., Halterman, D.A., Jia, Y. and Martin, G.B. 1996. Initiation of plant disease resistance by physical interaction of *AvrPto* and *Pto* kinase. *Science* 274, 2060-2063.

The Arabidopsis Genome Initiative. 2000. Analysis of the genome sequence of the flowering plant *Arabidopsis thaliana*. *Nature* 408, 796-815.

Thomzik, J.E., Stenzel K., Stocker, R., Schreier, P.H.,Hain, R. and Stahl, D.J. 1997. Synthesis of a grapevine phytoalexin in transgenic tomatoes (*Lycopersicon esculentum* Mill.) conditions resistance against *Phytophthora infestans*. *Physiol. Mol. Plant Path.* 51, 265-278.

Thordal-Christensen, H., Brandt, J., Cho, B.H., Rasmussen, S.K., Gregersen, P.L., Smedegaard-Petersen, L. and Collinge, D.B. 1992. cDNA cloning and characterization of two barley peroxidase transcripts induced differentially by the powdery mildew fungus *Erysiphe graminis*. *Physiol. Mol. Plant Pathol.* 40, 395-409.

Thordal-Christensen, H., Gregersen, P.L. and Collinge, D.B. 2000. "The barley/*Blumeria* (syn. *Erysiphe*) *graminis* interaction: a case study." In: *Mechanisms of Resistance to Plant Diseases*. eds. A.J. Slusarenko, R.S.S. Fraser, and L.C. van Loon, pp. 77-100. Kluwer Academic Publishers, Dordrecht.

Thordal-Christensen, H., Zhang, Z.G., Wei, Y.D. and Collinge, D.B. 1997. Subcellular localization of H_2O_2 in plants. H_2O_2 accumulation in papillae and hypersensitive response during the barley-powdery mildew interaction. *Plant J.* 11, 1187-1194.

Thulke, O. and Conrath, U. 1998. Salicylic acid has a dual role in the activation of defence-related genes in parsley. *Plant J.* 14, 35-43.

Tropf, S., Lanz, T., Rensing, S.A., Schroder, J. and Schroder, G. 1994. Evidence that stilbene synthases have developed from chalcone synthases several times in the course of evolution. *J. Mol. Evol.* 38, 610-618.

Tsuji, J., Gage, D.A., Hammerschmidt, R., and Somerville, S.C. 1992. Phytoalexin accumulation in *Arabidopsis thaliana* during the hypersensitive reaction to *Pseudomonas syringae* pv. *syringae*. *Plant Physiol.* 98, 1304-1309.

Uknes, S., Mauch-Mani, B., Moyer, M., Potter, S., Williams, S., Dincher, S., Chandler, D., Slusarenko, A.J., Ward, E. and Ryals, J. 1992. Acquired-resistance in *Arabidopsis*. *Plant Cell* 4, 645-656.

van Loon, L.C. 1970. Polyacrylamide disc electrophoresis of the soluble leaf proteins from *Nicotiana tabacum* var. "Samsun" and "Samsun NN". *Virology* 40, 199-211.

van Loon, L.C., Bakker, P.A. and Pieterse, C.M.J. 1998. Systemic resistance induced by rhizosphere bacteria. *Ann. Rev. Phytopath.* 36, 453-483.

van Loon, L.C., Pierpoint, W.S., Boller, T. and Conejero, V. 1994. Recommendations for naming plant pathogenesis-related proteins. *Plant Mol. Biol.Rep.* 12, 245-264.

van Loon, L.C. and van Strien, E.A. 1999. The families of pathogenesis-related proteins, their activities, and comparative analysis of PR-1 type proteins. *Physiol. Mol. Plant Path.* 55, 85-97.

van Peer, R., Niemann, G.J. and Schippers, B. 1991. Induced resistances and phytoalexin accumulation in biological control of fusarium wilt of carnation by Pseudomonas sp. Strain WCS417r. *Phytopathology* 91, 728-734.

Vanacker, H., Harrison, J., Ruisch, J., Carver, T.L.W. and Foyer, C.H. 1998. Antioxidant defences of the apoplast. *Protoplasma* 205, 129-140.

VanEtten, H. D., Mansfield, J. W. Bailey, J. A. and Farmer, E. E. 1994. Two classes of plant antibiotics - phytoalexins versus phytoanticipins. *Plant Cell* 6, 1191-1192.

VanEtten, H. D., Sandrock, R.W., Wasmann, C.C., Soby, S.D., McCluskey, K. and Wang, P. 1995. Detoxification of phytoanticipins and phytoalexins by phytopathogenic fungi. *Can. J. Bot.* 73, S518-S525.

Vera-Estrella, R., Barkla, B.J., Higgins, V.J. and Blumwald, E. 1994. Plant defence response to fungal pathogens, activation of host-plasma membrane H^+-ATPase by elicitor-induced enzyme dephosphorylation. *Plant Physiol.* 104, 209-215.

Vernooij, B., Friedrich, L., Morse, A., Reist, R., Kolditz-Jawhar, R., Ward, E., Uknes, S., Kessmann, H. and Ryals, J. 1994. Salicylic acid is not the translocated signal responsible for inducing systemic acquired resistance but is required in signal transduction. *Plant Cell* 6, 959-965.

von Röpenack, E., Parr, A. and Schulze-Lefert, P. 1998. Structural analyses and dynamics of soluble and cell wall-bound phenolics in a broad spectrum resistance to the powdery mildew fungus in barley. *J. Biol. Chem.* 273, 9013-9022.

Wang, H., Li, J., Bostock, R.M. and Gilchrist, D.G. 1996. Apoptosis: A functional paradigm for programmed plant cell death induced by a host-selective phytotoxin and invoked during development. *Plant Cell* 8, 375-391.

Ward, E.R., Uknes, S.J., Williams, S.C., Dincher, S.S., Wiederhold, D.L., Alexander, D.C., Ahl-Goy, P., Métraux, J.-P. and Ryals, J.A. 1991. Coordinate gene activity in response to agents that induce systemic acquired-resistance. *Plant Cell* 3, 1085-1094.

Wei, G., Kloepper, J.W. and Tuzun, S. 1991. Induction of systemic resistance of cucumber to *Colletrotichum orbiculare* by select strains of plant-growth promoting rhizobacteria. *Phytopathology* 81, 1508-1512.

White, R.F. 1979. Acetylsalicylic acid (aspirin) induces resistance to tobacco mosaic virus in tobacco. *Virology*, 410-412.

Chapter 8

RESPONSES AND ADAPTATIONS OF PLANTS TO METAL STRESS

David Salt

Chemistry Department, Northern Arizona University, Flagstaff, AZ 86011, USA.
david.salt@nau.edu

INTRODUCTION

"Over 200 papers, 3 reviews, a small book, and now a big book. What is it about this curious and unimportant character that has merited such attention?" This is the opening sentence from "The Essential Qualities", an article discussing the "curious and unimportant character" of how plants adapt to elevated concentrations of trace metals in their environment (Bradshaw *et al.,* 1990). Bradshaw and co-authors found the study of plant metal tolerance to be a valuable tool for the investigation of natural selection in plants (Bradshaw *et al.,* 1990). However, in the proceeding ten years since the publication of this article, the "curious and unimportant character" of metal tolerance has attracted even more attention; and given rise to a new field of study termed phytoremediation. Because of the attractiveness of using plants to remove pollutant metals from the environment, numerous researchers have now begun to investigate phytoremediation, generating over 200 publications and millions of dollars of grant funding. However, at the center of all this new attention is still that "curious and unimportant character" highlighted by Bradshaw and coworkers. What do we know about its physiological, biochemical and molecular nature?

The last decade has seen an almost exponential growth in the molecular tools available to plant scientists. With the release of completed plant genomes and new tools in bioinformatics and functional genomics, the pace of these developments will only increase in the coming decades. The intent of this review is to highlight how these developments have impacted our understanding of the processes involved in how plants respond and adapt to

M.J. Hawkesford and P. Buchner (eds.),
Molecular Analysis of Plant Adaptation to the Environment, 159–179.

heavy metal stress. Let us first start with a brief synopsis of the basic processes involved in this "curious and unimportant character".

PHYSIOLOGY AND BIOCHEMISTRY OF PLANT ADAPTATIONS AND RESPONSES TO ELEVATED CONCENTRATIONS OF POTENTIALLY TOXIC TRACE METALS

To resist the potentially toxic effects of such metals as zinc, nickel, copper, cadmium, selenium and arsenic (strictly speaking selenium and arsenic are metalloid but will be consider here as potentially toxic trace metals), plants can adopt a combination of two different strategies; either exclude the metal, or allow the metal to accumulate in the tissues and evolve internal tolerance mechanisms (Baker, 1981).

Metal exclusion from the root

Metal exclusion from a plant can occur at one of two levels, exclusion from the root, or exclusion from the shoot. Exclusion from the root can occur by reduction of the concentration of free metal ion in the soil, available for uptake. This can be achieved by secretion of certain metal chelates into the rhizosphere. Alternatively, modification of metal transporters at the plasma membrane could lead to reduced metal influx and/or increased metal efflux. Chelation of metals has the effect of reducing the effective concentration of the toxic free ionic form of the metal in the soil solution, thereby reducing the overall toxicity of the metal. For example, aluminium resistance is achieved by the secretion of citrate or malate into the rhizosphere. Both organic acids are capable of chelating ionic aluminium, thereby reducing the effective soil solution concentration of the toxic free ionic form (Kochian, 1995). The release of these organic acids from roots is induced by aluminium (Miyasaka *et al.*, 1991; Delhaize *et al.*, 1993).

This phenomenon does not appear to be restricted to aluminium. *Thlaspi arvense* was found to increase the amount of histidine and citrate produced in root exudates in response to treatment with nickel (Salt *et al.*, 1999a). The *Arabidopsis thaliana* has also been observed to increase the amount of citrate produced in root exudates in response to copper (Murphy *et al.*, 1999). Interestingly, in contrast to these non-accumulator species, the nickel hyperaccumulator *T. goesingense* showed no increase in histidine or citrate concentrations in root exudate when exposed to nickel. An intriguing possibility is that in hyperaccumulator species, this metal-chelate facilitated metal exclusion system has been suppressed, thereby allowing for an increased uptake of metals from the soil by the hyperaccumulator.

To enter root cells, metal ions must cross the plasma membrane that surrounds all cells. Our understanding of how metal ions cross the plasma membrane in plants has increased dramatically with the recent cloning and characterisation of genes involved in both iron and zinc transport. The iron regulated transporter, *IRT1* was recently cloned from *Arabidopsis thaliana* (Eide *et al.*, 1996). When assayed for function in a yeast fet3/fet4 mutant background, the IRT1 protein was found to be specific for iron (II) uptake and had almost no affinity for iron (III). In addition to iron (II), IRT1 is also capable of transporting cadmium, cobalt, manganese (II) and zinc, with a lower affinity. IRT1 is highly expressed in roots of iron-deficient plants (Eide *et al.*, 1996). However, its expression is repressed in roots when plants are supplied with sufficient iron. By suppressing expression of this iron (II) transporter, the plant is able to exclude excess iron (II) from the root cell cytoplasm.

A related family of *IRT1*-like genes has recently been cloned from *A. thaliana* (Grotz *et al.*, 1998). This family includes *ZIP1*, *ZIP2*, *ZIP3* and *ZIP4*. The ZIP1-3 proteins have been shown to be able to transport zinc, and cadmium, copper (II) and possibly manganese (II) with a lower affinity (Grotz *et al.*, 1998). Expression of *ZIP1* and *ZIP3* is strongly induced in the roots of zinc-deficient plants, and suppressed in the presence of sufficient or excess zinc (Grotz *et al.*, 1998); again allowing the plant to exclude excess zinc from its roots.

Another family of plasma membrane metal transporters that warrants comment is the Nramp (Natural resistance associated macrophage protein) family. Nramp genes were originally characterised in mammals and yeast, but several representatives have now been cloned from plants. Three Nramp genes have been cloned from rice, *OsNramp1*, *OsNramp2* and *OsNramp3* (Belouchi *et al.*, 1997). Three Nramp genes have also recently been cloned from *A. thaliana*, *AtNramp1*, *AtNramp2* and *AtNramp3*. Individual expression of each of these genes in yeast has suggested that the Nramp proteins from *A. thaliana* can transport manganese (II) (Thomine *et al.*, 2000). These transporters appear to be suppressed by elevated concentrations of metals (Thomine *et al.*, 2000).

An alternative method of metal exclusion from roots is enhanced efflux of metals out of the root, as a way to maintain a low root metal concentration. Very little is known about the mechanisms of metal efflux in plants. Recently *ZAT1*, a homologue of the mammalian zinc efflux gene *ZnT1*, was cloned from *A. thaliana* (Van der Zaal *et al.*, 1999). Expression of this gene was not regulated by zinc, however, its overexpression in *A. thaliana* was found to confer zinc resistance. Based on its homology to ZnT1, it is possible that ZAT1 is a zinc efflux protein involved in effluxing zinc out of cells. A partial EST cDNA from *Brassica campestris* ssp. *pekinensis* also exists that shows homology to ZnT1 (Lim *et al.*, 1996). In our laboratory, a homologue of the gene represented by this EST was recently found to be up-regulated, at the mRNA level, in young seedlings of *Brassica juncea* after treatment with nickel, zinc or cadmium (Persans *et al.*, 1999a). This support the hypothesis that these ZnT1 homologues may be involved in effluxing excess metal out of plant cells, as part of a metal exclusion system. Clearly, further work is needed on this class of proteins to fully understand their role in metal detoxification.

Metal exclusion from the shoots

When exposed to elevated trace metal concentrations, most plants respond by accumulating metals in their roots, restricting the movement of metals to the shoot. One way to achieve this restricted movement of metals to the shoot is to store the metals in an immobile pool within the root. Plants exposed to cadmium achieve this by storing cadmium complexed to the thiol groups of phytochelatins (Rauser, 1999). Using X-ray absorbance spectroscopy, it has been shown that cadmium-phytochelatin complexes are not translocated in the xylem sap to the shoot (Salt *et al.*, 1995). Plants exposed to arsenic also appear to use a similar strategy to restrict the mobility of arsenic to the shoot. Arsenate entering roots is rapidly reduced to arsenite and co-ordinated with thiol groups (Pickering *et al.*, 2000); most likely from phytochelatins (Schmoger *et al.*, 2000).

Due to its chemical properties, nickel prefers oxygen and nitrogen ligands rather than sulfur ligands, as is the case for cadmium and arsenic. This ligand preference is reflected in how nickel is complexed in plants. X-ray absorbance spectroscopy of nickel exposed roots of *T. arvense*, a non-accumulating plant, reveal that approx. 30 – 50 % of root nickel is co-ordinated with oxygen and nitrogen ligands, modeled as free histidine (Persans *et al.*, 1999b). This suggests that in this plant nickel movement to the shoot is restricted by its storage as a nickel-histidine complex. Interestingly, both total nickel and nickel-histidine concentrations are significantly lower in roots of the nickel hyperaccumulator *Thlaspi goesingense*. In this plant, we hypothesis that any nickel exclusion mechanism would be suppressed, as this plant is known to accumulate high concentrations of nickel in its shoots (Krämer *et al.*, 1997a).

A comparison of the amount of zinc in root cell vacuoles from the zinc hyperaccumulator *T. caerulescens* and the non-accumulator *T. arvense,* suggests that vacuolar storage of zinc is also an effective mechanisms for the immobilization of metals in root cells. The accumulation of zinc in the vacuoles of roots cells of *T. arvense* was found to be approximately double that measured in the hyperaccumulator *T. caerulescens* after 24h exposure to zinc (Lasat *et al.,* 1998). Extended efflux studies on zinc loaded roots of these two species also revealed that roots of *T. arvense* retained 6-fold more zinc than *T. caerulescens* after an efflux period of 48h. This supports the hypothesis that in the non-accumulator, a significantly higher proportion of cellular root zinc is immobilized in the root by storage in the root cell vacuoles. This allows the non-accumulator to limit excess zinc from being translocated to the shoot.

Metal accumulation in shoots

Certain plants have adapted to the presence of elevated concentrations of metals in the soil, by allowing the metals to pass unhindered from the soil, through the root to the shoot. Once in the shoot, the metals are detoxified.

The zinc hyperaccumulator *Thlaspi caerulescens* has evolved a system to bypass the mechanisms that normally limit zinc uptake. In this hyperaccumulator, the ability to suppress expression of the zinc uptake transporter, ZNT1, when soil zinc concentrations are elevated, has been lost (Pence *et al.,* 2000). This leads to an over expression of the transporter and increased zinc uptake rates (Lasat *et al.,* 1996; Pence *et al.,* 2000). In contrast, the zinc non-accumulator *T. arvense* is able to suppress expression of this zinc transporter when soil zinc concentrations are high (Pence *et al.,* 2000).

Once taken up by roots from the soil, metal transport to the shoots is controlled by two main processes - movement into the xylem and volume flux through the xylem; the latter mediated by root pressure and transpiration. In order to enter the xylem, solutes have to be taken up into root cells for passage through the root endodermis. There is some evidence for a second, wholly apoplastic pathway for the entry of water and possibly cations into the xylem in certain regions of the root (Marschner, 1995). In general, however, plants are likely to have tight control over the solutes entering the shoot via the xylem, through solute release and absorption by xylem parenchyma cells. Xylem loading appears to be energised by a negative membrane potential generated in xylem parenchyma cells through the operation of proton pumping ATPases (Pitman, 1972; De Boer *et al.,* 1983; Clarkson and Hanson, 1986). It is therefore possible that xylem loading of metal ions could operate through proton antiports, ATPases, or ion channels (Wegner and Raschke, 1994; Roberts and Tester, 1995; 1997). In the nickel hyperaccumulator *Alyssum lesbiacum,* and the zinc hyperaccumulator *T. caerulescens*, xylem loading of the hyperaccumulated metal occurs at a high rate, with millimolar concentrations of nickel and zinc being found in the xylem sap (Krämer *et al.,* 1996; Lasat *et al.,* 1998; Salt *et al.,* 1999b). Xylem concentrations of the free amino acid histidine and nickel show a linear correlation in several nickel hyperaccumulators in the genus

Alyssum (Krämer *et al.*, 1996). This suggests that histidine may facilitate xylem loading of nickel, possibly by forming a nickel-histidine complex. In support of this, exposure to D-histidine in the hydroponic medium has been observed to reduce the amount of nickel translocated to the shoots in the non-accumulator *Thlaspi arvense* (Persans *et al.*, 1999b), suggesting that L-histidine may be involved in nickel translocation in both accumulator and non-accumulator species.

Once loaded into the xylem metal ions are transported to the shoot. Both direct and predictive evidence suggests that organic and amino acids are involved in transporting metal ion, as metal ions complexes, in the xylem sap. Direct analysis of the xylem exudates of the zinc hyperaccumulator *T. caerulescens*, using X-ray absorbance spectroscopy, revealed that at least 20% of the zinc in the sap was co-ordinated to a carboxylic acid, modelled as citrate (Salt *et al.*, 1999b). The remaining zinc appeared to be hydrated Zn^{2+} (Salt *et al.*, 1999b). A similar analysis of xylem sap isolated from *Brassica juncea* exposed to cadmium revealed that the majority of the cadmium in the sap was also co-ordinated by carboxylic acids (Salt *et al.*, 1995). In the Ni hyperaccumulating tree *Sebertia acuminata*, NMR analysis revealed that citrate was also a major ligand of nickel in the latex of this plant (Sagner *et al.*, 1998). Extensive modelling studies on the metal species present in both soybean and tomato xylem sap have predicted the major species for copper, iron and zinc to be $Cu(Asn)_2$, $Cu(Gln)_2$, $Cu(His)_2$, Fe-citrate, and Zn-citrate (White *et al.*, 1981). In the nickel hyperaccumulator, *Alyssum lesbiacum*, similar predictions suggested that nickel in the xylem sap is co-ordinated by histidine (19%), and glutamine (15%), with citrate and malate playing a minor role, and the remaining nickel (48%) being present as the hydrated cation (Krämer *et al.*, 1996). Studies of the speciation of arsenic in plants have revealed that arsenic is transported in the xylem as the oxyanions arsenate and arsenite (Pickering *et al.*, 2000). This is contrasted by arsenic's fate in the roots, where it is chemically reduced and co-ordinated by thiol ligands for storage (Pickering *et al.*, 2000).

Shoot metal tolerance

Because of the potentially toxic nature of most metal ions, on arrival in the shoot, they need to be "processed" in such a way as to allow their safe, long-term storage. Storage in the leaf tissue is the only option for most metal ions, other than certain volatile elements such as mercury and selenium.

The complexation of metal ions by specific high-affinity ligands, or precipitation as an insoluble compound, can reduce the solution concentration of the free metal ions (chemical activity), preventing metal toxicity. The ability to control and modify the chemical activity of metal ions in selected cellular compartments could provide a mechanism for the control of metal toxicity. Detoxification of cadmium and arsenic are good examples of such a mechanism. On entering cells, cadmium ions are rapidly co-ordinated through the thiol groups of phytochelatins (Grill *et al.*, 1985; Rauser, 1999). Recently, the enzyme that catalyses the final step in the biosynthesis of phytochelatins has been characterised, and the gene cloned (Clemens *et al.*, 1999; Ha *et al.*, 1999; Vatamniuk *et al.*, 1999). This enzyme is capable of transferring a γ-glutamylcysteine unit from glutathione to form phytochelatins. Expression of the enzyme, at the transcriptional level, appears to be regulated by Cd in wheat roots (Clemens *et al.*, 1999), but not in *A. thaliana* (Ha *et al.*, 1999; Vatamniuk *et al.*, 2000). However, enzyme activity is dependent on the presence of heavy metal ions, particularly cadmium. Interestingly, this metal ion dependence is not mediated by the direct binding of the metal ion to the enzyme, but rather by the formation of a metal-thiolate complex with the substrate, glutathione (Vatmaniuk *et al.*, 2000). Once formed, the cadmium-phytochelatin complex is then accumulated in the vacuole for long-term storage (Vögeli-Lange and Wagner, 1990; Salt and Rauser, 1995). Arsenic is also complexed with phytochelatins (Schmoger *et al.*, 2000) once it has been chemically reduced from As(V) to As(III) (Pickering *et al.*, 2000).

Vacuolar accumulation of metal ions also appears to be a common mechanism for their long-term storage in plants. As discussed, cadmium is stored in the vacuole as a cadmium-phytochelatin complex (Vögeli-Lange and Wagner, 1990). Both nickel and zinc have also been observed to be stored in the vacuole (Krämer *et al.*, 2000; Brune *et al.*, 1994; Vázquez *et al.*, 1992, 1994; Küpper *et al.*, 1999); most likely co-ordinated by carboxylic acids (Salt *et al.*, 1999b; Krämer *et al.*, 2000). A cadmium/proton antiport and a cadmium-phytochelatin ATPase activity have been identified on the tonoplast (Salt and Wagner, 1993; Salt and Rauser, 1995), however, at present, very little is known about the molecular biology of these tonoplast membrane transport activities (Persans and Salt, 2000).

Metals are also localised in specific tissues in the shoot, and this metal localisation may be assumed to be part of the metal detoxification process. Scanning proton microscopy and energy-dispersive X-ray microanalysis have been used to localise metals in the nickel hyperaccumulators *Senecio coronatus* (Mesjasz-Przybylowicz *et al.*, 1994), *Alyssum lesbiacum* (Krämer *et al.*, 1997b), and *Thalspi montanum* var. *siskiyouense* (Heath *et al.*, 1997), and the zinc hyperaccumulator *T. caerulescens* (Vázquez *et al.*, 1992, 1994; Küpper *et al.*, 1999). The highest nickel concentrations were found in the unicellular stellate trichomes covering the leaf surface in *A. lesbiacum*, and in the subsidiary cells that surround guard cells in *T. montanum* var. *siskiyouense*. Interestingly, several non-accumulator plant species have also been observed to accumulate cadmium (Salt *et al.*, 1995), copper, nickel and zinc (Neumann *et al.*, 1995), lead (Martell, 1974), and manganese (Blamey *et al.*, 1986) in leaf trichomes.

Chemical reduction and/or incorporation into organic compounds are other possible detoxification mechanism, as observed for metalloids such as selenium and arsenic. To convert arsenic to a form that is more easily complexed, arsenate (As^{V}) is rapidly reduced to arsenite (As^{III}) by plants (Pickering *et al.*, 2000), before being complexed to phytochelatins (Schmoger *et al.*, 2000). At present, the mechanism of As^{V} reduction in plants is not understood. Selenium is an analogue of sulfur and is therefore metabolised in plants to selenocysteine and selenomethionine (Shift, 1969), causing cellular toxicity. By funneling selenium into the non-protein amino acids methylselenocysteine and selenocystathionine, selenium accumulator species of *Astragalus* are able to reduce the amount of selenium incorporated into proteins, thereby tolerating elevated concentrations of selenium in shoots (Läuchi, 1993). Recently, the enzyme responsible for the methylation of selenocysteine in the selenium accumulator *Astragalus bisculatus* has been isolated and characterised, a first step in determining the molecular basis of selenium resistance in plants (Neuhierl and Böck, 1996; Neuhierl *et al.*, 1999).

MOLECULAR APPROACHES TO UNDERSTANDING PLANT ADAPTATIONS AND RESPONSES TO ELEVATED CONCENTRATIONS OF POTENTIALLY TOXIC TRACE METALS

In order to fully understand the basic mechanisms that underlie the biochemical and physiological processes outlined above, it is imperative that the genes that direct these processes are identified. Outlined below are a number of strategies that have been successfully used to clone genes involved in plant responses and adaptations to heavy metal stress.

Functional complementation cloning

In this approach, yeast or *E. coli* strains are identified which lack a functional homologue of a gene of interest. For this approach to work, it is important that these mutants have a visible phenotype, such as the inability to grow under certain nutrient limiting conditions. The yeast or *E. coli* mutant is then transformed with a cDNA expression library created from the plant species under investigation. Transformants are screened for plant cDNA's that restore the yeast or *E. coli*'s phenotype. This approach has been used to successfully clone genes encoding both ion transporters and enzymes from plants.

Because uptake of metal ions through the roots is the first plant associated rate-limiting step for metal ion uptake in plants, several research groups have focused on cloning genes involved in this process. For example, Zhao and Eide (1996) generated a yeast double mutant lacking genes encoding the plasma membrane zinc transporters, zrt1 and zrt2. This double mutant was unable to grow on medium containing low zinc concentrations. By screening an *A. thaliana* cDNA expression library for cDNA's able to restore the ability of the zrt1/zrt2 double yeast mutant to grow on low zinc medium, Grotz *et al.*, (1998) were able to clone three *A. thaliana* cDNA's encoding the plasma membrane zinc transport proteins ZIP1, ZIP2 and ZIP3 (Grotz *et al.*, 1998).

The *ZIP1-3* genes have been shown to functionally complement the zrt1/zrt2 yeast double mutant, and it was therefore concluded that they are able to transport zinc (Grotz *et al.*, 1998). This conclusion was confirmed by direct zinc transport assays in the zrt1/zrt2 yeast double mutant expressing the *ZIP1-3* genes. The *ZIP1-3* genes were found not to functionally complement a fet3/fet4 yeast double mutant unable to acquire iron (II) from the media, suggesting that proteins of the ZIP family are unable to transport iron (II) into cells. Ten-fold excess concentrations of manganese (II), nickel, iron (II), and cobalt were found to have no inhibitory effect on zinc transport by ZIP1. Only cadmium and copper were found to be able to partially inhibit zinc transport. ZIP1 is therefore a highly specific zinc transporter (Grotz *et al.*, 1998). ZIP2 zinc transport activity was strongly inhibited by cadmium and copper (II). ZIP3 has the broadest range of metal transport specificities of the ZIP protein family members. Its transport activity can be inhibited by manganese (II), iron, cobalt, cadmium and copper (II), but inhibition was found to be not greater than 50% for any of these metals. The results of the metal specificity assays for the ZIP transporters are similar to the IRT1 transporter in that the proteins have a high affinity for one particular metal (i.e. zinc or iron) but may also, with some lower affinity, transport other metals. Expression of *ZIP1* and *ZIP3* is strongly induced in the roots of zinc-deficient plants. However, these genes are not expressed in zinc-sufficient roots. Neither *ZIP1* nor *ZIP3* are expressed in shoot (Grotz *et al.*, 1998). The *ZIP2* mRNA was undetectable in all plant tissues investigated and zinc conditions tested (Grotz *et al.*, 1998).

As discussed previously, reduction of metal toxicity by chelation is an important mechanism in plants. Recently, free histidine has been identified as a putative molecule involved in nickel and possibly zinc co-ordination in plants (Krämer *et al.*, 1996, Salt *et al.*, 1999b; Krämer *et al.*, 2000). To further understand the role of histidine in nickel hyperaccumulation in *Thlaspi goesingense*, genes encoding enzymes involved in histidine biosynthesis have been cloned by functional complementation in *E. coli* (Persans *et al.,* 1999b). By screening a *T. goesingense* cDNA expression library for cDNA's able to restore the histidine auxotrophic phenotype of various *E. coli* histidine mutants, our laboratory was able to identify genes encoding ATP phosphoribosyl transferase, imidazole glycerol phophate dehydrogenase and histidinol dehydratase, enzymes involved in key steps in the histidine biosynthestic pathway (Persans *et al.,* 1999b). These cDNA's were used as tools to investigate the regulation of histidine biosynthesis by nickel in *T. goesingense.* These studies revealed that nickel appears not to transciptionally regulate histidine biosynthesis. Biochemical studies also revealed that nickel does not regulate histidine biosynthesis post-transciptionally in *T. goesingense* (Persans *et al.,* 1999b).

Phenotypic cloning

This approach allows for the identification of particular cDNA's based on the phenotype that they confer when over expressed in yeast or *E. coli.* Unlike functional complementation, this approach does not require the existence of particular yeast of *E. coli* mutants lacking functional homologues of the gene of interest. Genes involved in both cadmium and nickel resistance have been identified using this approach. By over expressing a wheat cDNA expression library in yeast and screening for yeast transformants with elevated levels of cadmium resistance a gene encoding phytochelatin synthase has been cloned (Clemens *et al.,* 1999). Using a similar approach, a homologous gene has also been cloned from *A. thaliana* (Vatamniuk *et al.,* 1999).

Taking a similar approach, our laboratory has cloned genes involved in nickel resistance from the nickel hyperaccumulator *T. goesingense.* By over expressing cDNA's from a *T. goesingense* cDNA expression library in *E. coli,* and screening for nickel resistant *E. coli* transformants, we identified serine acetyl transferase (SAT) as a putative gene involved in nickel resistance in *T. goesingense* (Persans *et al.,* 1999a). Over expression of SAT in *E. coli* was found to confer a 6-fold increase in resistant Ni and Co. The mechanism of this SAT induced resistance is not clear, however we have established that it is not due to the over production of sulfide or cysteine. At present, the actually role of SAT in nickel resistance in *T. goesingense* is unclear; however, in our laboratory we hypothesis that it may be involved in the biosynthesis of a nickel chelate molecule.

Homology cloning

There are a number of different ways that homology at the DNA and protein level can be used to clone particular genes of interest. By searching

DNA sequence databases, it may be possible to identify genes from other organisms that perform similar functions to those of interest in plants. These homologues can then be used to design oligonucleotide for use as primers in PCR, for the amplification of a homologous sequence from the plant of interest. A cDNA of the homologues gene could also be used as a probe to screen cDNA libraries for similar genes from plants. In our laboratory, we have used such a strategy to clone new members of the CDF (Cation Diffusion Facilitator) family of metal pumps from the nickel hyperaccumulator *T. goesingense* (Persans and Salt, 2000). By designing oligonucleotides with a sequence based on known members of the CDF family from yeast and *A. thaliana*, and using these oligonucleotides as primers we have been able to amplify 3 full length cDNA's from *T. goesingense* mRNA using RT PCR. These genes have been termed MTP1, 2 and 3 (Metal Transport Protein). Recently, our laboratory have shown that over expression of the MTP genes in certain metal-sensitive yeast strains confers resistance to Ni, Co, Cd and Zn. Preliminary data suggests that this resistance may be due to enhanced metal efflux from the yeast, suggesting that in yeast the *Thlaspi* MTP proteins are acting as plasma membrane metal efflux pumps. A present we are investigating the role of these MTP genes in nickel resistance in *T. goesingense*.

If it is possible to purify a protein of interest to homogeneity, then partial amino acid sequence data can be obtained. Based on the partial amino acid sequence of the target protein, it is possible to design specific oligonucleotides, which can be used to screen a cDNA library for the full-length cDNA clone. Based on the partial amino acid sequence, it may also be possible to identify the gene or its homologues using genome sequence data, or EST sequence data. In the selenium hyperaccumulator, *Astragalus bisulcatus*, the toxic product of selenium assimilation selenocysteine, is detoxified by methylation to form the non-toxic dead-end metabolite methylselenocysteine. By purifying the enzyme involved in methylating selenocysteine, selenocysteine methyl transferase, researchers were able to obtain amino acid sequence information, which they used to design an oligonucleotide. This oligonucleotide were used in a series of PCR reactions to amplify a full-length cDNA from *A. bisulcatus* mRNA. Biochemical studies confirmed that this cDNA encoded the selenocysteine methyl transferase enzyme (Neuhierl *et al.*, 1999).

Using a similar strategy, Higuchi *et al.*, (1999) were able to obtain DNA sequence information specific for a barley gene encoding nicotianamine synthase (NAS). This enzyme is involved in the biosynthesis of the mutagenic acid family of phytosiderophores. Graminaceous plants secrete phytosiderophores from their roots to solublise sparingly soluble iron in the rhizosphere. By searching the rice EST sequence database, using the partial barley nicotianamine synthase gene sequence, it was possible to identify a rice homologue of this gene. Degenerate oligonucleotides were design using this rice EST sequence, and these primers used to amplify a cDNA fragment from a barley cDNA library. This amplified cDNA fragment contained more sequence information than was obtained from the original amino acid sequence. Because of its increased size this PCR amplified cDNA fragment could be used to probe a cDNA library, and identify a putative, full-length cDNA. Using various biochemical methods, this cDNA sequence was confirmed to encode nicotianamine synthase (Higuchi *et al.*, 1999).

Differential cloning

In investigating the mechanisms of plant responses to the environment, it can be assumed that plants react by differentially expressing various genes. Identification and characterisation of these differentially expressed genes should lead to a better understanding of the basic biochemical and molecular mechanisms involved in the plant response. This approach has been used successfully to clone numerous genes from plants.

De Miranda *et al.*, (1990) prepared a cDNA library from copper exposed root tissue of a copper tolerant ecotype of *Mimulus gutattus* (yellow monkey flower). This library was screened with radiolabeled mRNA isolated from copper exposed and control tissue. Using this approach, they were able to identify a differentially expressed gene encoding the copper-binding protein, metallothionein. This was the first example of a metallothionein gene from plants, and its identification laid the foundation for the study of the role of metallothioniens in copper metabolism and resistance in plants.

By screening a soybean cell culture for genes that are expressed during senescence, Crowell and Amasino, (1991) identified a putative senescence regulated gene, *SAM45*. Recently, a full length *A. thaliana* EST was identified which has 97% identity to *SAM45*. This *A. thaliana* gene was characterised as encoding a copper chaperon protein, CCH (Himelblau *et al.*, 1998). Expression of the *CCH* gene is induced 7-fold during leaf senescence and the N-terminal region of the predicted protein sequence contains a heavy-metal-binding motif. Based on this and other evidence, it was suggested that this gene is involved in the mobilisation of copper from plant tissues during senescence (Himelblau *et al.*, 1998).

Mutant analysis

Identification of metal sensitive *A. thaliana* has proved to be a very useful approach to understanding some of the basic mechanisms involved in plant responses to heavy metal stress. Of particular import was the identification of the cadmium sensitive *cad1 A. thaliana* mutant (Howden *et al.*, 1992). By screening 10,000 M_2 mutagenized seeds on agar containing 3 μM $CdSO_4$, and scoring the plants for brown coloration of the roots, the authors identified an allelic series of *cad1* mutants. Further biochemical studies revealed that the *cad1* mutants lacked a functional phytochelatin synthase (Howden *et al.*, 1995), the enzyme involved in synthesising phytochelatins from glutathione. Phytochelatins had already been established to play a central role in the mechanism of cadmium resistance in plants (Rauser, 1999), and therefore the identification of this mutant provided an opportunity to clone this important enzyme. Using a positional cloning strategy, the *CAD1* gene was isolated and sequenced, and it was confirmed to encode phytochelatin synthase (Ha *et al.*, 1999). Similar work was also published at the same time by two other laboratories, who independently cloned the same gene from *A. thaliana* (Vatamniuk *et al.*, 1999) and wheat (Clemens *et al.*, 1999) using a phenotypic cloning strategy described previously (section 3.2).

Serendipity

"The faculty of accidentally making fortunate discoveries". Though unconventional, this strategy has produced some very important results. Recently, the discovery of a putative zinc pump, ZAT1, in *A. thaliana* is an example of such a serendipitous discovery. During a search for genes that were differentially expressed by auxin-treatment in *A. thaliana* root cultures, a particular clone was identified that contained two cDNA inserts (Van der Zaal *et al.*, 1999). One of the inserts was found to be auxin-regulated, but the other was not. Further investigations determined that this second unregulated cDNA showed high homology to known zinc transporters of the CDF family, and it was designated ZAT (Van der Zaal *et al.*, 1999).

DEVELOPMENT OF METAL RESISTANT AND METAL ACCUMULATING PLANTS

With the recent cloning of numerous genes involved in plant responses and adaptations to heavy metal stress, the stage is set to utilise these genes to generate transgenic plants with an enhanced ability to resist and accumulate metals. These plants would have utility in the phytoremediation of metal contaminated soils and waters (Salt *et al.*, 1998). Below are outlined several examples of transgenic plants with enhanced resistance to certain metals.

Selenium resistance

Based on current details of sulfate (Leustek and Saito, 1999) and selenium (Shrift, 1969; Brown and Shrift, 1982) metabolism in plants, it is clear that plants are able to metabolise selenate to form the selenium isologs of the primary end-products of the sulfate assimilation pathway; cysteine, methionine and various intermediates. An important part of this process is the reduction of selenate to selenite via the enzymatic couple ATP sulfurylase and APS reductase. In order to enhance the ability of plants to assimilate selenium, for phytoremediation of selenium enriched soils and waters, Pilon-Smits *et al.,* (1999) overexpressed ATP sulfurylase in *Brassica juncea.* Over expression of ATP sulfurylase caused increased reduction of selenate and incorporation of selenium into selenomethionine. Analysis of the transgenic plants also revealed them to be more tolerant to selenium and to have an increased ability to accumulate selenium, as selenomethionine (Pilon-Smits *et al.,* 1999).

Cadmium resistance

As discussed above, resistance of plants to cadmium is closely linked to sulfur metabolism via the production of glutathione, the substrate for phytochelatin synthesis (Rauser, 1999). In plants, glutathione is synthesised from its component amino acids via the sequential action of two enzymes, γ-glutamylcysteine synthetase and glutathione synthetase (Rauser, 1999). Over expression of these enzymes in *B. juncea* has been show to increase the levels of both glutathione and phytochelatins (Zhu *et al.,* 1999a; Zhu *et al.,* 1999b). Concomitant with increased synthesis of glutathione and phytochelatins, cadmium resistance and accumulation was also increased. Recently, phytochelatin synthase, the enzyme which catalyses the biosynthesis of phytochelatins from glutathione (see section 3.2 and 3.5) was cloned from *A. thaliana* (Vatamanuik *et al.,* 1999; Ha *et al.,* 1999) and wheat (Clemens *et al.,* 1999). Over production of this enzyme now offers the possibility of increasing still further the cadmium resistance and accumulation of these plants.

Aluminium resistance

Plants are known to secrete chelating compounds into the rhizosphere to reduce the uptake of potentially toxic metals. Aluminum resistance is known to be mediated by secretion of citrate or malate into the rhizosphere (see section 2.1). Recently, over expression of citrate synthase in tobacco was found to be sufficient to impart aluminum resistance in the transgenic plants (Manuel de la Fuente *et al.,* 1997). A *Pseudomonas* citrate synthase gene was fused to a cauliflower mosaic virus 35S promoter and constitutively over expressed in tobacco and papaya. The level of citrate synthase activity in the transgenic plants was found to be between 3 and 10-fold higher than wild type plants. Over production of citrate caused an increase in citrate efflux from the roots of between 2 and 4-fold. Roots of transgenic plant, over producing citrate, accumulated less aluminum than wild-type plants. This decreased accumulation was thought to be due to a decrease in aluminum availability in the rhizosphere induced by an increased concentration of extracellular citrate.

Zinc resistance

As discussed in section 3.6, a putative zinc pump, ZNT1 was recently cloned from *A. thaliana* (Van der Zaal *et al.,* 1999). Based on sequence homology it was thought this gene encodes a protein involved in effluxing zinc either out of cells or into internal membrane bound organelles such as the vacuole. Over expression of ZNT1 in *A. thaliana* lead to increased root resistance to zinc. Interestingly, this increased zinc resistance in roots was associated with increased root zinc concentrations, though no differences in shoot zinc concentrations were observed. The authors concluded that ZNT1 mediated zinc resistance in roots was therefore not due to a reduction in net zinc influx, but rather increased vacuolar compartmentalisation of zinc.

Nickel resistance

Recently, Arazi *et al.,* (1999) cloned a gene from tobacco, designated NtCBP4, which showed homology to previously cloned non-selective plasma membrane cation channels. Surprisingly, over expression of NtCBP4 in tobacco caused increased nickel resistance. Analysis of the shoot nickel content of the transgenic plants showed that over expression of NtCBP4 reduced the uptake of nickel by the plants, and it was concluded that this reduced nickel uptake was responsible for the enhanced nickel resistance. It would appear paradoxical that over expression of a putative cation channel would reduced uptake of a particular cation, in this case nickel. The authors suggest two possible resolutions to the paradox. By binding nickel directly at the plasma membrane NtCBP4 might act to reduce the effective concentration of nickel at the membrane surface, thereby reducing the uptake of nickel. Alternatively, the NtCBP4 protein might interact with and suppress other plasma membrane channels that are selective for nickel.

Mercury resistance

Mercury resistance in Gram-negative bacteria is mediated in part by the activity of an organomercury lyase involved in protonolysis of carbon mercury bonds (MerB), and a mercuric ion reductase (MerA) that catalyses the reduction of Hg^{2+} to Hg^0. The co-ordinated action of both these enzymes leads to the efficient conversion of highly toxic organomercury compounds to the less toxic, volatile, Hg^0. In a series of elegant experiments, *MerB* and *MerA* have been over expressed in plants. Both *A. thaliana* and yellow poplar over expressing a modified *MerA* gene have been shown to have enhanced resistance to ionic mercury (Rugh *et al.,* 1996; Rugh *et al.,* 1998). This resistance was found to be correlated with an increased capacity to volatilise Hg^0. The authors concluded that mercury resistance in these transgenic plants was mediated by enhanced conversion of toxic ionic mercury to Hg^0, followed by loss of Hg^0 from the plant by volatilisation. Transgenic *A. thaliana* were also generated that over express *MerB* (Bizily *et al.,* 1999). These transgenic plants showed a remarkable level of resistance to highly toxic monomethylmercuric chloride and phenylmercuric acetate, compared to wild-type *A. thaliana*. By crossing separate *A. thaliana* lines over expressing either *MerA* or *MerB*, (Bizily *et al.,* 2000) recently produced crosses that over expressed both *MerA* and *MerB*. These plants efficiently detoxify monomethylmercury by first converting monomethylmercury to Hg^{2+} using the organomercury lyase (MerB). Ionic mercury was then rapidly reduced to elemental mercury, through the action of mercury reductase (MerA). Elemental mercury was then removed from the plant by volatilisation. The creators of these mercury resistant plants propose that transgenic plants over expressing *MerA* and *MerB* could be used to remove mercury from polluted soils and waters, though questions still remain as to the desirability of removing mercury via volatilisation into the atmosphere (Salt, 1998).

CONCLUSION

Over the past 10 years, we have learnt a lot about how plants respond and adapt to heavy metal stress. However, only in the last couple of years have we started to gain understanding of these phenomena at the molecular level. In this review, we have briefly discussed some of these recent discoveries, and provided some examples of how, using transgenic technologies, some of these discoveries have been used to generate plants with enhanced resistance to various metals. However, this is just the tip of the iceberg. In the coming years, as our understanding of the molecular processes involved in heavy metal stress responses and adaptations increases, we should be able to design transgenic plants to perform ever more sophisticated tasks. We might expect to see transgenic plants with the capacity to solubilise pollutant metals, such as lead, from soils and accumulate them in their shoots for harvest and removal. This type of technology should help remediate some of the heavy metal pollution generated by various industrial processes. We might also expect to see the generation of food crops with an enhanced capacity to accumulate essential micronutrients such as iron and zinc from soils. These micronutrient efficient crops should help to alleviate some of the mineral deficiencies that plague a large proportion of the world's human populations. So, now that we know more about this "curious and unimportant character", we find that it is far from unimportant, and its understanding may lead to large advances in environmental biotechnology and human mineral nutrition.

REFERENCES

Arazi, T., Ramanjulu, S., Kaplan, B. and Fromm, H. 1999. A tobacco plasma membrane calmodulin binding transporter confers Ni^{2+} tolerance and Pb^{2+} hypersensitivity in transgenic plants. *Plant J.* 20, 171-182.

Baker, A.J.M 1981. Accumulators and excluders – strategies in the response of plants to heavy metals. *J. Plant Nutr.* 3, 643-654.

Belouchi, A., Kwan, T. and Gros P. 1997. Cloning and characterization of the OsNramp family from *Oryza sativa*, a new family of membrane proteins possibly implicated in the transport of metal ions. *Plant Mol. Biol.* 33, 1085-1092.

Bizily, S.P., Rugh, CL. and Meagher, R.B. 2000. Phytodetoxification of hazardous organomercurials by genetically engineered plants. *Nature Biotechnol.* 18, 213-217.

Bizily, S.P., Rugh, C.L., Summers, A.O. and Meagher, R.B. 1999. Phytoremediation of methylmercury pollution: merB expression in *Arabidopsis thaliana* confers resistance to organomercurials. *Proc. Natl. Acad. Sci. USA* 96, 6808-6813.

Blamey, F.P.C., Joyce, D.C., Edwards, D.G. and Asher C.J. 1986. Role of trichomes in sunflower tolerance to manganese toxicity. *Plant Soil* 91, 171-180.

Bradshaw, A.D., McNeilly, T. and Putwain, P.D. 1990. "The essential qualities". In: *Heavy Metal Tolerance in Plants: Evolutionary Aspects*, ed. A.J. Shaw, pp. 323-334, CRC Press, Inc., Boca Raton.

Brown, T.A. and Shrift, A. 1982. Selenium: toxicity and tolerance in higher plants. *Biol. Rev.* 57, 59-84.

Brune, A., Urbach, W. and Dietz, K-J. 1994. Compartmentation and transport of zinc in barley primary leaves as basic mechanisms involved in zinc tolerance. *Plant. Cell. Environ.* 17, 153-162.

Clarkson, D.T. and Hanson, J.B. 1986. Proton fluxes and the activity of a stelar proton pump in onion roots. *J. Exp. Bot.* 37, 1136-1150.

Clemens, S., Kim, E.J., Neumann, D. and Schroeder, J.I. 1999. Tolerance to toxic metals by a gene family of phytochelatin synthases from plants and yeast. *EMBO J.*18, 3325-3333.

Crowell, D. and Amasino, R.M. 1991. Induction of specific mRNA's in cultured soybean cells during cytokinin or auxin starvation. *Plant Physiol.* 95, 711-715.

De Boer, A.H., Prins, H.B.A. and Zanstra, P.E. 1983. Bi-phasic composition of trans-root electrical potential in roots of *Plantago* species: involvement of spatially separated electrogenic pumps. *Planta* 157, 259-266.

De Miranda, J.R., Thomas, M.A., Thurman, D.A. and Tomsett, A.B. 1990. Metallothionein genes from the flowing plant *Mimulus guttatus*. *FEBS Lett.* 260: 277-280.

Delhaize, E., Ryan, P.R. and Randall, J. 1993. Aluminum tolerance in wheat (*Triticum aestivum* L.) II. Aluminum-stimulated excretion of malic acid from root apicies. *Plant Physiol.* 103, 695-670.

Eide, D., Broderius, M., Fett, J. and Guerinot, M-L. 1996. A novel iron-regulated metal transporter from plants identified by functional expression in yeast. *Proc. Nat. Acad. Sci. USA* 93, 5624-5628.

Grill, E., Winnacker, E-L. and Zenk, M.H. 1985. Phytochelatins: the principle heavy metal complexing peptides of higher plants. *Science* 230, 674-676.

Grotz, N., Fox, T., Connolly, E., Park, W., Guerinot, M-L. and Eide, D. 1998. Identification of a family of zinc transporter genes from *Arabidopsis* that respond to zinc deficiency. *Proc. Nat. Acad. Sci. USA* 95, 7220-7224.

Ha, S-B., Smith, A.P., Howden, R., Dietrich, W.M., Bugg, S., O'Connell, M.J., Goldsbrough, P.B. and Cobbett, C.S. 1999. Phytochelatin synthase genes from *Arabidopsis* and the yeast *Schizosaccharomyces pombe*. *Plant Cell* 11, 1153-1163.

Heath, S.M., Southworth, D. and D'Allura, J.A. 1997. Localization of nickel in epidermal subsidiary cells of leaves of *Thlaspi montanum* var. *siskiyouense* (Brassicaceae) using energy-dispersive x-ray microanalysis. *Int. J. Plant. Sci.* 158, 184-188.

Higuchi, K., Suzuki, K., Nakanishi, H., Yamaguchi, H., Nishizawa, N-K. and Mori, S. 1999. Cloning of nicotianamine synthase genes, novel genes involved in the biosynthesis of phytosiderophores. *Plant Physiol.* 119, 471-479.

Himelblau, E., Mira, H., Lin, S-J., Culotta, V.C., Peñarrubia, and Amasino, R.M. 1998. Identification of a functional homolog of the yeast copper homeostasis gene *ATX1* from *Arabidopsis*. *Plant Physiol.* 117, 1227-1234.

Howden, R. and Cobbett, C.S 1992. Cadmium-sensitive mutants of *Arabidopsis thaliana*. *Plant Physiol.* 99, 100-107.

Howden, R., Goldsbrough, P.B., Andersen, C.R. and Cobbett, C.S. 1995. Cadmium-sensitive, *cad1* mutants of *Arabidopsis thaliana* are phytochelatin deficient. *Plant Physiol.* 107, 1059-1066.

Kochian, L.V. 1995. Cellular mechanisms of aluminum toxicity and resistance in plants. *Ann. Rev. Plant Physiol. Plant Mol. Biol.* 46, 237-260.

Krämer, U., Cotter-Howells, J.D., Charnock, J.M., Baker, A.J.M. and Smith, J.A.C. 1996. Free histidine as a metal chelator in plants that accumulate nickel. *Nature* 379, 635-638.

Krämer, U., Smith, R.D., Wenzel, W., Raskin, I. and Salt, D.E. 1997a. The role of nickel transport and tolerance in nickel hyperaccumulation by *Thlaspi goesingense* Hálácsy. *Plant Physiol.* 115, 1641-1650.

Krämer, U., Grime, G.W., Smith, J.A.C., Hawes, C.R. and Baker, A.J.M. 1997b. Micro-PIXE as a technique for studying nickel localization in leaves of the hyperaccumulator plant *Alyssum lesbiacum*. *Nucl. Instr. Meth. Phys. Res. B.* 130, 346-350.

Krämer, U., Pickering, I.J., Prince, R.C., Raskin, I. and Salt, D.E. 2000. Subcellular localization and speciation of nickel in hyperaccumulator and non-accumulator *Thlaspi* species. *Plant Physiol.* 122, 1343-1353.

Küpper, H., Zhao, F.J. and McGrath, S.P. 1999. Cellular compartmentalization of zinc in leaves of the hyperaccumulator *Thlaspi caerulescens*. *Plant Physiol.* 119, 305-311.

Lasat, M.M., Baker, A.J.M. and Kochian, L.V. 1996. Physiological characterization of root Zn^{2+} absorption and translocation to shoots in Zn hyperaccumulator and nonaccumulator species of *Thlaspi*. *Plant Physiol.* 112:1715-22

Lasat, M.M., Baker, A.J.M. and Kochian, L.V. 1998. Altered zinc compartmentation in the root symplasm and stimulated zinc absorption into the leaf as mechanisms involved in Zn hyperaccumulation in *Thlaspi caerulescens*. *Plant Physiol.* 118, 875-883.

Läuchli, A. 1993. Selenium in plants: uptake, functions, and environmental toxicity. *Bot. Acta* 106, 455-468.

Leustek, T, and Saito, K. 1999. Sulfate transport and assimilation in plants. *Plant Physiol.* 120, 637-643.

Lim, C.O., Kim, H.Y., Kim, M.G., Lee, S.I., Chung, W.S., Park, S.H., Hwang, I. and Cho. M..J. 1996. Expressed sequence tags of chinese cabbage flower bud cDNA. *Plant Physiol.* 111, 577-588.

Manuel de la Fuente, J., Ramírez-Rodríguez, V., Cabrera-Ponce, J.L., Herrera-Estrella, L. 1997. Aluminum tolerance in transgenic plants by alteration of citrate synthesis. *Science* 276, 1566-1568.

Marschner, H. 1995. *Mineral Nutrition of Higher Plants*, 2nd edition, Academic Press, London.

Martell, E.A. 1974. Radioactivity of tobacco trichomes and insoluble cigarette smoke particles. *Nature* 249, 215-217.

Mesjasz-Przybylowicz, J., Balkwill, K., Przybylowicz, W.J. and Annegarn, H.J. 1994. Proton microprobe and X-ray fluorescence investigations of nickel distribution in serpentine flora from South Africa. *Nucl. Instr. Meth. B.* 89, 208-212.

Miyasaka, S.C., Buta, J.G., Howell, R.K. and Foy, C.D. 1991. Mechanism of aluminum tolerance in snapbeans. Root exudation of citric acid. *Plant Physiol.* 96, 737-743.

Murphy, A.S., Eisinger, W.R., Shaff, J.E., Kochian, L.V. and Taiz. L. 1999. Early copper-induced leakage of K^+ from *Arabidopsis* seedlings is mediated by ion channels and coupled to citrate efflux. *Plant Physiol.* 121, 1375-1382.

Neuhierl, B. and Böck, A. 1996. On the mechanism of selenium tolerance in selenium-accumulating plants. Purification and characterization of a specific selenocysteine methyltransferase from cultured cells of *Astragalus bisculatus. Eur. J. Biochem.* 239, 235-238.

Neuhierl, B., Thanbichler, M., Lottspeich, F. and Böck, A. 1999. A family of S-methylmethionine-depenent thiol/selenol methyltransferases. Role in selenium tolerance and evolutionary relation. *J. Biol. Chem.* 274, 5407-5414.

Neumann, D., zur Nieden, U., Lichtenberger, O. and Leopold, I. 1995. How does *Armeria maritima* tolerate high heavy metal concentrations? *J. Plant Physiol.* 146, 704-717.

Pence, N.S., Larsen, P.B., Ebbs, S.D., Letham, D.L., Lasat, M.M., Garvin, D.F., Eide, D. and Kochian, L.V. 2000. The molecular physiology of heavy metal transport in the Zn/Cd hyperaccumulator *Thlaspi caerulescens. Proc. Nat. Acad. Sci. USA* 97, 4956-60.

Persans, M. and Salt, D.E. 2000. Possible molecular mechanisms involved in nickel, zinc and selenium hyperaccumulation in plants. *Biotech. Gen. Eng. Rev.* 17, 385-409.

Persans, M.W., Albrecht, C., Nieman, K.S., Shaffer, I.N., Motley, P.L. and Salt, D.E. 1999a. Molecular dissection of the cellular mechanisms involved in nickel hyperaccumulation. *Abstracts of the American Society of Plant Physiology*, Annual Meeting, Baltimore, MD., USA.

Persans, M., Xiange, Y., Patnoe, J.M.M.L., Krämer, U. and Salt, D.E. 1999b. Molecular dissection of histidine's role in nickel hyperaccumulation in *Thlaspi goesingense* (Hálácsy). *Plant Physiol.* 121, 1-10.

Pickering, I.J., Prince, R.C., George, J.M., Smith, R.D., George, G.N. and Salt, D.E. 2000. Reduction and coordination of arsenic in Indian mustard. *Plant Physiol.* 122, 1171-1177.

Pilon-Smits, E.A.H., Hwang, S., Lytle, C.M., Zhu, Y., Tai, J.C., Bravo, R.C., Chen, Y., Leustek, T. and Terry, N. 1999. Overexpression of ATP sulfurylase in Indian mustard leads to increased selenate uptake, reduction, and tolerance. *Plant Physiol* 119, 123-132.

Pitman, M.G. 1972. Uptake and transport of ions in barley seedlings. II. Evidence for two active stages in transport to the shoot. *Aust. J. Biol. Sci.* 25, 243-257.

Rauser, W.E. 1999. Structure and function of metal chelators produced by plants. *Cell Biochem. Biophys.* 31, 19-48.

Roberts, S.K. and Tester, M. 1995. Inward and outward K^+-selective currents in the plasma membrane of protoplasts from maize root cortex and stele. *Plant J.* 8, 811-825.

Roberts, S.K. and Tester, M. 1997. Permeation of Ca^{2+} and monovalent cations through an outwardly rectifying channel in maize root stelar cells. *J. Exp. Bot.* 48, 839-846.

Rugh, C.L., Wilde, H.D., Stack, N.M., Thompson, D.M., Summers, A.O. and Meagher, R.B. 1996. Mercuric ion reduction and resistance in transgenic *Arabidopsis thaliana* plants expressing a modified bacterial merA gene. *Proc. Nat. Acad. Sci. USA* 93, 3182-3187.

Rugh, C.L., Senecoff, J.F., Meagher, R.B. and Merkle, S.A. 1998. Development of transgenic yellow poplar for mercury phytoremediation. *Nature Biotechnol.* 16, 925-928.

Sagner, S., Kneer, R., Wanner, G., Cosson, J-P., Deus-Neumann, B. and Zenk, M.H. 1998. Hyperaccumulation, complexation and distribution of nickel in *Sebertia acuminata*. *Phytochemistry* 47, 339-347.

Salt, D.E. and Wagner, G.J. 1993. Cadmium transport across tonoplast of vesicles from oat roots. Evidence for a Cd/H antiport activity. *J. Biol. Chem.* 268, 12297-12302.

Salt, D.E. and Rauser, W.E. 1995. MgATP-dependent transport of phytochelatins across the tonoplast of oat roots. *Plant Physiol.* 107, 1293-1301.

Salt, D.E., Prince, R.C., Pickering, I.J. and Raskin, I. 1995. Mechanisms of cadmium mobility and accumulation in Indian mustard. *Plant Physiol.* 109, 1427-1433.

Salt, D.E. 1998. Arboreal alchemy. *Nature Biotechnol.* 16, 905.

Salt, D.E., Smith, R.D., Raskin, I. 1998. Phytoremediation. *Ann. Rev. Plant Physiol. Plant Mol. Biol.* 49, 643-668.

Salt, D.E., Kato, N., Krämer, U., Smith, R.D. and Raskin, I. 1999a. "The role of root exudates in nickel hyperaccumulation and tolerance in accumulator and non-accumulator species of *Thlaspi*". In: *Phytoremediation of Contaminated Soil and Water*, eds. N. Terry and G.S. Bañuelos, pp. 191-202. CRC Press LLC, Boca Raton.

Salt, D.E., Prince, R.C., Baker, A.J.M., Raskin, I. and Pickering, I.J. 1999b. Zinc ligands in the metal hyperaccumulator *Thlaspi caerulescens* as determined using X-ray absorption spectroscopy. *Environ. Sci. Tech.* 33, 713-717.

Schmoger, M.E., Oven, M. and Grill, E. 2000. Detoxification of arsenic by phytochelatins in plants. *Plant Physiol.* 122, 793-802.

Shrift, A. 1969. Aspects of selenium metabolism in higher plants. *Ann. Rev. Plant Physiol.* 20, 475-495.

Thomine, S., Wang, R., Ward, J.M., Crawford, N.M. and Schroeder, J.I. 2000. Cadmium and iron transport by members of a plant metal transporter family in *Arabidopsis* with homology to nramp genes. *Proc. Nat. Acad. Sci. USA* 97, 4991-4996.

Van der Zaal, B.J., Neuteboom, L.W., Pinas, J.E., Chardonnes, A.N., Schat, H., Verkleji, J.A.C. and Hooykaas, P.J.J. 1999. Overexpression of a novel *Arabidopsis* gene related to putative zinc-transporter genes from animals can lead to enhanced zinc resistance and accumulation. *Plant Physiol.* 119, 1047-1055.

Vatamaniuk, O.K., Mari, S., Lu, Y-P. and Rea, P.A. 1999. AtPCS1, a phytochelatin synthase from *Arabidopsis*: isolation and *in vitro* reconstitution. *Proc. Nat. Acad. Sci. USA* 96, 7110-7115.

Vatamaniuk, O.K., Mari, S., Lu, Y.P., Rea, P.A. 2000. Mechanism of heavy metal ion activation of phytochelatin (PC) synthase: blocked thiols are sufficient for PC synthase-catalyzed transpeptidation of glutathione and related thiol peptides. *J. Biol. Chem.* 275, 31451-32459.

Vázquez, M.D., Barceló, J., Poschenrieder, Ch., Mádico, J., Hatton, P., Baker, A.J.M. and Cope, G.H. 1992. Localization of zinc and cadmium in *Thlaspi caerulescens* (Brassicaceae), a metallophyte that can hyperaccumulate both metals. *J. Plant Physiol.* 140, 350-355.

Vázquez, M.D., Poschenrieder, Ch., Barceló, J., Baker, A.J.M., Hatton, P. and Cope, G.H. 1994. Compartmentation of zinc in roots and leaves of the zinc hyperaccumulator *Thlaspi caerulescens* J & C Presl. *Bot. Acta* 107, 243-250.

Vögeli-Lange, R. and Wagner, G.J. 1990. Subcellular localization of cadmium and cadmium-binding peptides in tobacco leaves. *Plant Physiol.* 92, 1086-1093

Wegner, L.H. and Raschke, K. 1994. Ion channels in the xylem parenchyma of barley roots. *Plant Physiol.* 105, 799-813.

White, M.C., Baker, F.D., Chaney, R.F. and Decker, A.M. 1981. Metal complexation in xylem fluid. *Plant Physiol.* 67, 301-310.

Zhao, H. and Eide, D. 1996. The ZRT2 gene encodes the low affinity zinc transporter in *Saccharomyces cerevisiae*. *J. Biol. Chem.* 271, 23203-23210.

Zhu, Y.L., Pilon-Smits, E.A., Tarun, A.S., Weber, S.U., Jouanin, L. and Terry, N. 1999a. Cadmium tolerance and accumulation in Indian mustard is enhanced by overexpressing γ-glutamylcysteine synthetase. *Plant Physiol.* 121, 1169-1177.

Zhu, Y.L., Pilon-Smits, E.A., Jouanin, L. and Terry, N. 1999b. Overexpression of glutathione synthetase in Indian mustard enhances cadmium accumulation and tolerance. *Plant Physiol.* 119, 73-79.

Chapter 9

MOLECULAR RESPONSES TO OXIDATIVE STRESS

John G. Scandalios

Department of Genetics, North Carolina State University, Raleigh, NC 27695-7614, USA. jgs@unity.ncsu.edu

INTRODUCTION

The evolution of oxygenic photosynthesis altered the Earth's atmosphere and enabled the development and sustenance of aerobic life. Thus, molecular oxygen is essential for life on Earth. However, the incomplete reduction of dioxygen (O_2) to water (H_2O) during normal aerobic metabolism generates reactive oxygen species (ROS) that pose a serious threat to all aerobic organisms (Fridovich, 1975). At the same time, it has become increasingly clear that ROS, in addition to their role as toxic agents, are also used as second messengers in various signal transduction pathways, and in many types of biological defense mechanisms (Scandalios, 1997). Thus, in addition to the historic perception of ROS as toxic and lethal agents, recent evidence indicates that they play beneficial roles as well.

ROS such as singlet oxygen ($^1O_2$), superoxide anion ($O_2^{\bullet-}$), hydrogen peroxide (H_2O_2), hydroxyl radical ($\cdot OH$), and nitric oxide (NO) are crucial for many physiological processes, and usually exist in the cell in a balance with biochemical antioxidants. However, excess ROS resulting from exposure to environmental oxidants, toxicants, radiation, or numerous biostressors, perturbs cellular redox balance (to a more oxidised state) and disrupts normal biological functions. This condition is referred to as "oxidative stress" and may be detrimental to the organism and contribute to the pathogenesis of disease and aging, and numerous physiological dysfunctions leading to cell death.

M.J. Hawkesford and P. Buchner (eds.),
Molecular Analysis of Plant Adaptation to the Environment, 181–208.

To counteract the oxidant effects of ROS and to restore a state of redox balance, cells must reset critical homeostatic parameters

Changes associated with oxidative damage and restoration of cellular homeostasis often lead to "activation" or "silencing" of genes encoding regulatory transcription factors, antioxidant defence genes and enzymes, and structural proteins.

Thus, the evolution of organisms that use molecular oxygen as a terminal electron acceptor for respiration (i.e., aerobic organisms) has given rise to a paradox (often referred to as "the oxygen paradox"). On the one hand, aerobic metabolism provides a much higher energy yield than anaerobic respiration, giving the aerobic organism a huge competitive advantage. Alternatively, the respiratory reduction of oxygen to water leads to the stepwise formation of highly reactive oxygen intermediates (ROS). In the presence of light, photosensitizer-mediated production of the reactive singlet form of oxygen ($^1O_2$) may also occur.

Essentially the opposite process, the four-electron oxidation of water and the associated formation of oxygen radicals, occurs during photosynthesis. Normally, these ROS remain bound during oxidation/reduction reactions involved in respiration and photosynthesis. Occasionally, however, they are released and unless dealt with immediately are capable of causing extensive cell damage. In addition, $O_2^{\bullet-}$ and H_2O_2 can act as substrates for the transition-metal-catalyzed Haber-Weiss reaction leading to the production of the hydroxyl radical (·OH). This is the most reactive of the oxygen radicals that directly and indiscriminately attacks all biomolecules.

Normal metabolic processes also produce ROS. The $O_2^{\bullet-}$, for example, can be generated by various enzymatic reactions in the cell (e.g., xanthine oxidase and uricase). Hydrogen peroxide may be produced by the dismutation of two superoxide anions, by the β-oxidation of fatty acids in the course of normal lipid metabolism, and in response to stress. The production of ROS in response to pathogen attack is also well documented (Halliwell, 1996; Scandalios, 1996).

To minimize the damaging effects of ROS, whether of biotic or abiotic origin, aerobic organisms have evolved both nonenzymatic and enzymatic antioxidant defense systems. Nonenzymatic systems include compounds with intrinsic antioxidant properties such as glutathione, β-carotene, ascorbic acid or vitamin C, α-tocopherol or vitamin E, and flavonoids. These are often linked with enzymes such as ascorbate peroxidase and glutathione-reductase, which generate the antioxidant capacity of the quenching molecule.

Purely enzymatic systems, such as superoxide dismutases (SOD), catalases (CAT), and peroxidases (Px) protect the organism by directly scavenging $O_2^{\bullet-}$ and H_2O_2 and converting them to less reactive species (reactions 1-3):

(1) $$O_2^{\bullet-} + O_2^{\bullet-} + 2H^+ \xrightarrow{\textbf{SOD}} O_2 + H_2O_2$$

(2) $$2H_2O_2 \xrightarrow{\textbf{CAT}} 2H_2O + O_2$$

(3) $$H_2O_2 + R(OH)_2 \xrightarrow{\textbf{Px}} 2H_2O + R(O)_2$$

Since the rate constants of reactions involving •OH radical attack are as fast as the fastest reactions catalyzed by the antioxidant enzymes such as SOD, no specific enzyme systems are known that directly scavenge the •OH. It is hypothesised that, instead, the superoxide dismutases, coupled with the catalases and peroxidases, act to limit the availability of substrate for the reactions leading to the formation of the hydroxyl radical.

OXIDATIVE STRESS

Oxidative stress occurs when antioxidants are depleted and/or if the formation of ROS increases beyond the ability of the defences to cope. Such stress can happen when severely adverse environments overwhelm biological systems. One rapid and clear indicator of oxidative stress is the induction of antioxidant defences and/or increases in endogenous ROS levels. The formation of ROS can be accelerated as a consequence of various environmental stress conditions, including UV-radiation, high light intensities, low CO_2 concentrations, exposure to herbicides, extreme temperatures, pathogens, fungal toxins such as cercosporin and aflatoxin, ozone and other air pollutants, metals, wounding, and various xenobiotics. Many inducers of oxidative stress in various organisms examined are known carcinogens, mutagens, and toxins. It has become evident from numerous studies by several laboratories, using a variety of organisms, that ROS production is a common denominator in many diseases and environmental insults which can lead to cell death in virtually all aerobes. It is also becoming clear that a variety of different biotic and abiotic stresses cause their deleterious effects via ROS generation, directly or indirectly (Scandalios, 1993; 1997).

XENOBIOTICS, ENVIRONMENTAL POLLUTANTS, AND ROS

All living organisms encounter naturally occurring xenobiotics. However, more recently, the most important challenge with xenobiotics derives from the manipulation of cultivars with large quantities of chemical herbicides, pesticides, and insecticides, as well as from environmental pollution, which, in part, is due to the use of these compounds. An inconsistency that becomes immediately apparent is that vigorous and high yielding cultivars need to be established in the adverse environments these xenobiotics create for every living organism. Whether and how organisms can avoid the deleterious effects of xenobiotics is a major agricultural, societal, health, and environmental concern. The most effective way for the affected organism to escape the deleterious consequences of a xenobiotic is to launch its defensive responses that will lead to detoxification; that is, metabolic processing and inactivation of the harmful compound. It has become apparent that the antioxidant defence system in aerobic organisms plays a key role in protecting against such compounds.

Xenobiotic metabolism in higher plants

Plants encounter a chemical environment made up of nutrients and xenobiotics (i.e., natural or synthetic substances that cannot be utilised by plants for energy-yielding processes). In most cases xenobiotics are toxic compounds; chemicals that are commonly classified as xenobiotics include pesticides and air pollutants. Higher plants are exposed to xenobiotics either deliberately, i.e. pesticide application, or accidentally, as the result of industrial emissions or agricultural uses. In order to survive, plants detoxify xenobiotics by an array of biotransformations following strikingly similar pathways with those of animals (Kreuz *et al.,* 1996). Biotransformations of xenobiotics in higher plants are grouped into three main phases: *Phase I* (conversion), *Phase II* (conjugation), and *Phase III* (compartmentation). Phase I reactions include nonsynthetic processes such as oxidations catalysed by cytochrome P450 monooxygenases (Barrett, 1995; Frear, 1995), reductions, and hydrolyses. Phase I metabolism often results in the formation of metabolites with reduced or modified phytotoxicity, increased polarity, or enhanced susceptibility to further processing. Phase II conjugation of xenobiotics with glutathione (GSH), catalysed by glutathione S-transferases (GST), sugars (glucosyltransferases) or amino acids are synthetic reactions. Phase II metabolism results in the formation of metabolites with greatly reduced or no phytotoxicity, higher water solubility and limited mobility. In phase III, xenobiotic conjugates are converted to secondary conjugates or insoluble, bound residues, deposited in the vacuole, or other compartments of plant cells. Export of xenobiotic conjugates from the cytosol to the vacuole or other compartments is a prerequisite for cellular detoxification occurring in Phase III. Membrane-bound transporters have been recently shown to mediate the energy-dependent export of xenobiotic conjugates into plant vacuoles. N-malonylation of xenobiotic glutathione or glucoside conjugates appears to be another common reaction in this phase.

Xenobiotic metabolism generates ROS

Generation of ROS during normal metabolism does not normally impose oxidative stress. This source of ROS encompasses such mechanisms as "leakage" of electrons to O_2 from mitochondrial electron transport chains, microsomal cytochromes P450 and their electron donating enzymes (Halliwell, 1995). At several stages in the catalytic cycle of cytochrome P450, direct and indirect evidence has been obtained for the generation of free radical intermediates of carbon as well as oxygen and other heteroatoms (Dolphin, 1987). ROS are also by-products of cytochrome P450 peroxygenase activity, as well as its reductive reactions (Vaz *et al.,* 1987). Cytochrome P450 is the first enzyme in the pathway to detoxification of numerous plant xenobiotics, mostly herbicides (Gronwald, 1994). Thus, during xenobiotic Phase I metabolism involving cytochrome P450 oxidases, ROS are generated. The Phase II detoxification process involves conjugation of xenobiotics with GSH catalysed by GST. The tripeptide glutathione is the major low molecular weight thiol in plants (Hatzios, 1997). The redox chemistry of GSH endows the molecule with antioxidant function mediated by the sulfhydryl group of cysteine, which, upon oxidation, forms a disulfide bond with a second molecule of GSH to form oxidised glutathione (GSSH). GSH protects cells against oxidative stress by scavenging free radicals and keeping the free radical scavenger ascorbic acid in its reduced active form. After oxidation, GSSH can then be regenerated to GSH with the action of glutathione reductase (GR), which catalyses the NADPH-dependent reduction of oxidised glutathione. The balance of GSH oxidation-reduction and biosynthesis-depletion is of major importance in the protection of organisms from oxidative stress. Depletion of GSH from the cell or an increased GSSH/GSH ratio can unbalance the active oxygen scavenging cycle and result in oxidative stress. Xenobiotic metabolism actively depletes cellular resources of GSH, which are utilised by GST in xenobiotic conjugation reactions. In addition, induction of ROS generation by the cytochrome P450 mediated oxidations of xenobiotic substrates could result in oxidative stress, which deplete intracellular GSH after oxidation by hydroxyperoxides (Forman *et al.,* 1995). The combined results of ROS formation, GSH depletion and high GSSH/GSH ratio during xenobiotic metabolism may impose severe oxidative stress.

Xenobiotic action, directly or indirectly, produces ROS

Xenobiotics that interfere with chloroplastic or mitochondrial electron transfer systems produce, directly or indirectly, ROS. Based on this mode of action herbicides have been developed that interact with photosystems (PS) I and II of the chloroplast. Herbicides that block photosynthetic electron transport, such as monuron, ioxynil, and atrazine, allow excitation energy to be transferred from chlorophyl to the carotenoids, which will be damaged progressively. Once they are destroyed, the light energy may be transferred to oxygen, generating ROS that can initiate lipid peroxidation. Similarly, herbicides that act by inhibiting carotenoid synthesis, such as aminotriazole, metflurazone, fluridone, norflurazon, and pyrichlor, eliminate an important quencher of excitation energy, thus leading to formation of ROS (Knox and Dodge, 1995). Atrazine (S-chloro-triazine) is a herbicide that binds to the D1 protein and blocks the electron transfer of PS II. However, the herbicidal activity of atrazine is not due to the interruption of photosynthesis, but to oxidative stress generated when photosynthetic electron transport is blocked, resulting in the destruction of the PS II reaction center and the photo-oxidation of lipid and chlorophyll molecules (Gronwald, 1994). Chloroplasts have multiple mechanisms to scavenge ROS, but in this case, they are overwhelmed by the magnitude of oxidative stress generated. Maize is resistant to S-chloro-triazines because of multiple mechanisms for detoxification. The main mechanism seems to be GST-catalysed conjugation of triazines with GSH. Atrazine can also be metabolised via N-acetylation catalysed by cytochrome P450, but this has not been demonstrated in vitro. Whether other ROS scavenging enzymes (e.g., CAT) in other cell compartments contribute to atrazine resistance in maize has not been examined. There are also chemicals that interact with the reducing site of PS I (due to its low redox potential). One category of chemicals able to accept electrons from PS I are the viologens (paraquat, diquat, benzyl-viologen). These bipyridyl compounds have been commercialised as non-selective contact herbicides (Preston, 1994). Paraquat, or methyl-viologen (MV), is a redox-active compound that interacts with PS I, acting as an electron acceptor and interrupting electron transfer. MV, by accepting electrons from one of the iron-sulfur centers of PS I, forms the bipyridyl cation radical, which is unstable and reacts rapidly with O_2 to form superoxide, regenerating the bipyridyl cation. The plant is able to detoxify $O_2^{\bullet-}$ via superoxide dismutase (SOD), producing H_2O_2 and O_2. H_2O_2 can be further detoxified by enzymes of the ascorbate-glutathione cycle in the chloroplast, or it can diffuse to the cytosol and be scavenged by other enzymatic (e.g.,

CAT) or non-enzymatic systems. ROS produced following MV action can attack double bonds in the fatty acid side chains of lipids, disrupting membranes, and resulting in cell death. MV is also able to interact with other electron transfer systems and hence is toxic to animals. It can also be toxic to plants grown in the dark but the herbicidal action is slower. The plant mitochondrial electron transport system is probably not the site of MV action in the dark, as the redox potential of MV is too low to accept electrons from the NADH-NAD$^+$ couple. Benzyl viologen (BV; 1,1'-dibenzyl-4,4'di-pyridinium dichloride), also a bipyridyl herbicide, acts as an electron carrier, bypassing part of the normal photosynthetic electron transport chain, but, due to its chemical properties, it is about 100-times less inhibitory than paraquat in CO_2 fixation (Lewinsohn and Gressel, 1984).

Mitochondrial respiratory pathway-specific inhibitors

Mitochondria can generate copious amounts of ROS at two sites: the flavoprotein region of the internal NADH dehydrogenase and the ubiquinone pool. Mitochondria exposed to biotic, or abiotic stress, produce $O_2^{\bullet-}$ and H_2O_2 when electron transport through the cytochrome pathway is restricted due to physical changes in membrane components. In addition, cytochrome pathway inhibitors (e.g., antimycin-A) that inhibit cytochrome bc_1 of the electron transport chain can accelerate production of $O_2^{\bullet-}$ and H_2O_2 in mitochondria. However, this process is sensitive to KCN (inhibitor of cytochrome oxidase), which inhibits ROS production (Boveris and Cadenas, 1982).

Plant mitochondria are unique in having additional pathways for oxidation of NAD(P)H, and an alternative pathway for electron transfer from reduced ubiquinone to O_2 that bypasses cytochrome c oxidase (Wagner and Krab, 1995). This alternative pathway is associated with an oxidase found in the inner membrane of mitochondria from all higher plants, and is distinguished from the normal cytochrome c oxidase by its insensitivity to cyanide, azide, and carbon monoxide; all of which inhibit the latter oxidase. However, several compounds, including SHAM and n-propyl gallate can specifically inhibit electron flow through the alternative pathway. A characteristic feature of this pathway is that as electrons are shunted off of the standard electron transfer pathway at the level of the ubiquinone pool, they bypass the second and third energy conservation and ATP formation sites. Thus, during the engagement of the alternative pathway, electrons are transferred to O_2, producing water as the reduced product and energy, which is lost as heat. The physiological role of the alternative pathway is still uncertain (except in *Arum lilies* where, it has been linked to thermogenic metabolism). Several lines of evidence suggest that the alternative pathway may be related to prevention of oxidative stress (Polidoros and Scandalios,

1999). Induction of alternative oxidase by $O_2^{\bullet-}$ has been observed in the yeast *Hansenula anomala* (Minagawa *et al.,* 1992). It has also been suggested that oxidation of ubisemiquinones by the alternative oxidase prevents their interaction with O_2 to generate $O_2^{\bullet-}$ (Purvis and Shewfelt, 1993). Diversion of e^- flow from the cytochrome pathway, where the sites of ROS production are present, could lower ROS production. Thus, inhibition of the alternative pathway could result in induction of other protective mechanisms for ROS scavenging.

Inorganic soil compounds generate ROS, or are transformed to free radicals in the cell

Arsenite/ Arsenate

Arsenic compounds are naturally present in the environment and mainly used in agriculture and forestry as pesticide or insecticide components. The major anthropogenic sources of environmental pollution with arsenic are industrial metal smelting, the burning of coal and, more recently, the semiconductor industry. Due to their different modes of action, trivalent arsenic As (III), *arsenite,* is more toxic than the pentavalent species As (V), *arsenate.* The primary biological targets of arsenite are thiol-containing molecules, e.g. in active sites of enzymes. Arsenate, on the other hand, is better tolerated (detoxified) at low concentrations, but at high levels competes with phosphate, a ubiquitous biological anion. The toxicity of pentavalent arsenic *in vivo* may be partly due to its intracellular reduction to the trivalent form. In aqueous media, the arsenical must chelate with some portion (including oxygen and sulfur atoms) of a biomolecule in order to inhibit biological function. Although there are some reports indicating that arsenic may bind to oxygen as a ligand, sulfur has always been assumed to be the ligand of choice for arsenical ions. The effects of arsenicals have been measured on many complex enzyme systems with thiol-groups at the active sites. These include pyruvate dehydrogenase, adenylate cyclase, lipoamide dehydrogenase, glutathione reductase, urease, pyridine nucleotide dehydrogenase and thioredoxin. Studies in animal mitochondria have shown that inorganic as well as organic arsenicals inhibit the coupling Factor B (F_B) activity. F_B is an essential component of the F_o segment of the H^+-ATPase of mitochondria (Joshi and Hughes, 1981). The results indicate that uncoupling of oxidative phosphorylation occurs and can be reversed by addition of dithiol groups, since arsenic ions inhibit F_B activity by binding to thiol and dithiol groups of the biomolecule. Moreover, arsenicals can also inhibit ATPase because of their oxidative abilities. In the presence of divalent ion (Ca^{++} and/or Mg^{++}) arsenic is reduced while important sulfhydryl groups of proteins are oxidised (Beeler, 1990). During this process $O_2^{\bullet-}$ can be formed depending on the valence status of arsenic.

Nitrate/ Nitrite

Nitrate and nitrite are normal plant metabolites and not xenobiotics in the strict sense. Due to the use of nitrogen fertilisers and to environmental pollution with nitrogen oxides and other nitrogen containing molecules, their concentration far exceeds normal soil concentrations. Nitrate assimilation also produces hydroxide ions that can produce a temporary imbalance of the redox status in root cells (Crawford, 1995). In addition, both nitrate and nitrite can affect, enzymatically or non-enzymatically, production of nitric oxide (NO), which can inhibit mitochondrial cytochrome c oxidase and consequently oxidative phosphorylation (Millar and Day, 1996).

Consequently, nitrate and nitrite can create an imbalance in ROS and in turn trigger the plant's antioxidant defences.

ENVIRONMENTAL PERTURBATIONS, HORMONES, AND ROS

In addition to xenobiotics, a variety of environmental conditions such as drought, light and temperature fluctuations, wounding, pathogens, hypoxia and hyperoxia, changes in osmoticum, salinity, as well as normal development and senescence have been found to affect antioxidant gene expression in various plant species. It has also been demonstrated that one mechanism by which such factors affect antioxidant gene expression is via induction of ROS through various signal transduction pathways (Guan and Scandalios, 2000). Such pathways often involve changes in endogenous levels of hormones as intermediaries.

Thus in plants, ROS are produced by an enormous variety of causes and sources in every compartment of the cell. Each organelle has potential targets for oxidative stress, as well as for eliminating and/or modulating these noxious intermediates of oxygen metabolism (Table 1).

ACTIVATION AND MOBILIZATION OF ANTIOXIDANT DEFENSES

It has become apparent that many environmental stresses exert their effect, at least in part, by causing oxidative damage and its ensuing consequences on the affected organism. Consequently, the mechanisms by which the antioxidant defence system is triggered in all aerobic organisms has attracted considerable interest in recent years (Scandalios, 1997) with the goal of developing organisms with increased tolerance or resistance to environmental stresses. To this end, attempts to over-express any one antioxidant defence gene in transgenic plants have proven useful and instructive, but have not provided long-term protection against oxidative stress. A more promising approach might be to engineer organisms to activate the entire complex battery of their antioxidant defences. Thus, individual plant responses to stress do not occur in isolation but are components of an integrated defence with each part being necessary, but often insufficient by itself, to provide comprehensive protection. Hence, in addition to deciphering the mechanisms by which any antioxidant gene(s) responds to oxidative stress, it is necessary to examine more global responses of various other genes and their products that might directly or indirectly interact to effect the total antioxidant response. To these ends, progress is currently being made in understanding antioxidant gene structure and the perception of signals that trigger the expression of such and related genes in both prokaryotes and eukaryotes including plants (Scandalios, 1997; Valentine *et al.*, 1998; Storz and Imlay, 1999).

Table 1. Partial summary of antioxidant systems in higher plants.

subcellular location	**type of active oxygen species**	**source of active oxygen species**	**enzymatic scavenging systems**	**products**	**nonenzymatic scavenging systems**
Chloroplast	Superoxide H_2O_2	photosystem II enzymatic	SOD peroxidases	H_2O_2 glutathione, $NADP^+$, dihydroascorbate	ferredoxin carotenoids, xanthophylls
Mitochondria	Superoxide H_2O_2	electron transport enzymatic	SOD Peroxidases catalase	H_2O_2 H_2O_2 + oxidised donor $H_2O_2 + O_2$	
Cytosol	Superoxide H_2O_2	enzymatic enzymatic	SOD catalase peroxidase	H_2O_2 $H_2O_2 + O_2$ H_2O_2 + oxidised donor	
Glyoxysomes	H_2O_2 H_2O_2	β-oxidation photorespiration	catalase catalase	$H_2O_2 + O_2$ $H_2O_2 + O_2$	
Extracellular	Superoxide H_2O_2	enzymatic enzymatic	none known peroxidase	none known lignin,suberin, hydroxyproline	

ANTIOXIDANT DEFENSES

The antioxidant defence systems that have evolved have proven to be far more complex (and interesting) in plants than in other eukaryotes, perhaps as a consequence of the fact that plants are unable to physically alter their environment, and must respond to changing environmental conditions by being more biochemically agile. This has resulted in responses that are far more complex at the cellular level than animal responses to the same stimuli. Each response is usually part of an interconnected, layered response by the whole organism to a complex stimulus; such interconnections are now beginning to be elucidated.

Herein, I will discuss the maize enzymatic antioxidant defence system as a paradigm towards our understanding of such complex systems in plants, and in the context of strategies towards engineering for stress tolerance.

TRANSCRIPTIONAL ACTIVATION OF STRESS-RESPONSIVE GENES

In order to consider strategies towards engineering stress-resistant plants, it is essential to first identify and characterise the genes encoding enzymes involved in the scavenging, neutralising, and/or modulating ROS levels in the cell (Scandalios *et al.,* 1997). To date, cDNAs coding for superoxide dismutases, catalases, and various peroxidases have been isolated from various species. Unfortunately, however, only a few antioxidant genes have been isolated and characterised in plants, among these being the maize catalases and superoxide dismutases. Knowledge of the architecture of these genes is essential in any efforts to decipher the mechanisms by which they perceive environmental signals and respond to trigger the antioxidant defences (Scandalios, 1997; Guan and Scandalios, 1996; Guan and Scandalios, 1995).

The maize *Cat* and *Sod* antioxidant defence gene families

Catalases and superoxide dismutases and their encoding genes have been extensively examined in maize (Scandalios, 1997), and provide an ideal model system to investigate the underlying regulatory mechanisms for the expression of these important genes. Unlike animals, plants posses multiple genes encoding multiple, but functionally distinct, isozymes (Scandalios, 1965) allowing for precise identification of responses to a variety of signals. In maize, three unlinked structural genes; *Cat1*, *Cat2*, and *Cat3* encode three distinct CAT isozymes (CAT-1, CAT-2, CAT-3). Expression of each of the

Cat genes is highly regulated spatially, temporally, and in response to various environmental signals. Maize SOD exists as nine distinct isozymes encoded by the unlinked nuclear genes *Sod1*, *Sod2*, *Sod3.1*, *Sod3.2*, *Sod3.3*, *Sod3.4*, *Sod4*, *Sod4A*, and *Sod5*. SOD-2, SOD-4, SOD-4A, and SOD-5 are cytosolic enzymes; SOD-1 is a chloroplast enzyme, while SOD-3.1, SOD-3.2, SOD-3.3, and SOD-3.4 are compartmentalised in mitochondria (Scandalios, 1997a). Recent detailed studies with both catalases (McClung, 1997) and superoxide dismutases (Kliebenstein *et al.*, 1998) in *Arabidopsis* clearly indicate that the complexity of these systems found in maize is paralleled in *Arabidopsis* and is likely common to most if not all plant species. In fact, these systems have been examined to varying degrees in numerous other species, such as tobacco, rice, peas, potato, etc. (Havir and McHale, 1987; Higo and Higo, 1995; Bowler *et al.*, 1992), and in each case found to be similar to maize and far more complex than non-plant aerobes.

Ascorbate peroxidases (APXs), believed to be the most critical H_2O_2 scavengers in chloroplasts, use ascorbic acid as the reducing substrate and form part of the "ascorbate-glutathione cycle". This antioxidant system is also complex, and has been examined in significant detail by several laboratories. APXs will not be dealt with further here, as they have been competently reviewed elsewhere (Asada, 1992; Gadea *et al.*, 1999).

Cat and *Sod* gene responses to environmental stress

In order to determine the role(s) of the different SOD and CAT isozymes and their encoding genes in protecting plants against oxidative stress, the levels of their steady-state mRNA and protein were determined in various maize tissues at different developmental stages from plants exposed to various physical (Boldt and Scandalios, 1995), chemical (Mylona *et al.*, 1998), and biological stresses (Williamson and Scandalios, 1993; Becana *et al.*, 2000). The overall picture that has emerged from such studies is that each gene responds differently to such signals in different tissues and at different developmental stages. Most transcripts are upregulated to differing degrees while some are downregulated. This complex picture suggests that these genes and their products are involved in most environmental stresses encountered by plants during their development. For example, fungi of the genus *Cercospora sp.* produce a light-induced, photoactivated polyketide toxin, cercosporin. Upon photoactivation, cercosporin attains an excited triplet state and reacts directly with oxygen to produce $^1O_2$ and $O_2^{\bullet -}$. Using fungal extracts and purified cercosporin, we demonstrated that total CAT activity, protein, and transcripts changed in parallel in response to the applied toxin (Willamson and Scandalios, 1992a). The responses were not identical for each of the *Cat* genes and their products, and varied with tissue and developmental stage. In contrast, while *Sod* transcript levels changed dramatically in response to the toxin, total SOD activity and individual isozyme protein levels did not change, suggesting that protein turnover is also an important aspect of the SOD response to oxidative stress.

MOLECULAR MECHANISMS IN RESPONSE TO ROS

The mechanisms by which cells sense ROS and activate their antioxidant defences are not well understood, but a number of transcriptional factors that regulate the expression of antioxidant genes are well characterised (Table 2). In *E. coli* and other prokaryotes, the transcription factor *OxyR* activates a number of genes inducible by H_2O_2, while the transcription factors *SoxR/SoxS* mediate responses to $O_2^{\bullet-}$ (Jamieson and Storz, 1997). In yeast, there also exist two distinct adaptive stress responses, one towards H_2O_2 and one towards $O_2^{\bullet-}$ (Ruis and Koller, 1997).

Table 2. Transcription regulators of antioxidant-defence genes.

Regulator	Inducer	Regulated Defence Genes
E. Coli:		
OxyR	Hydrogen peroxide	*katG, ahpCF, dps, gorA*
SoxR	Superoxide	*soxS*
SoxS	SoxR	*sodA, nfo, zwf, fumC, micF*
MarA	Antibiotics, Redox-Cycling Agents	*sodA, zwf, fumC, micF*
Rpos	Stationary phase	*katE, xthA, dps, katG*
ArcA	Anaerobiosis	*sodA*
Fnr	Anaerobiosis	*sodA*
Fur	Iron	*sodA*
S. cerevisiae:		
ACE1	Copper	*SOD1*
MAC1	Hydrogen peroxide	*CTT1*
YAP1	Hydroperoxides, thioloxidants	*TRX2*
HAP1	Oxygen	*CTT1*
HAP2/3/4	Oxygen, carbon sources	*SOD2*
Mammalial cells:		
NF-κB	Hydrogen peroxide, UV, etc.	*Cat, Sod* (?)
AP-1	Hydrogen peroxide, antioxidants, UV	*Cat, Sod* (?)
ARE	Hydrogen peroxide	*Cat*

In higher eukaryotes, oxidative stress responses are more complex, and are modulated by several different regulators. In mammalian systems, two classes of transcription factors, nuclear factor κB (*NF-κB*) and activator protein-1 (*AP-1*) are involved in the regulation of the oxidative stress response (Angel and Karin, 1991; Meyer *et al.*, 1993). Antioxidant-specific gene induction has been reported for a number of enzymes involved in xenobiotic metabolism, mediated by a regulatory motif common in the promoter region of these genes. This motif, the "antioxidant responsive element" (*ARE*), is present in the promoter of mammalian glutathione S-transferase (GST), metalothioneine-I, and Mn*Sod* genes (Kahl, 1997; Rushmore at al, 1991; Nguyen and Pickett, 1990). An ARE motif has not been identified in any plant *Gst* gene (Marrs, 1996). However, ARE-like motifs are present in the promoter region of the three maize *Cat* genes (Scandalios *et al.,* 1997). In plants, ROS have been implicated in the damaging effects of various environmental stresses. Many plant defence genes, including the three maize *Cat* genes, are activated by stressors such as light, radiation, xenobiotics, ozone, drought, and pathogen attack (Scandalios, 1993; 1997). The presence of such motifs as *ARE, NF-κB*, and *AP-1* in the maize *Cat* and *Sod* (Kernodle and Scandalios, 1996) gene promoters renders these gene-enzyme systems suitable for the further characterisation of their potential regulatory roles in plants, and may help to uncover the signal transduction pathways involved in a more global regulation of the antioxidant response in plants.

For brevity, we will focus our discussion herein on the results with the maize catalase genes (*Cat1, Cat2*, and *Cat3*), but will refer to the *Sod* and other genes as necessary.

PROMOTER STRUCTURE OF THE THREE *CAT* GENES

Although there is no extensive similarity among the three *Cat* gene promoters, several regulatory motifs related to oxidative stress are present in each (Figure 1). The ARE motif (PuGTGACNNNGC) has been found in the promoter region of all three *Cat* genes. The ARE responds to H_2O_2 and phenolic antioxidants that undergo redox cycling, generating ROS (Rushmore *et al.,* 1991; Polidoros and Scandalios, 1999). The AP-1 motif (TGANTCA) that serves as the binding site of the mammalian antioxidant transcription factor AP-1 is also present in the *Cat1* and *Cat3* promoters. NFκB is a major transcription factor of defensive responses mounted by cells against diverse environmental challenges. It binds to a consensus sequence GGGPuNNPyPyCC located within promoters, or enhancers, of these genes where it functionally interacts with other transcription factors to regulate expression (Siebenlist *et al.;* 1995). The NFκB binding site is present twice in the *Cat3* promoter within 500bp from the start of transcription. It is also found in the promoters of *Cat1* and *Cat2*, but is located beyond 1500bp from the start of transcription. Xenobiotic responsive elements (XRE) with the core sequence GCGTG are present in mammalian cytochrome P450 gene promoters and are activated in response to xenobiotics (Pabo and Sauer, 1992). XRE motifs are repeatedly present in the *Cat2* and *Cat3* promoters.

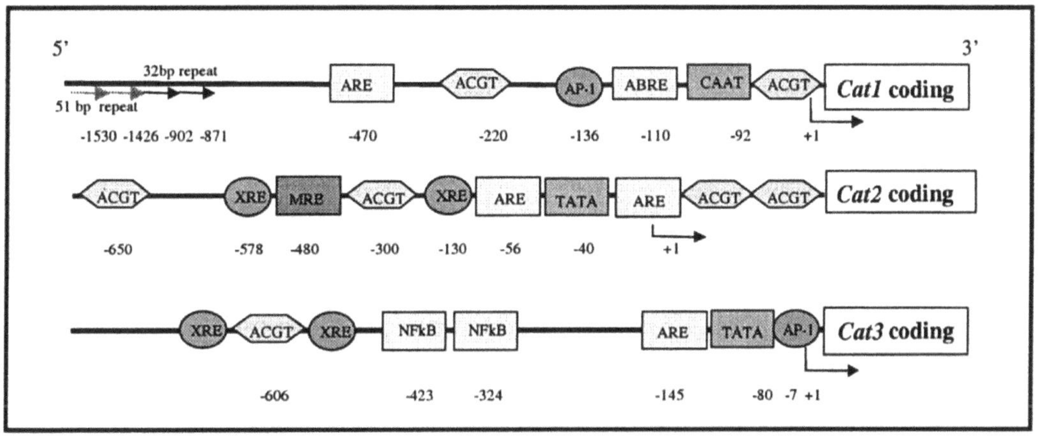

Figure 1. Schematic representation of the motifs located in the promoter region of each of the three maize *Cat* genes. The location of each motif is relative to the transcription start site of each gene. ABRE, ABA responsive element; ARE, antioxidant responsive element; ACGT core, or leucine zipper protein binding site; AP-1, AP-1 binding site; NFκB, NFκB binding site; XRE, xenobiotic responsive element; MRE, metal responsive element.

No XRE-like sequence is found in the *Cat1* promoter. ACGT-core regulatory elements recognised by basic leucine zipper transcription factors (Armstrong *et al.,* 1992) are found in diverse gene promoters such as the Em1a motif (ABRE, 'CCACGTGG', ABA responsive element) of the ABA regulated wheat *Em* gene (Guiltinan *et al.,* 1990). The ACGT-core is present in the promoter region of all three *Cat* genes, but is more frequent in the *Cat1* and *Cat2* promoters. This is in agreement with the fact that *Cat1* is highly induced by ABA (Williamson and Scandalios, 1992b).

Cat transcript fluctuations in response to abscisic acid (ABA), osmotic stress, and dehydration

The expression of many plant genes is affected by changes in hormone levels in various tissues at different developmental stages. Abscisic acid (ABA) mediates physiological processes in response to osmotic stress. ABA levels increase in tissues subjected to osmotic stress because of high osmoticum, salt, dehydration, and cold. The effect of ABA and high osmoticum on *Cat* gene expression was examined in developing and germinating maize embryos, and in leaves (Guan and Scandalios, 1998). Steady-state levels of *Cat1* transcript increased with all applied ABA doses, with a maximum at 10^{-4} M ABA. *Cat1* transcript also increased in response to osmotic stress (mannitol) in immature embryos and in young leaves. The *Cat2* and *Cat3* transcripts were downregulated by ABA and osmotic stress. These data suggest that *Cat1* mRNA accumulation in response to ABA is not developmental stage-dependent, and might represent a general stress response in which the *Cat1* gene product increases to protect plants from ABA-mediated stress. We also examined the effect of dehydration on *Cat* gene expression in maize, in order to determine if ABA may be involved in such a response. *Cat1* was induced after 4h of dehydration and continued increasing with a maximum at 12-24h. However, *Cat2* and *Cat3* transcripts decreased after 4h and are almost undetectable after 24h. The pattern is similar to that of the ABA response in W64A (standard inbred line for CAT) leaves.

Carotenoid-deficient viviparous mutants of maize have great potential for providing information concerning the regulation of gene expression in response to ABA and osmotic stress. Seeds of viviparous mutants germinate precociously while still attached to the ear, before maturity. Some viviparous mutants (i.e., *vp5*) are deficient in ABA (Neill *et al.,* 1986). Seedlings rescued from viviparous kernels contain negligible levels of ABA compared to wild-type seedlings, and the ABA concentration does not increase in response to water deficit (Moore and Smith, 1984). This ABA deficient mutant has been used to determine if ABA is involved in the response of *Cat* genes to osmotic stress and dehydration. Like W64A, *Cat1* is induced while *Cat2/Cat3* are repressed in response to dehydration, in *Vp5/-* wild-type leaves. In *vp5/vp5* mutant leaves, *Cat1* is induced in response to dehydration while *Cat2* is repressed; however, the pattern of *Cat3* transcript accumulation is different from that in the *Vp5* wild-type leaves. *Cat3* transcript is greatly induced in *vp5* mutant leaves in response to dehydration, while in wild-type leaves, *Cat3* transcript accumulation is repressed, implying that ABA may mask the effects of dehydration on *Cat3*. That is, when ABA is present in high levels as in W64A and in *Vp5* wild-type leaves (or when ABA is induced by dehydration), then *Cat3* transcript is repressed by dehydration. However, when ABA levels are low as in the *vp5* mutant (deficient in ABA), then *Cat3* mRNA is induced by dehydration. This suggests that *Cat* gene responses to dehydration occur via at least two pathways: an ABA-dependent and an ABA-independent pathway. *Cat* transcript accumulation in *Vp5* wild-type leaves is similar to that of the W64A leaves in response to ABA and mannitol, with *Cat1* increasing, while *Cat2* and *Cat3* transcripts are downregulated. In *vp5* mutant leaves, *Cat1* mRNA also accumulates in response to ABA and mannitol.

TRANSFORMATION AND IDENTIFICATION OF RELEVANT *CIS*-REGULATORY ELEMENTS

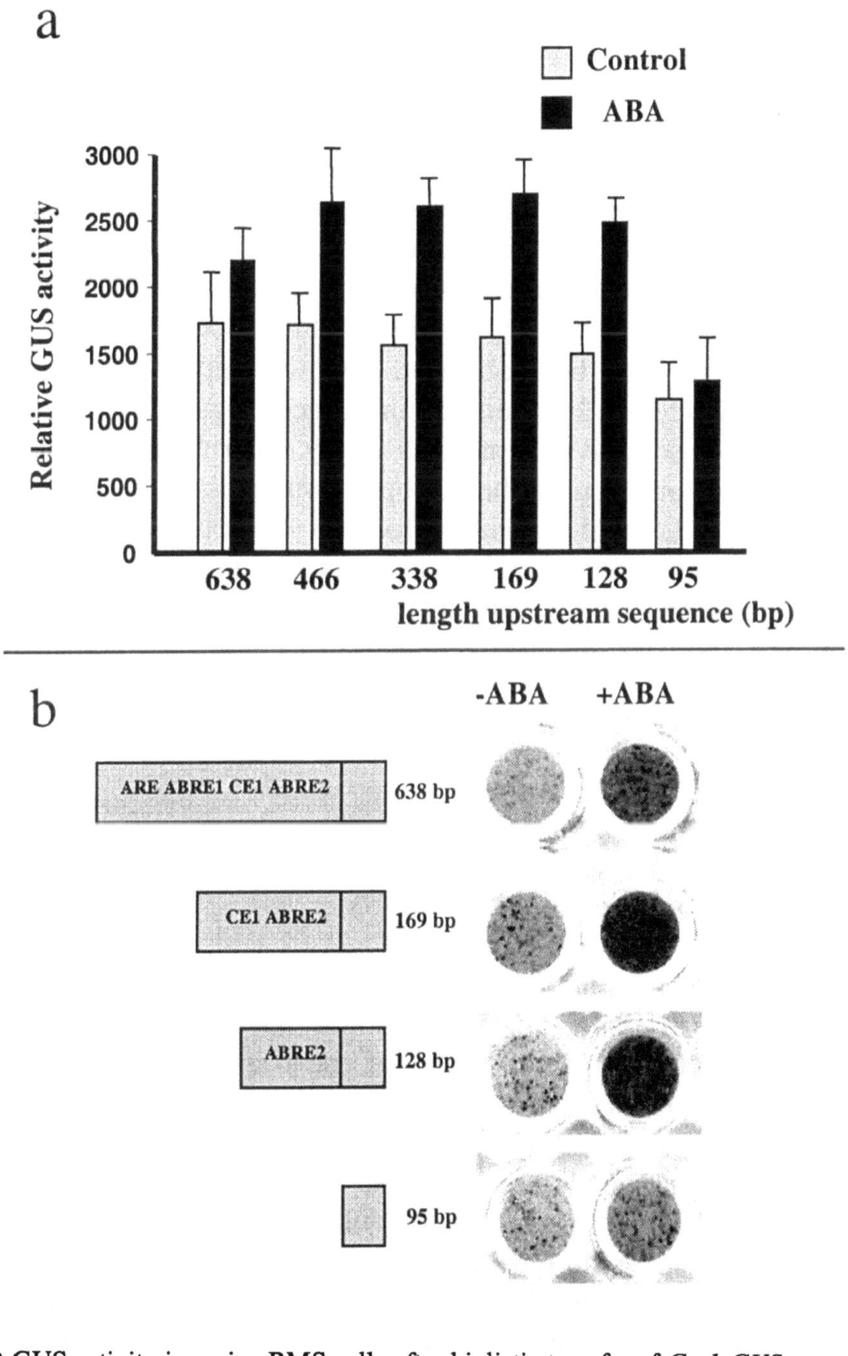

Figure 2. **(a)** GUS activity in maize BMS cells after biolistic transfer of *Cat1-GUS* constructs and treatment with ABA for 24h. GUS activity is the average of three independent experiments ± S.E. **(b)** *Cat1* promoter fragments used for transient assays. Constructs were introduced into maize BMS cells by particle bombardment. GUS activity was determined by color staining after ABA treatment.

Several deletion fragments of the *Cat1* promoter were obtained by either restricted digestion or PCR amplification with the 5'-end at -638bp, -338bp, -169bp, -128bp and -95bp, relative to the transcription start site. The only difference in sequence between the fragments of 128bp and 95bp is that a putative ABRE element is deleted from the 95bp fragment. Each deletion fragment was then fused to the upstream region of the 35S minimal promoter (-46 bp)::intron::Gus construct (pIG46), generating serial *Cat1* promoter deletion-Gus reporter plasmids (Guan *et al.*, 2000).

To locate any *cis*-elements in the *Cat1* promoter responsible for induction by ABA, the plasmids were individually delivered by particle bombardment into maize (var. Black Mexican Sweet; BMS) cells, for transient expression assays. Cell extracts were prepared from bombarded BMS cells, treated with or without ABA after bombardment. β-Glucuronidase activities in the cell extracts were detected by chemiluminescence (Figure 2a) and calibrated by internal luciferase activities, as well as by colorimetric detection of GUS (Figure 2b). Expression of the 638bp construct in cells treated with ABA was increased 1.3-fold compared to cells without ABA; the 169bp and 128bp constructs were also induced, but the induction decreased as the deletion size increased. In contrast, induction of the 95bp construct by ABA was almost abolished due to the deletion of the ABRE element, suggesting that the ABRE is at least one of the *cis*-acting elements responsible for the ABA induction of *Cat1*.

DNA BINDING PROTEINS INTERACT WITH "ABRE" (G-BOX) AND "ARE" MOTIFS

Nuclear proteins binding to the ABRE motif of the *Cat1* promoter were identified by gel retardation. Two oligos covering the ABRE motif (5'-GAAGTCCACGTGGAGGTGG-3') of the *Cat1* promoter and the mutant ABRE (5'-GAAGTaacatgttcGGTGG-3') core of *Cat1* were used. At least four binding complexes formed in the interaction between the ABRE probe and the nuclear protein extract prepared from 21 days post pollination (dpp) embryos of the *vp5* mutant or its wild-type *Vp5* sibling. Binding was competed by cold, "wild-type" ABRE, but not by mutated ABRE, indicating that the binding is specific for ABRE. The binding protein complex-1 (CBF1) appears to be the major protein complex in 21dpp embryos (Figure 3).

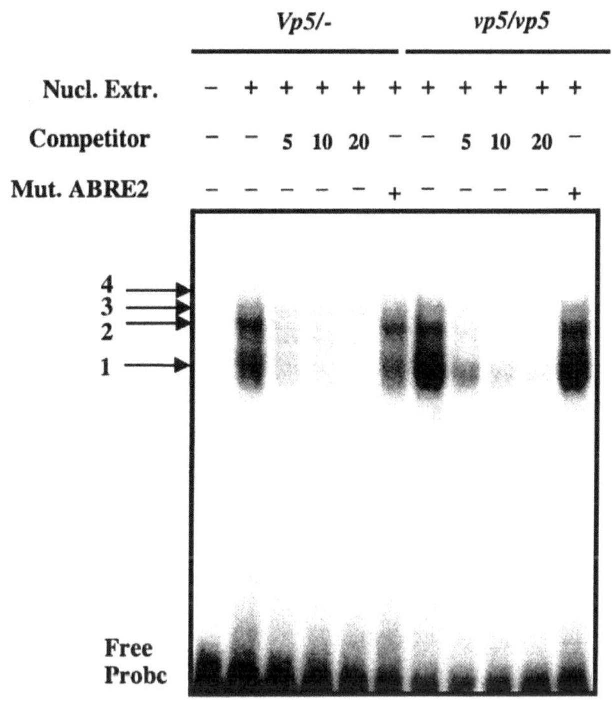

Figure 3. Gel retardation assay of binding activities to the ABRE motif of *Cat1* in the nuclear protein extracts from 21dpp embryos of the *vp5/vp5* mutant and its *Vp5/-* siblings. Competitor: cold, wild type ABRE at 5, 10 and 20 fold in excess; Mut. ABRE: mutated ABRE.

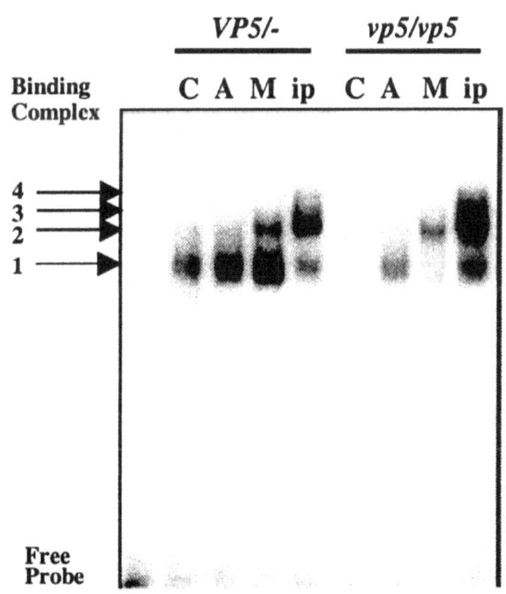

Figure 4. Gel retardation assay of the binding activities to the ABRE motif of *Cat1* in nuclear protein extracts from 17dpp embryos of *vp5/vp5* and its Vp5/- siblings. Embryos were cultured for 24h on MS media with ABA (A), or mannitol (M). ip: embryos isolated directly from seeds. C: control.

No major difference is observed in binding intensity of the 4 binding complexes between wild type (*Vp5/*) and mutant (*vp5/vp5*) embryos, in agreement with previous observations that endogenous ABA does not play a major role in *Cat1* expression via ABRE in 21dpp embryos. We also examined the ABRE motif binding to proteins isolated from 17dpp embryos.

In untreated embryos, the protein complex distribution is not the same as in 21dpp embryos (Figure 4). Complex-2 (CBF2) is the major binding complex in 17dpp embryos while CBF1 is the major binding complex in 21dpp embryos.

Upon ABA treatment, CBF1 was induced in the nuclear extracts from embryos of both the *vp5* mutant and its wild-type siblings, compared to its absence in the control embryos of the *vp5* mutant and less amount in the control *VP5/-* embryos. CBF1 was significantly induced in the embryos of *VP5/-* genotype in the presence of 11% mannitol in the MS media compared to the control, but was not in the embryos of the *vp5* mutant (Figure 4). However, CBF2 was only induced in mannitol treated embryos but not in ABA treated embryos.

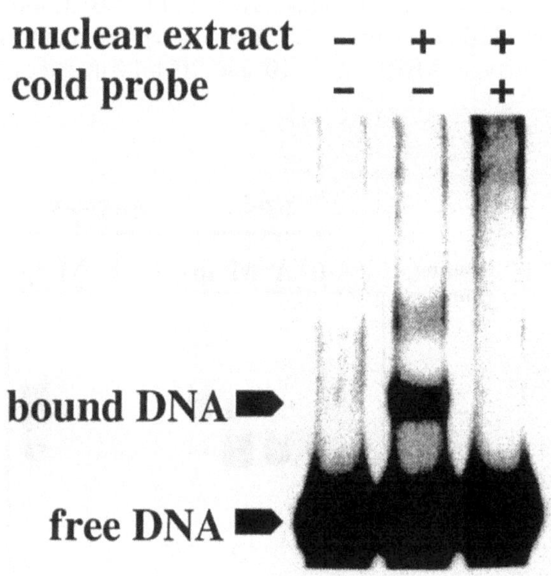

Figure 5. Protein-DNA interactions in the *Cat1* promoter region harbouring the ARE motif. Electrophoretic mobility shift assay was probed with a 158bp *Sau3*A *Cat1* promoter fragment containing the ARE motif. Specificity of the interaction was determined by addition of 30X cold probe as specific competitor in the assay. Lanes are: Probe alone control (-,-), probe with nuclear extract and no specific competitor (+,-), and probe with nuclear extract and 30X cold probe as specific competitor (+,+). The bands of free probe and bound probe are indicated by arrowheads. Note the very strong interaction (+,-) revealed by the retardation of the whole amount of probe used in the assay.

These results suggest that there are two different pathways for *Cat1* expression in response to mannitol, an ABA-dependent pathway likely through the CBF1 complex, and an ABA-independent pathway likely through the CBF2 complex (Guan *et al.*, 2000). The possible regulatory role of ARE for the expression of the *Cat1* gene was examined in the scutellum during germination. Interactions of nuclear proteins, isolated from 10 days post imbibition (dpi) scutella, with the promoter region of *Cat1* were examined by gel retardation assays (Polidoros and Scandalios, 1999).

Results show that a 158bp fragment, containing the ARE core sequence, interacts strongly with nuclear extracts (Figure 5).

H_2O_2-MEDIATED *CAT* GENE EXPRESSION IN RESPONSE TO WOUNDING

The effect of wounding on catalase expression was examined in embryos and leaves of maize (Guan and Scandalios, 2000). All three *Cat* genes are upregulated in response to wounding in immature embryos.

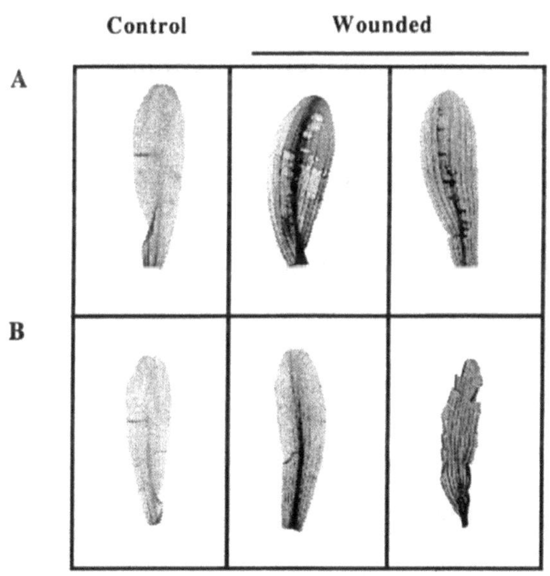

Figure 6. The production of H_2O_2 in leaves of maize plants in response to wounding. Plants (7 dpi) were excised at the base of the stems and supplied with DAB (3,3-diaminobenzidine) for 6h. The plants then were wounded and continuously supplied with DAB for 4h. The production of H_2O_2 can be visualised by the deposition of the brown-red (dark areas) colour products in the leaves. (a) Control and wounded leaves (wounding was conducted near the main vein). (b) Control and wounded leaves (wounding was conducted at the main vein and cut at the leaf edges).

Cat expression also increased in response to jasmonic acid (JA), raising the possibility that JA and wounding may share a common signal transduction pathway in up-regulating *Cat* mRNA in immature embryos. In young leaves, only *Cat1* and *Cat3* transcripts increase in response to wounding, but JA does not play a role. *Cat1* and *Cat3* transcript accumulation also increases in response to wounding in both wild-type and ABA-deficient mutant leaves, implying that *Cat1* and *Cat3* induction in response to wounding is not mediated by abscisic acid in leaves. Transient assays using the *Cat1* promoter fused with the reporter gene *Gus*, showed that the DNA sequence motif responsible for *Cat1* up-regulation by wounding overlaps with the ABA-responsive element (ABRE, G-box) in the *Cat1* promoter. In addition, production of H_2O_2 is significantly induced in wounded leaves. H_2O_2 was detected in tissue surrounding the wound site, and then appeared in the major and minor veins throughout the leaves (Figure 6). Thus, ROS is the likely mediator for the observed wounding-induced *Cat* gene expression

EFFECT OF H_2O_2 ON *CAT* TRANSCRIPT LEVELS

The direct effect of H_2O_2 on *Cat* gene expression was examined in 9dpi-leaves hydroponically treated with increasing H_2O_2 (0 to 50 mM) concentrations for 24h. *Cat1* transcript increased with increasing H_2O_2 concentrations, with maximum induction at 50mM H_2O_2, whereas both *Cat2* and *Cat3* transcripts decreased with increasing H_2O_2 concentrations (Figure 7).

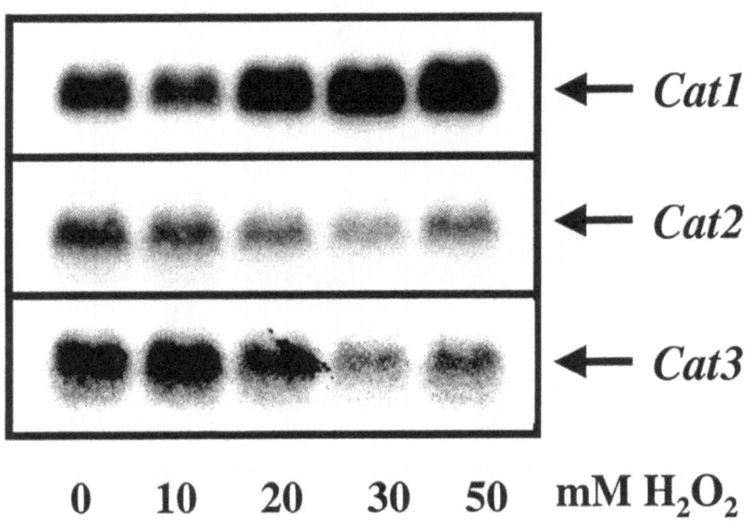

Figure 7. Cat mRNA accumulation in response to increasing concentration of H_2O_2. W64A seedlings (9dpi) were treated with increasing concentrations of H_2O_2 for 24h. Total RNA was isolated from treated leaves and *Cat* transcript levels determined by RNA-blots, with *Cat* gene specific probes.

The accumulation pattern of the three *Cat* transcripts in response to H_2O_2 is similar to previously reported ABA effects on *Cat* expression (Guan and Scandalios, 1998). Thus, we hypothesised that ABA may cause changes in endogenous H_2O_2 levels. Consequently, we examined H_2O_2 levels in cultured maize cells after ABA treatment. Maize BMS cells were treated with 10^{-4}M ABA for 0.5, 1, 2, 3, and 4h. After treatment, cells were collected and H_2O_2 levels determined. H_2O_2 levels increase in response to ABA between 0.5h to 3h. At 4h after ABA treatment, H_2O_2 levels are almost the same between the control cells and ABA treated cells (Figure 8). The reduced H_2O_2 levels after 4h ABA treatment may be due to activation of the *Cat1* gene and protein.

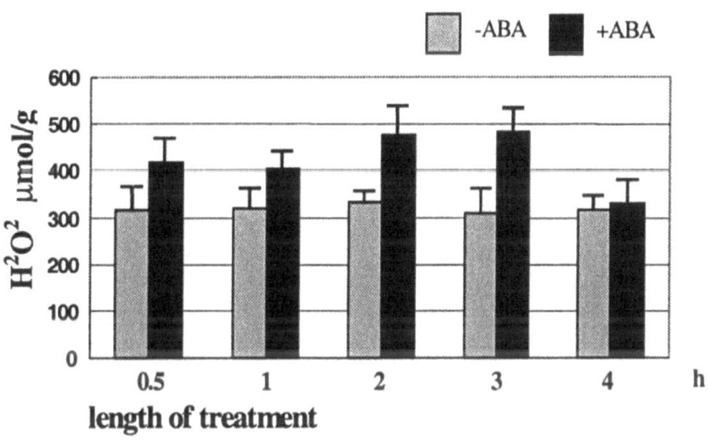

Figure 8. Hydrogen peroxide concentration in maize BMS cells after ABA treatment. BMS cells were treated with 10^{-4} M ABA for 0.5, 1, 2, 3 and 4h. After treatment, cells were collected and H_2O_2 levels determined using the luminol assay. H_2O_2 levels are based on the average of three independent experiments.

CONCLUDING REMARKS AND FUTURE PROSPECTS

Plants, because of their growth under high intensities of sunlight and a high cellular concentration of dioxygen, are subjected to the most severe oxidative stress, relative to other organisms, which can result in membrane leakage, senescence, chlorophyll destruction, and impaired photosynthetic capacity. Aerobes have evolved efficient antioxidant defences among which, are the enzymes superoxide dismutase (SOD) and catalase (CAT). All aerobes contain CAT to modulate and degrade H_2O_2. No aerobes are known to completely lack catalase and survive in an O_2 environment, underscoring the importance of CAT for living in an aerobic environment.

The correlation between elevated *Cat* and *Sod* gene expression and oxidative stress suggests that enzyme levels are optimally maintained to provide better protection. Whether all stresses examined induce CAT and SOD via the same mechanisms is as yet unclear. The changes in CAT and SOD expression under these environmental stresses or during normal development may be caused, in part, by changes in ROS levels, or redox potentials. Thus, these maize antioxidant defence systems provide an excellent opportunity to decipher the mechanisms involved in the regulation of these important defence genes. Elucidation of *Cat* and *Sod* gene structure and regulation during normal development and under stress may provide important insights into the molecular mechanisms for the differential expression of these and other antioxidant genes, and the functional roles of their encoded proteins in eukaryotes. The data presented here are helping to unravel the roles of antioxidants and the possible signaling roles of ROS in modulating antioxidant gene expression and other metabolic activities. Deciphering the mechanisms by which these genes are regulated during development and under various stressful environments is critical in determining the various physiological roles these important genes play. We have established, in addition to other factors, that the *Cat* and *Sod* genes differentially respond to biotic and abiotic environmental signals related to oxidative stress. Our results also show that each *Cat* gene may respond differently to a certain signal, or all three *Cat* genes may respond similarly to several different stresses.

Our results support the hypothesis that the observed *Cat* gene responses are mediated by oxidative stress signals, directly or indirectly. Further investigation of *Cat* gene responses to stress in different tissues and developmental stages will serve as a guide to define the precise molecular mechanisms by which the *Cat* genes perceive and respond to various signals.

Sequence comparisons show that each *Cat* and *Sod* gene promoter is unique with no homology to other *Cat* or *Sod* promoters. Each *Cat* promoter contains a unique set of motifs known to induce gene expression in response to oxidative stress in other eukaryotes. Use of promoter deletions, gel retardation, southwestern and DNase footprinting analyses are helping to identify *cis*-acting elements in each *Cat* promoter responsible for increased transcription under oxidative stress. Once relevant *cis*-elements have been identified (as with ABRE and ARE), use of gel-retardation will lead to identification of *trans*-acting factors that interact with the DNA motifs to trigger expression of these and other antioxidant genes, in tandem, leading to a more global response to oxidative stress. This will likely be a more effective strategy than merely overexpressing any one gene at a given time and place.

Our results are providing insights into the underlying mechanisms utilised by plant genomes to perceive stress signals and trigger the appropriate defences against environmental oxidative stress. These findings are relevant not only for maize, but also for all other eukaryotes. Knowledge of the architecture of these genes and use of transgenics to decipher the underlying regulatory mechanisms for their developmental expression will provide a sound basis for efficiently engineering stress tolerant organisms. It has been amply demonstrated by several groups, that overexpression of any one antioxidant gene may provide some protection under specified conditions, but is not sufficient to provide permanent effective tolerance. The work discussed herein is aimed at identifying transcription factors that are able to turn on all of the antioxidant genes simultaneously, or in tandem, to maximise the defences on a sustained basis.

Acknowledgements

I thank my colleagues, past and present, for their significant contributions to the work described in this article. I especially thank Lingqiang Michael Guan, Stephanie Ruzsa, and Sheri Kernodle for both their contributions and for their comments on this manuscript. My apologies to those, whose work could not be cited due to space limitations. Research from the author's laboratory was supported in part by grants from the U.S. Environmental Protection Agency, the National Science Foundation, and the Department of Agriculture.

REFERENCES

Angel, P. and Karin, M. 1991. The role of *Jun, Fos* and the AP-1 complex in cell-proliferation and transformation. *Biochim. Biophys. Acta* 1072, 129-157.

Armstrong, G. A., Weisshaar, B. and Hahlbrock, K. 1992. Homodimeric and heterodimeric leucine zipper proteins and nuclear factors from parsley recognize diverse promoter elements with ACGT cores. *Plant Cell* 4, 525-537.

Asada, K. 1992. "Production and scavenging of active oxygen in chloroplasts". In: *Molecular Biology of Free Radical Scavenging Systems*, ed. J. G. Scandalios, pp. 173-192. Cold Spring Harbor Laboratory Press, Cold Spring harbor.

Barrett, M. 1995. Metabolism of herbicides by cytochrome P450 in corn. *Drug Metab. Drug Interact.* 12, 299-315.

Becana, M., Dalton, D.A., Moran, J.F., Ormaetxe, I.I., Matamoros, M.A. and Rubio, M.C. 2000. Reactive oxygen species and antioxidants in legume nodules. *Physiol. Plant.* 109, 372-381.

Beeler, T. 1990. Oxidation of sulfhydryl groups and inhibition of the (Ca^{++} and Mg^{++})-ATPase by arsenoazoIII. *Biochim. Biophys. Acta.* 1027, 264-267.

Boldt, R. and Scandalios, J.G. 1995. Circadian regulation of the *Cat3* catalase gene in maize (*Zea mays* L.): Entrainment of the circadian rhythm of *Cat3* by different light treatments. *Plant J.* 7, 989-999.

Bowler, C., Van Montague, M. and Inze, D. 1992. Superoxide dismutase and stress tolerance. *Ann. Rev. Plant Physiol. Plant Mol. Biol.* 43, 83-116.

Boveris, A. and Cadenas, E. 1982. "Production of superoxide radicals and hydrogen peroxide in mitochondria". In: *Superoxide Dismutase* Vol II, ed. L.W. Oberley, pp. 15-30. CRC Press, Boca Raton.

Crawford, N.M. 1995. Nitrate: Nutrient and signal for plant growth. *Plant Cell* 7, 859-868.

Dolphin, D. 1987. "The generation of radicals during the normal and abnormal functioning of cytochromes P450". In: *Oxygen Radical in Biology and Medicine*, eds. M.G. Simic, K.A. Taylor, J.F. Ward and C. von Sonntag, pp. 491-500. Plenum Press, New York.

Forman, H.S., Liu, R.M. and Shi, M.M. 1995. "Glutathione synthesis in oxidative stress". In: *Biothiols in Health and Disease*, eds. L. Packer and E. Cadenas, pp. 189-212. Marcel Dekker, New York.

Frear, D.S. 1995. Wheat microsomal cytochrome P450 monooxygenases: characterization and importance in the metabolic detoxification and selectivity of wheat herbicides. *Drug Metab. Drug Interact* 12, 329-357.

Fridovich, I. 1975. Oxygen: boon and bane. *Am. Sci.* 63, 54-59.

Gadea, J., Conejero, V. and Vera, P. 1999. Developmental regulation of a cytosolic ascorbate peroxidase gene from tomato plants. *Mol. Gen. Genet.* 262, 212-219.

Gronwald, J.W. 1994. "Resistance to photosystem II inhibiting herbicides". In: *Herbicide Resistance in Plants*, eds. S.B. Powels and J.A.M. Holtum, pp. 27-60. CRC Press, Boca Raton.

Guan, L. and Scandalios, J.G. 1995. Developmentally related responses of the maize catalase genes to salycilic acid. *Proc. Natl. Acad. Sci. USA* 92, 5930-5934.

Guan, L. and Scandalios, J.G. 1996. Molecular evolution of maize catalases and their relationship to other eukaryotic and prokaryotic catalases. *J. Mol. Evol.* 42, 570-579.

Guan, L. and Scandalios, J.G. 1998. Effects of the plant growth regulator abscisic acid and high osmoticum on the developmental expression of the maize catalase genes. *Physiol. Plant.* 104, 413-422.

Guan, L.M., Zhao, J. and Scandalios, J.G. 2000. Characterization of *cis*-acting elements and *trans*-acting factors that regulate expression of the maize *Cat1* gene in response to ABA and osmotic stress. *Plant J.* 22, 87-95.

Guan, L.M. and Scandalios, J.G. 2000. Hydrogen peroxide mediated catalase gene expression in response to wounding. *Free Rad. Biol. & Med.* 28, 1182-1190.

Guiltinan, M.J., Marcotte, W.R. Jr. and Quatrano, R.S. 1990. A plant leucine zipper protein that recognizes an abscisic acid response element. *Science* 250, 267-271.

Halliwell, B. 1995. "The biological significance of oxygen-derived species". In: *Active Oxygen in Biochemistry*, eds. V.J. Selverstone, C.S. Foote, A. Greenberg and J.F. Liebman, pp. 313-335. Blackie Academic and Professional, New York.

Halliwell, B. 1996. "Free radicals in biology and medicine". In: *The Encyclopedia of Molecular Biology and Molecular Medicine,* Vol. 2, ed. R.A. Meyers, pp.330-337. VCH Publishers, New York.

Hatzios, K, ed. 1997. "Regulation of enzyme systems detoxifying xenobiotics in plants". *NATO-asi series, vol. 37.* Kluwer Acad. Publ., Dordrecht.

Havir, A.E. and McHale, N.A. 1987. Biochemical and developmental characterization of multiple forms of catalase in tobacco leaves. *Plant Physiol.* 84, 450-455.

Higo, K. and Higo, H. 1995. Cloning and characterization of three catalase genes from rice. *Plant Physiol.* (*suppl.*) 108, 74.

Jamieson, D.J. and Storz, G. 1997. "Transcriptional regulators of oxidative stress responses". In: *Oxidative Stress and the Molecular Biology of Antioxidant Defenses*, ed. J.G. Scandalios, pp. 91-115. Cold Spring Harbor Laboratory Press, Plainview.

Joshi, S. and Hughes, J.B. 1981. Inhibiting of coupling factor B activity by cadmium ion and arsenite-2,3-dimercaptopropanol and phenylarsine oxide and preferential reactivation by dithiols. *J. Biol. Chem.* 256, 11112-11116.

Kahl, R. 1997. "Phenolic antioxidants: Physiological and toxicological aspects". In: *Handbook of Synthetic Antioxidants*, eds. L. Packer and E. Cadenas, pp 177-224. Marcel Dekker, New York.

Kernodle, S.P. and Scandalios, J.G. 1996. A comparison of the structure and function of the highly homologous maize antioxidant Cu/Zn superoxide dismutase genes, *Sod4* and *Sod4A*. *Genetics* 144: 317-328.

Kliebenstein, D.J., Monde, R.A. and Last, R.L. 1998. Superoxide dismutase in *Arabidopsis*: an eclectic enzyme family with disparate regulation and protein localization. *Plant Physiol.* 118, 637-650.

Knox, J.P. and Dodge, A.D. 1985. Singlet oxygen and plants. *Phytochem.* 24, 889-896.

Kreuz, K., Tommasini, R. and Martinoia, E. 1996. Old enzymes for a new job. Herbicide detoxification in plants. *Plant Physiol.* 111, 349-353.

Lewinsohn, E. and Gressel, J. 1984. Benzyl viologen-mediated counteraction of diquat and paraquat phytotoxicities. *Plant Physiol.* 76, 125-130.

Marrs, K.A. 1996. The functions and regulation of glutathione S-transferases in plants. *Ann. Rev. Plant Physiol. Plant Mol. Biol.* 47, 127-158.

Meyer, M., Schreck, R. and Baeuerle, P.A. 1993. Hydrogen peroxide and antioxidants have opposite effects on activation of NF-κB and AP-1 in intact cells: AP-1 as secondary antioxidant-responsive factor. *EMBO J.* 12, 2005-2015.

Millar, H.A. and Day, D.A. 1996. Nitric oxide inhibits cytochrome oxidase but not alternative oxidase of plant mitochondria. *FEBS Lett.* 398, 155-158.

Minagawa, N., Koga, S., Nakano, M., Sakajo, S., and Yoshimoto, A. 1992. Possible involvement of superoxide anion in the induction of cyanide resistant respiration in *Hansenula anomala*. *FEBS Lett.* 302, 217-219.

Moore, R. and Smith, J.D. 1984. Growth graviresponsiveness and abscisic acid content of *Zea mays* seedlings treated with fluoridone. *Planta* 162, 342-844.

McClung, C.R. 1997. The regulation of catalase in *Arabidopsis*. *Free Rad. Biol. Med.* 23, 489-496.

Neill, S.J., Horgan, R. and Parry, A.D. 1986. The carotenoid and abscisic acid content of *viviparous* kernels and seedlings of *Zea mays* L. *Planta* 169, 87-96.

Nguyen, T. and Pickett, B. 1990. Transcriptional regulation of the rat glutathione transferase Ya subunit. *J. Biol. Chem.* 265, 14648-14653.

Pabo, C.O. and Sauer, R.T. 1992. Transcription factors: structural families and principles of DNA recognition. *Ann. Rev. Biochem.* 61, 1053-1095.

Polidoros, A.N. and Scandalios, J.G. 1999. Role of hydrogen peroxide and different classes of antioxidants in the regulation of catalase and glutathione S-transferase gene expression in maize (*Zea mays* L.). *Physiol. Plant.* 106, 112-120.

Preston, C. 1994. "Resistance to photosystem I disrupting herbicides". In: *Herbicide Resistance in Plants*, eds. S.B. Powels and J.A.M. Holtum, pp: 61-82. CRC Press, Boca Raton.

Purvis, A.C. and Shewfelt, R.L. 1993. Does the alternative pathway ameliorate chilling injury in sensitive plant tissues? *Physiol. Plant.* 88, 712-718.

Rich, P.R., and Bonner, W.D. Jr. 1978. The sites of superoxide anion generation in higher plant mitochondria. *Arch. Biochem. Biophys.* 188, 206-213.

Ruis, H. and Koller, F. 1997. "Biochemistry, molecular biology, and cell biology of yeast and fungal catalases". In: *Oxidative Stress and the Molecular Biology of Antioxidant Defenses*. ed. J.G. Scandalios, pp. 309-342. Cold Spring Harbor Laboratory Press, Plainview.

Rushmore, T.H., Morton, M.R. and Pickett, C.B. 1991. The antioxidant responsive element. *J. Biol. Chem.* 266, 11632-11639.

Scandalios, J.G. 1965. Subunit dissociation and recombination of catalase isozymes. *Proc. Natl. Acad. Sci. USA* 53, 1035-1040.

Scandalios, J.G. 1993. Oxygen stress and superoxide dismutases. *Plant Physiol.* 101, 7-12.

Scandalios, J.G. 1996. "Genomic responses to environmental stress II". In: *The Encyclopedia of Molecular Biology and Molecular Medicine,* vol. 2, ed. R. A. Meyers, pp. 216-222. VCH Publishers, New York.

Scandalios, J.G. 1997a. "Molecular genetics of superoxide dismutases in plants". In: *Oxidative Stress and the Molecular Biology of Antioxidant Defenses*, ed. J.G. Scandalios, pp. 527-568. Cold Spring Harbor Laboratory Press, Cold Spring Harbor.

Scandalios, J.G. ed. 1997b. *Oxidative Stress and the Molecular Biology of Antioxidant Defenses.* Cold Spring Harbor Laboratory Press, Plainview.

Scandalios, J.G., Guan, L. and Polidoros, A.N. 1997. "Catalases in plants: gene structure, properties, regulation, and expression". In: *Oxidative Stress and the Molecular Biology of Antioxidant Defenses,* ed. J.G. Scandalios pp. 343-406. Cold Spring Harbor Laboratory Press. Plainview.

Siebenlist, U., Brown, K. and Franzoso, G. 1995. "NF-κB: a mediator of pathogen and stress responses". In: *Progress in Gene Expression: Inducible Gene Expression. Vol.1 Environmental Stresses and Nutrients*, ed. P.A. Baeuerle, pp. 93-141. Birkhaeuser Boston, New York.

Storz, G. and Imlay, J.A. 1999. Oxidative stress. *Curr. Opin. Microbiol.* 2, 188-194.

Valentine, J.S., Wertz, D.L., Lyons, T.J., Liou, L.L. and Gralla, E.B. 1998. The dark side of dioxygen biochemistry. *Curr. Opin. Chem. Biol.* 2, 253-262.

Vaz, A.D.N., Roberts, E.S. and Coon, M.J. 1987. "Radical intermediates in the catalytic cycles of cytochrome P450". In: *Oxygen Radical in Biology and Medicine*, eds. M.G. Simic, K.A. Taylor, J.F. Ward and C. von Sonntag, pp. 501-507. Plenum Press, New York.

Wagner, A.M. and Krab, K. 1995. The alternative respiratory pathway in plants: role and regulation. *Physiol. Plant.* 95, 318-325.

Williamson, J.D. and Scandalios, J.G. 1992a. Differential response of maize catalases and superoxide dismutases to the photoactivated fungal toxin cercosporin. *Plant J.* 2, 351-358.

Williamson, J.D. and Scandalios, J.G. 1992b. Differential response of maize catalases to abscisic acid: Vp1 transcriptional activator is not required for abscisic acid-regulated *Cat1* expression. *Proc. Natl. Acad. Sci. USA* 89, 8842-8846.

Williamson, J.D. and Scandalios, J.G. 1993. Plant antioxidant gene responses to fungal pathogens. *Trends Microbiol.* 1, 239-245.

Chapter 10

RESPONSES TO LOW TEMPERATURE AND ADAPTATIONS TO FREEZING

Gareth J. Warren

School of Biological Sciences, Royal Holloway, University of London, Egham, Surrey TW20 0EX, United Kingdom.
g.warren@rhbnc.ac.uk

INTRODUCTION

The nature of chilling and freezing stresses

Low temperature can be stressful to plants in two fundamentally different ways. One way is through the direct effects of low temperature on the properties and behaviour of biological molecules. This is called "chilling stress" and is the theoretical counterpart of heat stress. *A priori* we might expect a number of potentially damaging results, for example cold denaturation of proteins, metabolic imbalances due to differential effects of temperature on different enzymes' function, and phase changes in membranes. In plants that are sensitive to chilling, evidence for such lesions has been sought and reported in an extensive literature (see reviews by Lyons, 1973; Graham and Patterson, 1982). There is broad agreement in the field that altered membrane behaviour is very important. Photoinhibition - a special case of metabolic imbalance - is also known to be an important component of chilling stress (Hayden and Baker, 1990), against which the photosynthetic cells of chilling-tolerant plants must be protected. Beside these two types of lesion, any other unifying principles of chilling stress - if they exist - are far from obvious.

The second way in which low temperature can be harmful is indirect: by causing water to freeze. The consequences of a phase change in the plant's most abundant component are, not surprisingly, profound. Of course, we can only ever observe the effects of "freezing stress" superimposed on chilling conditions. Freezing and chilling stresses are nonetheless distinguishable

M.J. Hawkesford and P. Buchner (eds.),
Molecular Analysis of Plant Adaptation to the Environment, 209–247.

because (i) freezing damage can be incurred on a much shorter timescale than chilling damage, because (ii) the expected lesions are not equivalent, and finally because (iii) supercooling (by prevention of ice nucleation in laboratory-grown plants) dramatically demonstrates the incremental effect of freezing over low temperatures *per se* (Lindow *et al.*, 1982). This chapter is concerned with plants' adaptations to freezing stress. Other recent reviews (Hughes and Dunn, 1996; Ingram and Bartels, 1996; Huner *et al.*, 1998; Thomashow, 1998; Pearce, 1999; Thomashow, 1999; Shinozaki and Yamaguchi-Shinozaki, 2000) are recommended for different approaches to the topic and alternative opinions.

Cold acclimation

Plant species that are "frost-hardy" are freezing-tolerant during the season when frost is likely (Graham and Patterson, 1982). At other times of year the same plants are freezing-sensitive. Low (non-freezing) temperature is the main environmental cue for their seasonal adaptation to freezing stress (Guy, 1990). (For some species, day length is an additional stimulus (Sakai and Larcher, 1987)). Therefore, the process of adaptation is called "cold acclimation". It takes place on a timescale of days or weeks.

Cold acclimation implies that freezing tolerance comes at a cost to the plant. If it was free, and compatible with tolerance to other environmental stresses, evolution would surely have dispensed with cold acclimation and produced plant species that were always freezing-tolerant. Constitutive tolerance would have a selective advantage on occasions of unseasonal (sudden) frost; its rarity argues for the existence of a counterbalancing disadvantage.

Cold acclimation provides an avenue for investigating how plant cells can tolerate freezing. The biochemical changes that occur during acclimation must include the changes that provide freezing tolerance. A variety of physiological and biochemical changes have been observed (Graham and Patterson, 1982), but direct evidence of their specific relevance to freezing-tolerance has been difficult to obtain (Steponkus, 1984), and it still is. A clearly important change is the increase in cellular concentrations of "compatible" osmolytes, which may be proline, betaines, and/or soluble carbohydrates (Pollock and Jones, 1979; Levitt, 1980; Delauney and Verma, 1993; Ristic and Ashworth, 1993; Kishitani *et al.*, 1994), according to the plant species. This appears to be a general phenomenon among hardy plants, and various experimental approaches have provided compelling evidence of its relevance to hardiness.

However, the biochemical changes that occur during acclimation are not necessarily all adaptive to freezing stress. Plants must also adapt to chilling stress even to survive during the acclimation process itself. Pearce (1999) lists several further plausible reasons for biochemical changes occurring during cold acclimation. There may be adaptive responses to other stresses associated with low temperatures, such as ice encasement or attack by snow moulds. Developmental responses such as vernalisation will also entail novel biochemistry. And certain changes may be non-adaptive - simply the mechanistic consequences of perturbing a complex system, rather than evolved responses. Thus, some changes that occur during cold acclimation will contribute positively to freezing tolerance and to our understanding of it; others may make a zero or even negative contribution and, unless we exercise scepticism, have corresponding effects on our understanding.

The biochemical changes occurring during cold acclimation include effects on gene expression (Guy *et al.*, 1985). At first glance, investigation of these effects does not seem to escape the problem of proving the relevance of correlative observations, but merely extends it to the molecular level. But the power of molecular genetics derives from its tools for testing hypotheses, and the molecular investigation of cold acclimation is beginning to produce empirical demonstrations of freezing-tolerance mechanisms. The molecular biology of low temperature responses is therefore the focus of this chapter.

Relationship of freezing to drought and osmotic stresses

During natural frosts, temperatures drop slowly and plant tissues freeze gradually. There is time for water to migrate out of protoplasts (down the gradient of chemical potential) to join the extracellular ice (Steponkus, 1984; Pearce, 1988; Pearce and Ashworth, 1992). Because protoplasts are never far from osmotic equilibrium, they do not become sufficiently supercooled for ice to nucleate within them. Mechanical penetration of protoplasts by ice spicules, which was once speculated to be a mode of damage, does not seem to occur either. In short, the aqueous phases of the protoplast remain liquid, but at greatly reduced volumes and concomitantly increased osmolalities.

The protoplast's situation during a frost episode thus resembles a state of advanced dehydration due to drought stress. It seems likely that the extreme changes in cell volume and osmolality are the major potential causes of damage in both freezing and drought stress. This notion is strongly supported by the observation that when hardy plants are cold-acclimated, they also show an elevated tolerance to dehydration, and vice versa.

However, although a close relationship between freezing and drought stresses is predicted and supported by experiment, their equivalence is not exact. The chilling stress that necessarily accompanies freezing may predispose cells to certain sorts of dehydrative damage that are not significant at warmer (i.e. drought-typical) temperatures. There might also be some types of freezing-dependent lesion that are unrelated to dehydrative stress, such as mechanical distortion by ice, prevention of gas exchange, generation of electric gradients, and so forth. These differences would create the need for additional mechanisms of protection against freezing. Conversely, low temperature might *prevent* some types of dehydrative damage that are incurred during drought, by slowing or preventing certain chemistries or by limiting gas exchange. Thus, the two types of stress are likely to overlap, substantially but not completely, in their mechanisms of damage and of tolerance, as illustrated in Figure 1. This seems to be borne out by the molecular biology of the response to each type of stress.

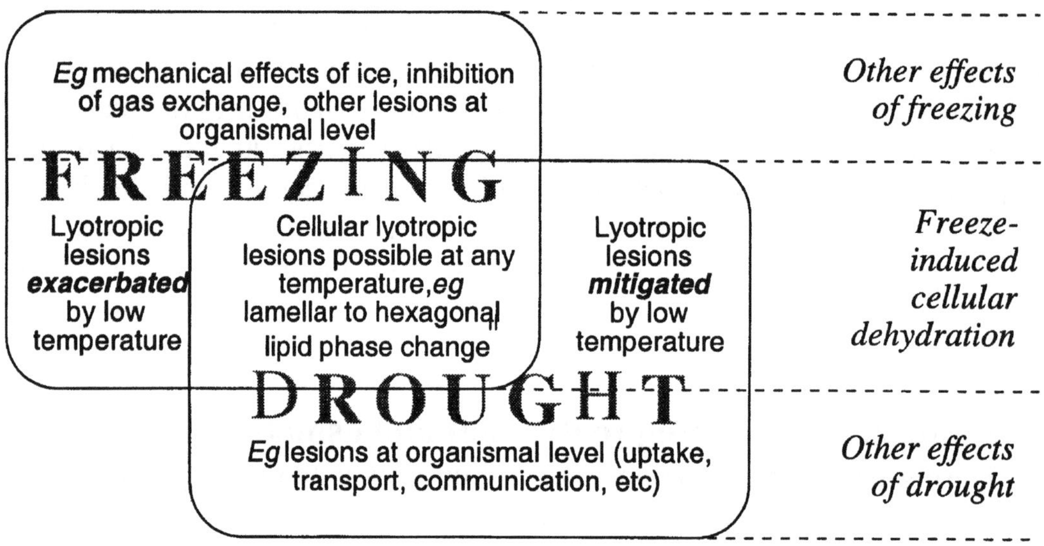

Figure 1. Overlap between the injurious effects of freezing and drought.

Cryopreservation has been investigated mainly for animal and microbial cells, but the same principles seem to apply to plant cells. In cryopreservation theory, high intracellular osmolality is not inherently damaging; rather, certain solutes are held to be "cryotoxic" and in cryopreservation their harmful effects are mitigated by replacing or diluting them (through human intervention) with "compatible" (harmless at high concentration) osmolytes. Some cryotoxic solutes - notably, the sodium cation - are a major component of salinity stress, damaging plants after their excess uptake and accumulation. Hence there is also a relationship between salinity and freezing stresses. An important difference is that freezing tolerance entails survival of a temporary extreme, whereas salinity tolerance usually means the ability to grow during prolonged exposure to less severe solute concentrations. Thus we might expect at least a minor overlap between salinity and freezing stresses in their mechanisms of tolerance.

The central role of membranes in freezing stress

What is the nature of the damage inflicted by lethal freezing events? Or to put it another way, what are the lethal cellular lesions that - after cold acclimation - a hardy plant is able to prevent or repair? Many possibilities have been suggested, but one type of lesion has found by far the greatest experimental support (Steponkus, 1984): the loss of membrane integrity.

Before having recourse to living systems, it can be observed that freeze-induced dehydration tends to cause the L_α-to-H_{II} phase transition in membrane lipids (Williams, 1990). H_{II} is a non-lamellar phase, so this physical phase transition will undoubtedly destroy membrane integrity. It is different from the phase transition that tends to be caused by chilling, whose product is another lamellar phase with less obviously destructive properties.

In planta, freeze-induced membrane destabilisation has been observed by electron microscopy and where studied has been attributed entirely to the dehydrative effects of freezing (Pearce, 1985; Pearce and Willison, 1985; Steponkus and Gordon-Kamm, 1985; Cudd and Steponkus, 1988). At different freezing temperatures, different membrane-associated lesions have been observed in rye (Gordon-Kamm and Steponkus, 1984; Dowgert *et al.*, 1987; Uemura and Steponkus, 1989; Steponkus *et al.*, 1993) and *Arabidopsis* (Uemura *et al.*, 1995; Steponkus *et al.*, 1998), and shown to be prevented or deferred by cold acclimation. In a classic series of experiments on rye protoplasts, the naturally-occurring changes in membrane lipid composition were directly shown to prevent freeze-induced membrane lesions (Steponkus *et al.*, 1988). The universal observation of compatible osmolyte accumulation during cold acclimation is easily reconciled with a central role of membranes, since compatible osmolytes may directly interact with and stabilise membranes during freeze-induced dehydration (Strauss and Hauser, 1986; Crowe *et al.*, 1992).

Freezing tolerance: alternative paradigms

With our currently incomplete understanding of freezing tolerance, what we look for in the evidence can determine what we see. In one paradigm it is a quantitative trait - it can certainly be measured quantitatively by minimum survival temperature. Mapping of quantitative trait loci (QTLs) for freezing tolerance uses a simplifying assumption that the contributing genes have additive effects, and yields estimates of the percent contribution of each identified locus. Alternatively, cold acclimation may be viewed as a developmental trait analogous to flower development. It does not seem meaningful to estimate the percent contribution of each gene involved in making a flower. Whichever model is preferred, most reviewers concur that freezing tolerance is likely to be a complex adaptation and thus to require the contributions of a number of genes.

COLD ACCLIMATION: GENES AND PHENOMENA

Discovery

Genes that are turned on during cold acclimation are prime candidates for providing the observed increase in freezing tolerance. Over a wide range of plant species, many studies have followed the work of Guy (Guy *et al.*, 1985) by identifying genes whose transcript levels are elevated in response to cold. In only a minority of cases has this been demonstrated to correlate with increased levels of the gene product, though it seems safe to assume that in most cases, increased transcript levels will in fact correspond to higher levels of protein synthesis. But not all cold-inducible transcripts' levels are controlled by the frequency of transcriptional initiation. In *Hordeum*, some cold-induced genes are known to be regulated by post-transcriptional mRNA stability (Dunn *et al.*, 1994; Phillips *et al.*, 1997); a similar phenomenon could easily have been overlooked in other species.

It is beyond the scope of this review to describe all the various cold-induced genes described in the plant kingdom. Instead, a survey is presented that is limited to the most intensively studied species, *Arabidopsis thaliana*, with references to other species where the identified sets of genes overlap (Table 1). The use of *Arabidopsis* as a model for plant freezing tolerance is not without controversy. Its chief shortcoming is that its tolerance level, while enough to be useful for survival in nature and potentially to genetic engineers, is only moderate by comparison with crop species such as barley and rye.

The survey in Table 1 is likely to be the most complete for strongly induced genes in a single plant species. Because it is based on the work of many groups, it can be seen as a result of a number of samplings of the real set of cold-inducible genes by a range of approaches. Multiple names for individual genes demonstrate oversampling for some genes. It is difficult to infer to what extent the real set has been saturated by the sampling; the observed set may omit many or most of the cold-inducible genes with lower levels of induction. When making comparisons to the sets of cold-inducible genes described in other plant species, it should be kept in mind that all other such observed sets are likely to be still less complete than that of *Arabidopsis*.

The more species in which homologs of a gene are cold-induced, the stronger the circumstantial evidence for that gene's importance. Conservation in species more distant from *Arabidopsis* (e.g. *Hordeum vulgare*) is more impressive since it implies a longer period of evolutionary conservation of the response. Induction by dehydration and/or ABA is consistent with the product of the gene protecting against freeze-induced dehydration (expected to be the major component of freezing stress). On both counts, as well the number of genes represented in Table 1, the dehydrins are particularly impressive.

Why are some cold-induced proteins unusually hydrophilic?

For obvious technical reasons, genes encoding the most abundantly cold-induced proteins tend to have been the earliest to be described, with publication dates preceding 1994. Several classes of abundantly cold-inducible proteins are unusually hydrophilic (Lin *et al.*, 1990). These are the dehydrins (also known as LEA Group 2 or LEA D11 proteins), as well as the protein families represented by the COR6.6, COR15a, COR78 and M17 proteins of *Arabidopsis*, the cas15 protein of alfalfa (*Medicago sativa*) (Monroy *et al.*, 1993a), and the HVA1 protein of barley (*Hordeum vulgare*) (Sutton *et al.*, 1992). Their inability to fold (or aggregate) via hydrophobic bonding allows their observable properties to remain unchanged after heat denaturation: hence they are sometimes called "boiling-soluble" proteins.

Dehydrins are cold-induced in a wide variety of species, and are also the class of protein most clearly associated with dehydration, being universally induced in response to both environmental and developmental dehydration. This argues powerfully for a role for dehydrins in protecting against the dehydrative effects of freezing. They are localised in the cytoplasm or the nucleus, according to their individual sequences (Close, 1996). Proteins of the COR6.6, COR15a and COR78 classes have also been observed in at least one other species, and are also induced by environmental dehydration stress (Table 1). These observations likewise argue for protective roles specifically in freeze-induced dehydration.

The hydrophilicity of such proteins is probably incompatible with any of them adopting a globular structure necessary for catalytic activity. In the absence of enzymatic functionality it is natural to infer that these proteins serve to stabilise other macromolecules and structures in the freeze-dehydrated protoplast. The shared property of hydrophilicity suggests that the protective action may be explicable at a physico-chemical level. Close (1996) suggests that they work by the preferential exclusion mechanism that was proposed by Carpenter and Crowe (1988) to explain the cryoprotection of proteins by certain solutes.

The stabilisation of proteins could in turn stabilise protein-containing membranes. However, hydrophilicity is not the only selected property of the proteins and therefore cannot be the sole requirement for their function. Dehydrins, for example, are recognised by their possession of several highly homologous motifs that have been conserved throughout the evolution of angiosperms (Close, 1996). This implies that hypotheses of action relying exclusively on nonspecific physico-chemical interactions can not provide the

Table 1. Cold-inducible genes in *Arabidopsis thaliana*.

Gene product	Features and/or putative function	Cold-induced homologs	Inducible by Drought	Inducible by ABA	Inducible by Salt	References
A1494	Cysteine thiol protease		yes	yes	?	53
ADH	Alcohol dehydrogenase	*Zea mays* Adh1, *O. sativa* Adh1	yes	yes	?	7, 21, 9, 3
ADS2	δ 9 acyl-lipid desaturase		?	?	?	11
AWI34	-		yes	?	yes	56
CBF1/DREB1B	AP2-domain, C-box-binding factor	CBF2, CBF3	no	no	?	46, 35
CBF2/DREB1C	AP2-domain, C-box-binding factor	CBF1, CBF3	no	no	?	14, 35, 38
CBF3/DREB1A	AP2-domain, C-box-binding factor	CBF1, CBF2	no	no	?	14, 35, 38
CBL1	Calcineurin B-like protein		yes	?	?	28
CCR1/GRP8	RNA-binding protein	CCR2, *Hordeum vulgare* blt801	no	no	?	41, 5, 10
CCR2/GRP7	RNA-binding protein	CCR1, *Hordeum vulgare* blt801	yes	no	?	41, 5, 10
CHS	Chalcone synthase; anthocyanin synthesis	*Zea mays* Chs	?	?	?	8, 33
COR6.6/KIN2	Short, alanine-rich, hydrophilic	KIN1, *Brassica napus* Bn28	?	yes	?	13, 30, 2
COR15a	Chloroplast-targeted, hydrophilic	COR15b, *Brassica napus* Bn115, Bn26,Bn19	yes	yes	?	34, 22
COR15b	Chloroplast-targeted, hydrophilic	COR15a, *Brassica napus* Bn115, Bn26, Bn19	no	yes	?	52, 51, 22
COR47/RD17	Dehydrin, hydrophilic	Many examples across species	yes	yes	yes	13
COR78/LTI78/RD29A	Very hydrophilic	*Spinacia oleracea* CAP160	yes	yes	yes	18, 42, 55, 23
CYP83A1	Cytochrome P45		?	?	?	1
ERD6	Sugar transporter		yes	?	?	27
ERD14	Dehydrin, hydrophilic	Many examples across species	yes	yes	?	25
FAD8	Chloroplast omega-3 desaturase		?	?	?	12
HK1	Histidine kinase		yes	?	yes	48
KIN1	Short, alanine-rich, hydrophilic	*Brassica napus* (Bn28)	?	yes	?	29, 2
LTI29/LTI45/ERD10	Dehydrin, hydrophilic	Many examples across species	yes	yes	?	13, 24, 50, 51
LTI30/XERO2	Dehydrin, hydrophilic	Many examples across species	yes	yes	?	50, 45
LTI65/RD29B	Very hydrophilic	*Spinacia oleracea* (CAP160)	yes	yes	yes	42, 55, 23
M17	Very hydrophilic	Many species	no	no	no	44
MEKK1	MAP kinase kinase kinase		?	?	yes	40
MPK3	MAP kinase		?	?	yes	40
P5CS1	δ-pyrroline-5-C synthase; proline synthesis	P5CS2	yes	yes	yes	57, 47
P5CS2	δ-pyrroline-5-C synthase; proline synthesis	P5CS1	yes	yes	yes	57, 47

Table 1. Cold-inducible genes in *Arabidopsis thaliana*

Gene product	Features and/or putative function	Cold-induced homologs	Inducible by Drought	ABA	Salt	References
PAL*	Phenylalanine ammonia lyase; anthocyanin synthesis	*Zea mays* (Pal)	?	?	?	8, 33
PDC1	Pyruvate decarboxylase		yes	?	?	6
PK19	Ribosomal protein S6-kinase	PK6	?	?	yes	39, 40
PK6/atpk1	Ribosomal protein S6-kinase	PK19	?	?	yes	39
PLC1	Phosphatidylinositol-specific phospholipase C		yes	?	yes	19
POX/ERD5	Proline oxidase; proline catabolism		yes	?	?	26, 49, 54
PUMP	Plant uncoupling mitochondrial protein		?	?	?	37
RAB18	Dehydrin	Many species	yes	yes	?	31, 32
RCI1A/GF14psi/GRF3	14-3-3 protein	RCI1B	no	no	?	21
RCI1B/GF14lambda	14-3-3 protein	RCI1A	no	no	?	21
RCI2A/LTI6A	Short, very hydrophobic	RCI2B, *H. vulgare* (blt101), *L. elongatum* (ESI3)	yes	yes	no	15, 16, 4
RCI2B/LTI6B	Short, very hydrophobic	RCI2A, *H. vulgare* (blt101), *L. elongatum* (ESI3)	yes	yes	no	15, 16, 4
RCI3A	Peroxidase		?	?	?	36
RPK1	Receptor-like protein kinase		yes	yes	yes	17
RR1	Two-component response regulator-like	RR2	yes	yes	yes	48
RR2/IBC6/ARR5	Two-component response regulator-like	RR1	yes	no	yes	48
TCH2	Calmodulin-related protein	TCH3	?	?	?	43
TCH3	Calmodulin-related protein	TCH2	?	?	?	43
TCH4	Xyloglucan endotransglycosylase		?	?	?	43

[1]Bilodeau *et al.*, 1999; [2]Boothe *et al.*, 1995; [3]de Bruxelles *et al.*, 1996; [4]Capel *et al.*, 1997; [5]Carpenter *et al.*, 1994; [6]Conley *et al.*, 1999 [7]Christie *et al.*, 1991; [8]Christie *et al.*, 1994; [9]Dolferus *et al.*, 1994; [10]Dunn *et al.*, 1996; [11]Fukuchi-Mizutani *et al.*, 1998; [12]Gibson *et al.*, 1994; [13]Gilmour *et al.*, 1992; [14]Gilmour *et al.*, 1998; [15]Goddard *et al.*, 1993; [16]Gulick *et al.*, 1994; [17]Hong *et al.*, 1997; [18]Horvath *et al.*, 1993; [19]Hirayama *et al.*, 1995; [20]Jarillo *et al.*, 1993; [21]Jarillo *et al.*, 1994; [22]Jiang *et al.*, 1996; [23]Kaye *et al.*, 1998; [24]Kiyosue *et al.*, 1993; [25]Kiyosue *et al.*, 1994; [26]Kiyosue *et al.*, 1996; [27]Kiyosue *et al.*, 1998; [28]Kudla *et al.*, 1999; [29]Kurkela and Franck, 1990; [30]Kurkela and Borg-Franck, 1992; [31]Lang and Palva, 1992; [32]Lang *et al.*, 1994; [33]Leyva *et al.*, 1995; [34]Lin and Thomashow, 1993; [35]Liu *et al.*, 1998; [36]Llorente *et al.*, 1998; [37]Maia *et al.*, 1998; [38]Medina *et al.*, 1999; [39]Mizoguchi *et al.*, 1995; [40]Mizoguchi *et al.*, 1996; [41]van Nocker and Vierstra, 1993; [42]Nordin *et al.*, 1993; [43]Polisensky and Braam, 1996; [44]Raynal *et al.*, 1999; [45]Rouse *et al.*, 1996; [46]Stockinger *et al.*, 1997; [47]Strizhov et al., 1997; [48]Urao *et al.*, 1998; [49]Verbruggen *et al.*, 1996; [50]Welin *et al.*, 1994; [51]Welin *et al.*, 1995; [52]Wilhelm and Thomashow, 1993; [53]Williams *et al.*, 1998; [54]Xin and Browse, 1998; [55]Yamaguchi-Shinozaki and Shinozaki, 1993; [56]Yang *et al.*, 1995; [57]Yoshiba *et al.*, 1995.

whole picture - the conserved motifs must contribute a specific aspect of structure or intermolecular interaction.

Genes of the *KIN1/COR6.6* family, which are cold-induced in *Arabidopsis* and *Brassica napus*, encode proteins that are not only boiling-soluble but also highly alanine-rich. At its first report, homology was noted between the predicted KIN1 protein and the alanine-rich Type I antifreezes of Arctic teleost fish (Kurkela and Franck, 1990), and this has sometimes been cited with the implication of a possible equivalence of function. Such a function is very unlikely: the *Brassica* homolog has been tested and lacks the recrystallization-inhibiting activity that is diagnostic of antifreezes (Boothe *et al.*, 1997), and the homology with Type I antifreezes is no less if the order of amino acids in KIN1 is randomised.

Genes and proteins with putative roles

The catalytic properties imputed to some of the other cold-induced genes' products give indications of their likely roles in promoting low temperature tolerance.

The *ADH* and *PDC1* genes encode enzymes of alcoholic fermentation, and are therefore adaptive for anoxic or hypoxic conditions. Root hypoxia can result from the waterlogging of soil that tends to accompany low temperatures; in this sense *ADH* and *PDC1* are protective against a winter stress that is technically distinct from both chilling and freezing. However, it is conceivable that freezing may also produce hypoxic conditions in aerial tissues, particularly where it develops into ice encasement.

The FAD8 and ADS2 desaturases presumably participate in the increase of membrane lipid unsaturation that almost always accompanies cold acclimation (Somerville and Browse, 1991; Palta *et al.*, 1993; Kodama *et al.*, 1995; Kwon and Markhart, 1997). It has been amply demonstrated that increased unsaturation of the fatty acids in membrane lipids is crucial in chilling tolerance (Murata *et al.*, 1992; Nishida and Murata, 1996; Yokoi *et al.*, 1998), probably in order to maintain membrane fluidity (Cossins, 1994). However, such changes may also be a necessary background for the specific changes in the molecular species composition of membrane lipids (Lynch and Steponkus, 1987; Uemura and Steponkus, 1994; Uemura *et al.*, 1995) that have been shown to contribute directly to freezing tolerance (Steponkus *et al.*, 1988). Similarly, the lipid transfer proteins that are cold-induced in barley (White *et al.*, 1994) might have some role in the adjustment of membrane composition, although the authors argue that this is unlikely because of the presence of presumptive export signals.

Antioxidant enzymes (such as the RCI3A peroxidase of *Arabidopsis*: Table 1) are cold-induced in various species (Hull *et al.*, 1997; Jahnke *et al.*, 1991; Prasad 1997; Walker and McKersie, 1993), and there is a concomitant increase in tolerance to oxidative stress (Bridger *et al.*, 1994). Is this likely to be an adaptation to chilling or freezing stress? It may be both. Oxidative stress has been proposed as a component of freezing stress, and the evidence for this, although indirect, is compelling: transgenesis with superoxide dismutase, an antioxidant enzyme, increased freezing tolerance in alfalfa (McKersie *et al.*, 1993). But antioxidants are also cold-induced in non-hardy species, where they have been associated with chilling-tolerance (Prasad, 1997).

Cold-induction of enzymes involved in proline biosynthesis (P5CS1 and 2: Table 1) is almost certainly responsible for the elevation of proline levels (Wanner and Junttila, 1999). Proline is a compatible osmolyte and its synthesis is therefore very likely to contribute to freezing tolerance.

Molecular chaperones induced during cold acclimation (Guy and Li, 1998) seem likely to be involved in refolding of denatured proteins. Evidence has been obtained for increased protein denaturation during chilling (Guy *et al.*, 1998), but the intracellular consequences of extracellular freezing are also likely to include denaturation of a subset of proteins.

Notable genes and proteins in search of roles

A number of studies have shown that the transcript levels of certain protein kinases are cold-inducible - including calcium-dependent kinases (Monroy and Dhindsa, 1995; Tahtiharju *et al.*, 1997), calcineurin-B-associated kinases, calmodulin-like proteins, receptor-like protein kinases, ribosomal protein S6 kinases, MAP kinases and MAP kinase kinase kinases (see Table 1). The cold-inducibility of such proteins has sometimes been invoked by their discoverers as evidence for their involvement in signaling of the low-temperature stimulus. This conclusion seems to presume that their transcriptional regulation is a step in signaling, and thus to imply that systems capable of extremely rapid phosphorylation responses have evolved to lie downstream of the much slower apparatus of transcription and translation in a number of signaling pathways. I find this implausible and suggest that the cold-inducibility of kinases has no bearing on whether they are likely to be involved in cold signal transduction. The cold-induced kinases are certainly likely to have roles in chilling or freezing tolerance - for example by altering homeostasis or the ability to respond to external stimuli before or after freezing - but without necessarily being links in a low temperature-response pathway. Unfortunately, except where their targets are other kinases (Ichimura *et al.*, 1998; Mizoguchi *et al.*, 1998), there is little specific information about which proteins' activities are regulated by the cold-induced kinases, still less of how such action may contribute to freezing or chilling tolerance.

Non-colligative antifreezes were first reported in rye (Griffith *et al.*, 1992; Griffith *et al.*, 1997) and subsequently in other hardy cultivated species (Chun *et al.*, 1998; Smallwood *et al.*, 1999) and a variety of hardy wild species (Duman and Olsen, 1993). Their presence was suggestive of a function in freezing tolerance, not only because of their specific ability to interact with ice, and their localisation in the apoplast where ice forms, but also because they are present only when plants are cold-acclimated. This correlative evidence has been extended by the observation that they were present in all the Antarctic species but only 25% of the overwintering temperate species that were surveyed (Doucet *et al.*, 2000). Thus, if they have a function in freezing tolerance it is likely to be required only toward more severe freezing conditions. Freezing-point depression can hardly be their function, since they are not present in sufficient quantities to depress freezing points by even 0.1°C. Neither does recrystallisation-inhibition, the property by which they are more sensitively assayed (Knight *et al.*, 1988), seem a useful activity since it is detectable only when water is frozen at least five orders of magnitude more quickly than happens during frosts. Their adaptive function might be to change the crystal habit of ice, or to reduce adhesions between membranes and ice crystals (Olien and Smith, 1977), but at present such hypotheses are purely speculative.

Their cold-inducibility in species as distantly related as *Arabidopsis* and barley (*Hordeum vulgare*) argues that the RNA-binding proteins are important for tolerance of chilling and/or freezing. It is difficult to rationalise this from current knowledge. However, it is noteworthy that the first two chilling-sensitive insertional mutations to be characterised both affected RNA modification (Tokuhisa *et al.*, 1997; Tokuhisa *et al.*, 1998). This hints that RNA structure or processing may be a point of particular vulnerability to chilling damage, and so tends to favour the hypothesis that the cold-induced RNA-binding proteins may be involved in chilling rather than freezing tolerance.

Phenomena in search of a genetic basis

While the accumulation of proline appears to have a clear molecular basis, we currently do not know exactly what causes sucrose and/or other sugars to accumulate during cold acclimation. One proposal has been that it results from an imbalance between photosynthesis and growth rate at low temperatures (Huner *et al.*, 1998). Consistent with this idea, the accumulation of sugars is lessened when the intensity or duration of light is reduced (Wanner and Junttila, 1999), suggesting that sugar level during cold acclimation could indeed be an unmediated consequence of photosynthetic output. Thus it does not seem necessary to postulate any low-temperature-specific control of sugar accumulation at the molecular genetic level.

However, I shall be surprised if a more exacting system has not evolved to govern sugar accumulation at low temperature, and look forward to learning of the proteins and genes likely to be involved in such control.

Superimposed on the increasing general level of fatty acid unsaturation are complex changes in the populations of individual molecular species of membrane lipids. Detailed descriptions of these changes have been published for several cereals (Lynch and Steponkus, 1987; Uemura and Steponkus, 1994) and for *Arabidopsis* (Uemura *et al.*, 1995). As discussed above (under *The central role of membranes in freezing stress*) certain of these changes are responsible for the membranes' increased resistance to freeze-dehydration-induced lesions. Currently it is not known how the composition of lipid species is controlled - whether it is by changes in gene expression at low temperature, or by temperature-dependence of the intrinsic activities of the synthetic and degradative enzymes, or by post-translational modification of enzyme activities.

BETWEEN COLD SIGNAL AND MOLECULAR RESPONSE

Calcium signaling

There is abundant evidence that fluxes of the calcium cation form a very early part of plants' signal pathways in responding to a wide variety of stimuli. Even before including the pathway from low temperature stimulus to the development of freezing tolerance, it is clear that there is a conundrum: how can calcium fluxes be coupled to the distinct downstream responses to the various stimuli? A second conceptual problem is that the apparent calcium fluxes are usually transient even when the downstream response is sustained. Both of these difficulties hang over the role of calcium in cold acclimation, also. But since invoking calcium signaling in the cold response does not create these difficulties (they are already well established), they do not argue against its involvement.

It is not straightforward to show that calcium flux is required for functional acclimation (acquisition of tolerance): the means of inhibiting calcium flux tend to be of broad spectrum and are likely to interfere with viability by the time freezing tolerance would have developed. However, in cell cultures of alfalfa (*Medicago sativa*) administration of lanthanum, which blocks calcium channels, was shown to prevent the acquisition of freezing tolerance in response to cold (Monroy *et al.*, 1993b). The use of cell cultures permitted uniform delivery of the inhibitor, and perhaps enabled cells to survive such drastic treatment to the point of testing. Corresponding evidence has been obtained for the acquisition of chilling-tolerance in roots of rice (*Oryza sativa*): caffeine (though not a universal inhibitor of calcium flux) blocked the cold-responsive calcium flux and the treated plants remained chilling-sensitive (Kitigawa and Yoshizaki, 1998).

The role of calcium in cold-responsive gene expression is more accessible to investigation: shorter response times can be examined, and interpretation is not clouded by longer-term effects on viability. Calcium chelators and calcium channel blockers (lanthanum and gadolinium) have been shown to block cold-induced gene expression in alfalfa (Monroy and Dhindsa, 1995) and in *Arabidopsis* (Knight *et al.,* 1996; Polisensky and Braam, 1996); Palva and coworkers additionally used ruthenium red to specifically inhibit calcium release from intracellular stores, and observed a partial reduction in cold-induced gene expression (Tahtiharju *et al.,* 1997). These experiments strongly support the involvement of calcium in cold-signaling in *Arabidopsis*, and comparable results have been obtained in other species (Berberich and Kusano, 1997). Two studies further supported this conclusion with a complementary approach: intoxication with calcium led directly to the expression of cold-inducible genes in *Arabidopsis* and alfalfa (Monroy and Dhindsa, 1995; Knight *et al.,* 1996).

Knight and colleagues have developed a system in which cytosolic calcium concentration can be monitored *in vivo*. Their method measures light emission from the calcium-dependent photoprotein aequorin, which is supplied by a cnidarian transgene. With this approach they have observed rapid elevations in cytosolic calcium in response to cold as well as a number of other stimuli (Knight *et al.,* 1991; Knight *et al.,* 1996). The calcium signatures are so transient that they give the appearance of being responsive to temperature change, rather than absolute temperature (Plieth *et al.,* 1999). By targeting aequorin to the cytosolic face of the vacuolar membrane, it was possible to examine the calcium signature in this microdomain within the cell (Knight *et al.,* 1996). The results concurred with those of ruthenium red intoxication, namely suggesting that a portion of the cold-induced calcium release occurs from the vacuole.

What causes calcium flux to respond to cold? There is some evidence to implicate phosphoinositide signaling (De Nisi and Zocchi, 1996; Knight *et al.,* 1996). Relying heavily on analogy to cyanobacteria, Murata and Los (Murata and Los, 1997) argue that the temperature sensor for cold measures membrane fluidity, which decreases towards lower temperature, or fluidity-dependent conformation. A fluidity-dependent calcium channel has been proposed (Monroy and Dhindsa, 1995). However, chilling-tolerant plants adjust their membrane composition homeostatically in response to temperature variation. Thus membrane fluidity is plausible as the signal for such adjustment, but it makes less sense as a long-term temperature sensor that could cause freezing tolerance to be maintained over periods of weeks to months.

Kinases

Calcium signaling would be expected to lead in turn to effects on protein phosphorylation, via the calcium-binding adaptor, calmodulin. In support of this, an inhibitor of calmodulin and calcium-dependent protein kinases blocks cold-induced gene expression (Monroy *et al.*, 1993b). Constitutively active variants of two calcium-dependent protein kinases (CDPKs), out of eight tested, cause constitutive activation of a normally stress-inducible promoter (Sheen, 1996). Low temperature was found to result in specific inhibition of one protein phosphatase in alfalfa; this could be mimicked at normal temperatures by treatments that induced calcium influx (Monroy *et al.*, 1998).

Independently of the case for their control by calcium, there is evidence that protein phosphorylation levels respond rapidly to low temperature, and that cold-induced gene expression is controlled by phosphorylation levels (Monroy *et al.*, 1997). Phosphorylation of a MAP kinase is rapidly induced in response to low temperature in plant cells (Jonak *et al.*, 1996; Meskiene *et al.*, 1998).

It has been pointed out that low temperature is a stimulus available to all parts of the cell, and that multiple sensors might exist in the different cellular compartments. This is supported by the observation of rapid cold-responsive changes in protein phosphorylation level in isolated nuclei of alfalfa (Kawczynski and Dhindsa, 1996).

In a previous section (*Cold acclimation: genes and phenomena*) I have argued, contrary to majority opinion, that the cold-inducibility of certain kinase genes neither supports nor refutes the involvement of their products in transduction of the low-temperature signal.

Promoter sequence motifs and transcription factors

A conserved, repeated motif is recognisable in the promoters of several strongly cold-induced *Arabidopsis* genes (Baker *et al.*, 1994) - the "C-repeat", CCGAC. By functional analysis, a 9 bp sequence that included this motif, taCCGACat, was found to be essential for the cold-inducibility of one of these genes (Yamaguchi-Shinozaki and Shinozaki, 1994). The taCCGACat sequence is also responsible for the same genes' induction in response to dehydration, and is therefore called the "CRT/DRE" (C repeat/drought-responsive element). A similar sequence with CCGAC at its core was recognised in a cold-inducible promoter in *Brassica napus* (Jiang *et al.*, 1996). The related sequence CCGAAA has been demonstrated to confer cold-inducibility on the blt4.9 gene in *Hordeum vulgare* (Dunn *et al.*, 1998).

The CRT/DRE was postulated to represent the binding site of a transcriptional activator; the activator gene was then sought by one-hybrid selection in yeast. The yeast screen duly yielded a gene, *CBF1*, encoding a protein with the AP2 domain characteristic of a subset of plant transcription factors. A surprising aspect of the screen was that *CBF1* was obtained as a *non-hybrid* gene: thus CBF1 protein had been able not only to bind the CRT/DRE but also to activate transcription in yeast, *without* the intended fusion to a yeast activation domain (Stockinger *et al.*, 1997). By transgenesis, it has been demonstrated that the CBF1 protein also functions as a CRT/DRE-specific transcription factor in *Arabidopsis* (Jaglo-Ottosen *et al.*, 1998); the phenotypic consequences of such manipulation are discussed in a later section (*transgenesis with transcription factors*).

The cold-responsivess of CRT/DRE promoters probably results from the fact that the *CBF1* gene is itself cold-induced, as are its two tandem homologs (Gilmour *et al.*, 1998; Liu *et al.*, 1998; Medina *et al.*, 1999). The early kinetics of the *CBF* genes' induction, and the degree of their up-regulation, seem sufficient to account for the cold-inducibility of genes dependent on the CRT/DRE. Interestingly, the initial failure to detect inducibility of the *CBF* genes (Stockinger *et al.*, 1997) can be attributed to inadvertent disturbance of the control plants; *CBF1* is, like *TCH2*, *3*, and *4* (Table 1), also induced by mechanical stimulation (Gilmour *et al.*, 1998). The fallacious idea that *CBF1* transcription was constitutive initially implied that the CBF1 protein itself might be temperature-sensitive. This was investigated and reversible cold-denaturation was observed in regions of the CBF1 protein (Kanaya *et al.*, 1999), but its relevance to activity is unknown; it is neither necessary nor sufficient to explain the cold-inducibility of CRT/DRE-containing promoters.

The expression of *CBF* genes can *not* account for the dehydration-responsiveness of CRT/DRE-containing promoters: *CBF* transcription appears to be unaffected by dehydration. However, there is another family of CRT/DRE-binding transcription factors, DREB2A and B; the genes encoding them *are* induced by drought and are *not* turned on at low temperature (Liu *et al.*, 1998). Thus the CRT/DRE is a point of integration of two distinct signal pathways, and this explains some of the overlap between the cold and drought responses.

There are also overlaps with responses to other stimuli. For example, some CRT/DRE-containing promoters (e.g. the promoter of *COR78*) also contain multiple copies of the "ABRE" (abscisic acid response element, aYACGTGgc) motif, and so are also activated when abscisic acid (ABA) levels rise (Shinozaki and Yamaguchi-Shinozaki, 1997). This is consistent with the observation that treatment with exogenous ABA can induce at least partial freezing tolerance in a variety of species (Robertson *et al.*, 1994; Wilen *et al.*, 1994; Churchill *et al.*, 1998), including in *Arabidopsis* (Mäntylä *et al.*, 1995). The *abi1* (abscisic acid insensitive) mutation, unlike *abi2* or *abi3*, blocks ABA-induced transcription of ABRE-containing promoters

(Gilmour and Thomashow, 1991), and correspondingly prevents ABA from inducing freezing tolerance (Mäntylä *et al.*, 1995).

Mutations of *Arabidopsis* that block ABA biosynthesis also depress freezing tolerance (Heino *et al.*, 1990; Gilmour and Thomashow, 1991; Mäntylä *et al.*, 1995), which might seem to suggest that ABA plays an indispensable role in cold acclimation. But such mutant plants lack vigour and their cells are more leaky than those of wild type: therefore they are compromised before freezing occurs, so that the meaning of their freezing-sensitivity is dubious. When induction of ABRE-containing promoters is prevented by the *abi1* mutation, low temperatures can still induce freezing tolerance (Gilmour and Thomashow, 1991), but its development may be delayed (Mäntylä *et al.*, 1995). This suggests that ABA, which is produced transiently in response to cold, may well serve to accelerate the development of freezing tolerance in the normal cold response. The *abi1* phenotype does not rule out ABA as an essential component of the cold response, since some ABA-responsive promoters depend on different, non-ABRE-recognising transcriptional activators (Abe *et al.*, 1997). However, ABA is not likely to have a major role in cold acclimation, not only because cold-induction of many genes occurs independently of ABA, but also because ABA levels rise only transiently (Lang *et al.*, 1994) whereas freezing tolerance is maintained indefinitely for as long as the plant experiences daily low temperatures (Wanner and Junttila, 1999).

A gene fusion between the promoter of *COR78* and *LUC*, a luciferase gene, has been used to examine the interaction of effects of the various inducing stimuli (Xiong *et al.*, 1999b). *COR78* was known to be inducible by multiple stresses (see Table 1) and CRT/DRE and ABRE sequences are recognisable in its promoter. *In vivo* expression of the *LUC* gene was estimated by measurement of the resulting luminescence. Cold and ABA had additive effects, which is consistent with the knowledge that these stimuli operate through signaling pathways that are independent all the way from perception to recognition of the target promoters. In contrast, the effects of cold and osmotic stress were *not* additive when applied simultaneously: induction was typical of cold alone, and less than due to osmotic stress alone. This is consistent with models in which the two stimuli share part of their signaling pathway - but it might instead be due to components of the osmotic stress pathway becoming less active at low temperature. In the control experiments, cold-induced expression from the *COR78* promoter was not detected until much later than has been reported by other investigators (Horvath *et al.*, 1993; Nordin *et al.*, 1993; Knight *et al.*, 1999).

Not all cold-inducible promoters of *Arabidopsis* contain a CRT/DRE - for example, the promoters of *CBF1-3*, *P5CS1*, and *RCI1A*. Gilmour and coworkers (Gilmour *et al.*, 1998) have proposed that such genes may be part of a higher-level regulon, controlled by a hypothetical "ICE" (inducer of cold expression) transcription factor. We can expect to see rapid progress in the identification of the "ICE box" (a phrase that, perhaps not coincidentally, is also an American term for refrigerator) and the isolation of transcription factors that recognise it.

Mutations affecting the cold response

The *sfr6* mutant was isolated on the basis of its freezing sensitivity (see *Genetics of freezing tolerance*, below), but has subsequently been found to inhibit the induction of some cold-inducible genes (Knight *et al.*, 1999). The recessiveness of *sfr6* implies that it is a loss-of-function mutation; therefore it appears that the wild type *SFR6* gene is essential for normal induction of CRT/DRE promoters. The *sfr6* mutation does not affect calcium-signaling (Knight *et al.*, 1999); neither does it prevent the cold-inducible expression of the *CBF* genes or of the other non-CRT/DRE gene that was tested (*P5CS1*). It was surprising that *sfr6* did not affect expression of the CBF transcription factors, whose binding to CRT/DRE promoters might have seemed to be the last step necessary for their activation.

sfr6 cannot represent a mutation in any of the *CBF* genes, since it maps to a different location (Knight *et al.*, 1999). An additional layer of regulation of CRT/DRE promoters had not been expected and there is no prior information to suggest how it might operate. The most plausible hypotheses are that SFR6 is a cofactor or adaptor for the CBF transcriptional activators, or that it is required for their maturation to a functional form, for example by controlling their phosphorylation. Either class of possibility would allow the SFR6 product to function in other pathways, consistent with the observation of pleiotropic effects of the *sfr6* mutation at normal growth temperature. The cofactor/adaptor hypothesis is perhaps slightly favoured by the observation that *sfr6* also inhibits ABA-induced and drought-induced expression from the *COR78* promoter (Knight *et al.*, 1999), in which two other transcriptional activators are involved.

The luciferase reporter system has been utilised to allow the identification of mutants in which expression from the *COR78* promoter is altered (Ishitani *et al.*, 1997). The screen allowed testing of large numbers of seedlings for constitutive expression (in the absence of stress) and then sequentially for altered expression in response to cold, ABA, and osmotic stress (created by sodium chloride). From the screening of luminescence in 300,000 M_2 plants, 103 mutants with strong phenotypes were selected for further analysis. Classified firstly by the level and responsiveness of reporter expression, these were called "*cos*" (showing constitutive expression of the reporter in the absence of stress), "*los*" (low expression in response to stress), and "*hos*" (high expression in response to stress).

The three mutant types were subclassified according to their responses to the various stimuli (cold, ABA, and NaCl). hos_{all} and los_{all} showed equivalent alterations in their expression to all three stresses, whereas hos_{cold} and los_{cold} had perturbations to expression only in response to cold. $hos_{cold/ABA}$ and $los_{cold/ABA}$ displayed altered expression in response to both cold and ABA, but *not* NaCl; $hos_{cold/NaCl}$ and $los_{cold/NaCl}$ showed altered expression after cold and osmotic stresses, but *not* in response to ABA (Ishitani *et al.*, 1997).

The phenotypes of hos_{cold} and los_{cold} mutations can be accommodated within existing models if they are assumed to be mutations in negative and positive regulators, respectively, of cold signal transduction up to the stage of CRT/DRE activation by the CBF family. The $hos_{cold/NaCl}$ and $los_{cold/NaCl}$ classes of mutants can also be rationalised simply: since both stresses lead to activation of the CRT/DRE, an interaction of control factors at that element is plausible. The other mutant classes are not easily accounted for by extrapolation from current knowledge.

Like *sfr6*, the mutants showing hos_{all} and los_{all} phenotypes imply the existence of a factor in common between ABA-induced, cold-induced, *and* drought-induced activation of *COR78*. This is unexpected, given that cold/drought and ABA act via different enhancer sequences, and that each stimulus can induce expression independently of the others. The authors suggested that there may be a set of transcription factors that interact with both the CRT/DRE and the ABRE elements. A transcriptional adaptor complex could fit this description (interacting with the promoter motifs indirectly, via its contacts with the cognate transcriptional activators); in yeast, for example, the SAGA transcriptional adaptor complex is involved in transcription from a small subset (approx. 5%) of yeast promoters (Holstege *et al.,* 1998). It is notable that genes exist in *Arabidopsis* which encode homologs of yeast adaptor complex components.

The existence of $hos_{cold/ABA}$ and $los_{cold/ABA}$ mutants implies that there is a factor involved in the cold and ABA signaling pathways but not the dehydration-signaling pathway. Such a commonality is not expected on the basis of genes' inducibility - whereas numerous genes are induced by all three stimuli, only one (*COR15B*) is known to be induced by both cold and ABA and yet not by drought (see Table 1). As was true for the hos_{all} and los_{all} mutants, the common factor can be hypothesised to be a transcriptional adaptor, this time postulating interactions with just two of the three transcriptional activators. Such hypotheses do not conflict with the observation that each stimulus alone is capable of inducing expression.

Two mutations of the hos_{cold} class, *hos1* and *hos2*, hypothesised to inactivate negative regulators of cold signal transduction, have been investigated further (Ishitani *et al.,* 1998; Lee *et al.,* 1999; Xiong *et al.,*

1999a). As well as the native *COR78* gene, several other cold induced genes including *ADH* were tested in *hos1* and *hos2* and all showed superinduction in response to cold. The *ADH* gene is notable because, alone among those tested, its promoter does not contain a recognisable CRT/DRE, suggesting that these lesions in cold-signaling lie in a pathway common to induction of CRT/DRE and other types of cold-inducible promoters.

Interestingly, the osmotic stress- and ABA-induced expression of several genes (though not *COR78*) was diminished in *hos1* (Ishitani *et al.,* 1998). The *hos1* mutant also flowered very early and the authors interpreted this as constitutive vernalisation. However, the *hos1* mutant appears to be under stress at normal temperatures - it is more freezing-sensitive than wild type *before* cold acclimation - which is sufficient to account for its early flowering without invoking a specific molecular interaction with the vernalisation pathway. The *hos2* mutant did not show either early flowering (Lee *et al.,* 1999) or a significant difference from wild type in its nonacclimated freezing sensitivity (my interpretation of the published results, which differs from the authors') after growth at normal temperature (Lee *et al.,* 1999).

In the *los*cold mutation, *los1*, *COR78-LUC* ceased to be inducible by the other stimuli when plants were chilled for 48 h (Xiong *et al.,* 1999b). This might mean that *los1* has a temperature-sensitive lesion in signaling, rather than a defect in the cold-signal transduction pathway.

Negative selection against expression of the *ADH* gene was used to isolate mutants which failed to induce *ADH* during germination (Conley *et al.,* 1999); the *aar1* and *aar2* mutants were also defective in the induction of both *ADH* and *PDC1* by cold, drought and anoxia. The inducibility of other genes has not yet been examined. Promoter analysis of *ADH* had previously led to the conclusion that its induction by anoxia was conferred by different cis-acting elements from those activated by the cold and drought stimuli. Thus the *aar* mutants mirror *sfr6* and a number of the *hos* and *los* mutants in the breadth of their effects, and the possibility of explaining them by transcriptional adaptors - a hypothesis that seems in danger of becoming overworked.

GENETICS OF FREEZING TOLERANCE

Natural genetic variation in freezing tolerance

Quantitative genetics has enabled the identification of loci showing allelic variation for freezing tolerance in crop species - in spite of the probable incorrectness of the assumption of additive effects. In wheat, a single locus (*Fr1*) has been found to show greater natural allelic variation in freezing tolerance than all others combined (Galiba *et al.*, 1995; Storlie *et al.*, 1998), when spring and winter varieties are crossed. Intriguingly, the degree of inducibility of a family of cold-induced genes has also been mapped to wheat chromosome 5A by the use of chromosome substitution lines (Limin *et al.*, 1997).

In barley, quantitative analysis of crosses between winter and spring types has similarly mapped the locus which makes the greatest contribution to natural variability in freezing tolerance, in this case to a region of barley chromosome 7 (Hayes *et al.*, 1993; Pan *et al.*, 1994; Van Zee *et al.*, 1995). Its relative contribution was estimated to be considerably less than that of *Fr1* in wheat, but this does not imply that the gene or genes at the identified locus are less potent in conferring freezing tolerance. For example, the observed result might happen because of a smaller allelic difference at the barley locus, or because of greater variability at other loci, or more subtly, because barley might lack the cofactors necessary for the gene to achieve its full potential. It is intriguing that the identified region of barley chromosome 7 is homologous to the portion of whcat chromosomc 5A that contains *Fr1* (see discussion in Thomashow (1999)), so it is conceivable that the same gene bears major responsibility for natural variation in freezing tolerance in both cereals.

Natural variation in freezing tolerance has also been mapped as a quantitative trait in *Brassica rapa* (Teutonico *et al.*, 1995). In the future it may become possible to select and test candidate genes in *Brassica*, based on its degree of synteny with *Arabidopsis* (Teutonico and Osborn, 1994).

The identification of loci contributing to natural variation in freezing tolerance is of obvious benefit to breeding programs. The identity of the genes at such loci would of course also be of considerable interest to fundamental understanding. But gene identification is not technically straightforward because the incomplete penetrance of quantitative trait loci greatly hinders fine mapping. However, even in theory the study of such natural variants is unlikely to supply a comprehensive understanding of freezing tolerance, because most of the genes contributing to freezing tolerance probably will not show allelic variation within natural populations.

Induced mutations affecting freezing tolerance

Some cold-inducible genes may not be detectable by most techniques of differential-expression screening (Hughes and Dunn, 1996); it may even be that some genes essential for freezing tolerance are constitutive rather than cold-inducible. Moreover, as noted above (*Cold-induced genes and proteins*), reverse genetic approaches have not been informative (although their attraction is being revitalised by the technology for identifying a rare insertion mutation in any gene of known sequence). These considerations have prompted the pursuit of a classical genetic approach: the isolation and study of mutations that affect freezing tolerance in *Arabidopsis*.

Mutations causing *deficiency* in freezing tolerance would be expected to identify non-redundant genes that are essential (though not necessarily sufficient) for hardiness. On the other hand such mutations would not show up in genes with redundant functions - for example, gene families with equivalent expression patterns among the members. Mutants deficient in freezing tolerance were isolated (Warren *et al.*, 1996; Warren *et al.*, 1997; Thorlby *et al.*, 1999) after eliminating those mutants that were compromised by chilling-sensitivity during the process of cold acclimation. The "*sfr*" (sensitive to freezing) mutations were identified in nine genes. Eight of these were represented by single mutant alleles: therefore the screen was far from saturating the genome with such mutations. The mutations have been mapped against cold-induced genes (and others likely to affect freezing tolerance, e.g. abscisic acid biosynthesis and perception genes) (Knight *et al.*, 1999; Thorlby *et al.*, 1999). Although this has greatly constrained the possibilities, it is too early to draw conclusions about the identity of *SFR* genes with other known genes: possible identities must be confirmed by other methods (e.g. sequencing or complementation).

In two of the *sfr* mutants, biochemical phenotypes have been observed that appear to be sufficient to account for their freezing-sensitivity. (i) The *sfr4* mutation causes depletion (in place of the normal accumulation) of sugars during cold acclimation (McKown *et al.*, 1996), depriving the mutant of the most important of compatible solutes, sucrose. (ii) The *sfr6* mutation, as noted above, prevents activation of CRT/DRE promoters and thus synthesis of the hydrophilic cold-induced proteins. Mutants in the other seven *SFR* genes do not have biochemical phenotypes that predict (or can be simply rationalised to explain) their freezing tolerance: thus we may expect to learn something new when the mechanisms of these genes' contribution to freezing tolerance are elucidated.

The gene expression phenotype of the *hos2* mutant is (by definition of the *hos* class) apparently the converse of *sfr6* - over-responsiveness of expression (see *Control of cold-induced gene expression: implications of mutants*, above). We might expect correspondingly greater tolerance but, far from it, *hos2* impairs freezing tolerance following cold acclimation (Lee *et al.*, 1999). If not for a convincingly contradictory result (see *Manipulating regulons with transcription factors*, below) it would be tempting to conclude that the expression of cold-induced proteins might be necessary but was not sufficient to provide freezing tolerance. In the *hos2* mutant, two frequently correlated parameters (expression and tolerance) are perturbed in opposite directions. This is reminiscent of interference with a pathway that is subject to feedback control.

Mutations causing an *elevation* in freezing tolerance were isolated by freezing selection of seedlings that had not undergone cold-acclimation (Xin and Browse, 1998). They might be expected to elucidate cold acclimation by demonstrating constitutive activity of its functional components. The first to be characterised, *esk1*, has proline and sugar levels elevated beyond what is normally seen during cold acclimation; most of its cold-inducible genes, by contrast, are normally-responsive and are *not* constitutively expressed. This suggests that compatible osmolytes alone can raise the level of freezing tolerance. There is an apparent paradox with the *sfr* mutants - all but one of which show normally-elevated sugar levels after cold acclimation, but have freezing tolerance limited by deficiencies in individual proteins. Two possible solutions to the paradox are suggested: (i) The proteins that are deficient in *sfr* mutants are not cold-inducible - and hence are already present in the nonacclimated *esk1* mutant. Alternatively, (ii) compatible osmolytes and cold-inducible proteins may contribute independently to freezing tolerance of the cold-acclimated plant, so that production of compatible osmolytes alone would lead to levels of freezing tolerance intermediate between the acclimated and non-acclimated state. To an approximation, intermediate levels are indeed observed in both the nonacclimated *esk1* mutant and in the cold-acclimated *sfr* mutants. When *esk1* plants are subjected to cold acclimation, their freezing tolerance increases further, surpassing that of the cold-acclimated wild type. This is consistent with the idea that compatible osmolytes can boost freezing tolerance independently of other mechanisms.

Manipulating the CRT/DRE regulon

Since the CBF transcription factors were understood to cause cold-induced transcription by virtue of themselves being produced in response to cold, it followed that their constitutive expression should cause constitutive expression of the CRT/DRE regulon - the set of genes controlled by the CRT/DRE promoter motif. Transgenic *Arabidopsis* plants were constructed that constitutively expressed CBF1 (Stockinger *et al.*, 1997), and constitutive expression of the downstream, normally cold-inducible, genes was confirmed. The transgenic plants thus mimicked cold acclimated plants in one respect. Remarkably, they also showed freezing-tolerance: in the absence of cold acclimation, one of the *CBF1* transgenic lines approached the level of freezing tolerance seen in cold-acclimated wild-type plants. Here was clear evidence that cold-inducible proteins were not only necessary but sufficient for freezing tolerance at the whole-cell and whole-plant level.

An independent group subsequently reported the constitutive expression of *CBF3* (a homolog of *CBF1*, and called by its synonym *DREB1A* in their work), in *Arabidopsis* (Liu *et al.*, 1998). They also saw constitutive freezing tolerance, but whereas Thomashow's group had not reported growth effects, Shinozaki's group observed that all such transgenic plants were dwarf: they appeared to be paying a growth penalty for their freezing tolerance. The cause of this difference - one of obvious significance for the engineering of freezing tolerance - was not apparent; it might be ascribed to differences in the level or timing of expression, or to a subtle difference between the target specificity of the two closely-related transcription factors. The earlier result - constitutive tolerance without a growth penalty - is the more surprising and its replication, especially in other species, is eagerly anticipated.

The *CBF3* gene has more recently been expressed under the control of a CRT/DRE-containing promoter, producing a positive feedback in transcriptional activation (Kasuga *et al.,* 1999). As intended, expression of the CRT/DRE regulon was hyper-responsive to cold or osmotic stress, without high levels of constitutive transcription from CRT/DRE-containing promoters. Although most such plants exhibited slight growth retardation under control conditions, it was less than seen previously. The freezing tolerance of these plants was reported to be high without prior cold acclimation. Presumably this must mean that the plants were cooled slowly enough to allow induced expression to take place - in other words, the plants received a brief period of "just-in-time" cold acclimation that sufficed for protein production, given their hyper-responsiveness. It is nonetheless remarkable that this approach was successful, because (i) in a positive feedback system there is only a narrow band of gain within which significant amplification can be achieved before a "permanently on" state results, and (ii) a high degree of amplification would seem necessary in order to compress the acclimation of days into at most several hours.

Manipulating individual traits

A number of attempts have been made at elevating the levels of compatible osmolytes. In tobacco (*Nicotiana tabacum*), introduction of genes encoding various biosynthetic enzymes has succeeded in generating transgenic lines that accumulate trehalose (Holmstrom *et al.,* 1996), proline (Kishor *et al.,* 1995), fructans (Pilon-Smits *et al.,* 1995), mannitol (Tarczynski *et al.,* 1993) or glycine betaine (Holmstrom *et al.,* 2000). These studies demonstrate the possibility for such manipulation but they were employing tobacco as a model system for *drought* stress; effects on freezing tolerance were not reported. Tobacco does not acclimate and is highly freezing-sensitive, so it would have been an interesting but very demanding system in which to test the potential of these osmolytes for raising freezing tolerance. However, two studies have deliberately manipulated compatible osmolyte levels in *Arabidopsis* and assessed their effect on freezing tolerance.

Glycine betaine was synthesised in *Arabidopsis* by introduction of a choline oxidase gene, which catalyses the conversion of choline into glycine betaine. Glycine betaine is well-established as a compatible osmolyte (Gorham, 1995) that is present naturally in some plants, but not in *Arabidopsis*. The transgenic lines accumulated glycine betaine while the expression pattern of their cold-inducible genes remained normal (Sakamoto *et al.*, 2000). In the absence of cold acclimation, these plants showed a greater tolerance to freezing than the wild type - a result comparable to that seen in the proline-accumulating *esk1* mutant. After cold acclimation, freezing tolerance was indistinguishable from that of wild type. This is in marked contrast to the situation in *esk1* mutant plants: it seems to suggest that, unlike proline, the freezing tolerance mechanism of glycine betaine can not work additively with the normal tolerance mechanisms of cold acclimation.

Proline levels in *Arabidopsis* were elevated by a novel approach: antisense suppression of the degradative enzyme proline oxidase (Nanjo *et al.*, 1999). This raised the freezing tolerance of nonacclimated plants, but to a level below that of the *CBF3*-hyperinducing plants described in the previous section. Freezing tolerance after cold acclimation was not assessed.

Several groups have constructed transgenic plant lines expressing antifreeze proteins. In two of these studies, antifreeze activity was detected in the apoplast (by means of recrystallisation inhibition or thermal hysteresis) but no effect could be discerned on organismal or tissue freezing tolerance (Hightower *et al.*, 1991; Kenward *et al.*, 1999). Both studies employed solanaceous species with very little intrinsic freezing tolerance, so it may be argued that the influence of antifreeze was tested in unfavourable circumstances, i.e., in the absence of other freezing tolerance mechanisms. This begs the question of how tolerance mechanisms may be expected to interact, and underlines the fallacy that a tolerance mechanism can necessarily be measured in terms of a differential in survival temperatures or a percent contribution to survival. A third study detected an enhancement in tissue freezing tolerance (Wallis *et al.*, 1997), but suffered the shortcoming of being based on comparisons between a small number of primary transformants and an untransformed control: the likelihood of interference from tissue culture effects undermines the inference of a protective effect.

Transgenic overexpression of a hydrophilic chloroplast protein, COR15a, has been found to cryoprotect chloroplasts *in vivo* and *in vitro* (Artus *et al.*, 1996). A protective effect on isolated protoplasts (though not at the level of whole plants or even intact cells) was also observed. These results were exciting and prompted research into the lesion which COR15a prevents or protects against. This appears to be the lamellar to hexagonal$_{II}$ phase transition, a lesion which the investigators suggest may be nucleated when multiple bilayers are appressed by freeze-induced dehydration (Steponkus *et al.*, 1998).

It is reasonable to suppose that there have been many unreported failures to detect any plant protection following individual overexpression of cold-inducible proteins. Such failures would be explicable if there must be co-operative action of different cold-induced proteins in order to produce freezing tolerance. (A framework for how they may interact is elaborated in the next section). However, one would still expect the complementary approach, of preventing normal cold-induction by antisense or cosuppression methods, to be able to confirm the proteins' contribution to freezing tolerance. Such an approach has been technically feasible for ten years yet there has, to date, been no convincing demonstration that any individual cold-induced gene is essential for freezing tolerance.

DISCUSSION

How do tolerance mechanisms interact?

In discussing the implications of much experimental work, it would have been highly desirable to know how tolerance mechanisms interact - to understand whether the action of one tolerance mechanism, or component thereof, would add to, be synergistic with, or possibly be redundant with another. At present such knowledge does not exist, but we can constrain the possibilities from what is known about freezing stress and freezing tolerance mechanisms.

Freezing stress is actually a combination of multiple potentially lethal effects (lesions). Freezing tolerance results from the combined action of whatever tolerance mechanisms prevent or mitigate, collectively, the entire set of potential lesions. This lesion-based view of freezing tolerance has been championed by Steponkus (Steponkus, 1984). Its premises are uncontroversial, yet it debunks the common assumption that effects on tolerance combine additively - why should the prevention of one lesion necessarily affect every other?

It is possible to construct a quantitative framework for freezing tolerance in place of the simple assumption of additivity. Each lesion will have a temperature threshold below which the severity of that type of damage becomes lethal. On the simplest interpretation, which neglects synergy between lesions, the highest of these thresholds will be the temperature limit at which freezing becomes lethal (Warren, 1998) - the highest threshold will thus determine the level of freezing tolerance. The crucial question is that of how multiple mechanisms of freezing tolerance will affect the temperature thresholds of different lesions.

Figure 2 illustrates two extremes. A tolerance mechanism that is *specific* to one type of lesion may, acting alone, have no effect on organismal freezing tolerance (comparing ***b*** with the nonacclimated situation at ***a***). I suggest that this corresponds to the failure of transgenics expressing COR15a or antifreeze to show improved tolerance. However, a combination of such mechanisms will be able to shift the highest threshold downward (***c***), and thus increase the level of freezing tolerance: this may represent the situation in plants overexpressing the CBF regulon. The opposite extreme would be a tolerance mechanism that is *general* to all lesions: this would necessarily improve tolerance whenever it operated (***d***). The elevation of compatible osmolytes seems likely to be a mechanism of general tolerance for all lesions caused by freeze-induced dehydration. Our understanding of protection by compatible osmolytes suggests that they should be nonspecific, and in practise the manipulation of their levels usually leads to increased freezing and drought tolerance.

The combined effects of a specific and a general mechanism acting on the same lesion might or might not be additive. Additivity is illustrated in Figure 2 (***e***), the outcome being a greater level of tolerance. Additivity might be expected between the effects of compatible osmolytes, which reduce the concentration of cryotoxic solutes at any given temperature, and a specific mechanism, which prevents a particular kind of injury by cryotoxic solutes. This broadly agrees with the effect of combining acclimation (causing induction of the CBF regulon) with the high osmolyte content of the *esk1* mutant, and also with the incremental effects of increasing sucrose levels on the freezing tolerance of cold-acclimated *Arabidopsis* (Wanner and Junttila, 1999). On the other hand, glycine betaine did not appear to act additively with the other mechanisms embodied in normal cold acclimation (see above, *Manipulating individual traits*). However, that inference can not be definitive because it was not established, for example, that glycine betaine had not interfered with the accumulation of other compatible solutes.

Because plants engineered to overexpress the CBF regulon do not show elevated levels of compatible osmolytes, their increased freezing tolerance seems attributable to the combined action of multiple lesion-specific tolerance mechanisms. However, two other results are in apparent conflict with this interpretation. (i) When wild type plants are cold-treated in the dark, they induce the CBF regulon normally but show no improvement in freezing tolerance (Wanner and Junttila, 1999). (ii) When *hos2* plants are cold-acclimated, they overexpress the CBF regulon and yet acquire less freezing tolerance than the wild type (Lee *et al.*, 1999). Crucially, we do not know the behaviour of compatible osmolyte levels in either case, although they seem likely to have fallen from their pre-acclimation levels in the light-deprived plants. Without this knowledge, it may be premature to draw conclusions from these apparently contradictory experiments.

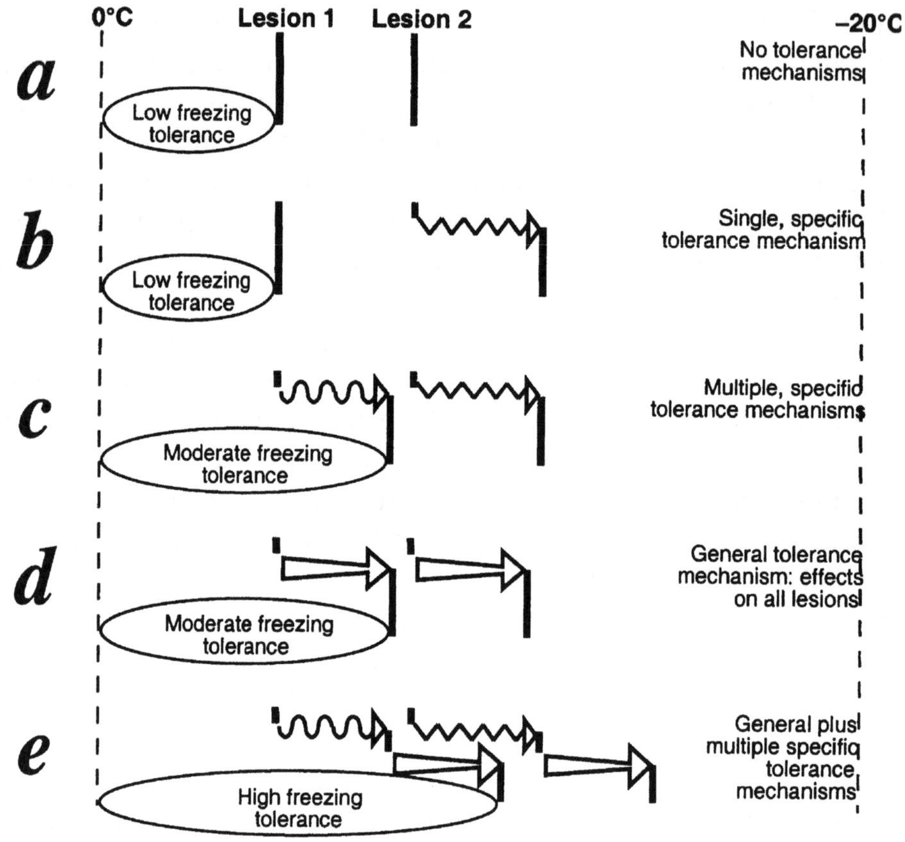

Figure 2. A framework for the interaction of multiple tolerance mechanisms. Temperature thresholds for two imaginary lesions (vertical lines) are moved rightward (toward lower temperatures) by tolerance mechanisms (rightward-pointing arrows). The tapered arrows represent the effects of a general tolerance mechanism on the thresholds of both lesions.

Outlook

We do not have clear evidence to indicate which types of injury are pre-empted by most of the cold-induced proteins; in fact, for most we have no evidence that they even contribute to freezing tolerance. Genetic and reverse-genetic approaches are poised to provide such evidence in the near future.

The ability to simulate cold acclimation by expression of transcription factors immediately suggests applications in crop improvement. Prospects are excellent for controlling the timing and tissue-specificity of freezing-tolerance in hardy species. This would have application in crops that are subject to frost during seasons in which freezing-tolerance is not naturally induced. A second application would be in enabling plant parts produced in summer to be shipped under mild freezing temperatures. The storage life and so the distribution range of fresh vegetables is highly temperature-dependent, and thus an ability to withstand mild freezing would add post-harvest value.

The prospects for conferring tolerance on non-hardy species by means of transcriptional activators are less clear. It will work only if non-hardy plants possess the appropriate structural genes to prevent freezing injury, under the control of appropriately responsive promoters. There are sharply contrasting views on the likelihood of this situation. Analysis of cold-induced genes and their promoters has explained how drought and/or abscisic acid (ABA) can induce hardy species to become freezing-tolerant by inducing substantially overlapping sets of genes. However, dehydration and ABA do not produce freezing tolerance in non-hardy plants. This indicates that non-hardy plants may lack more than an equivalent of CBF1 protein.

Rye is much more freezing-tolerant than *Arabidopsis*, surviving temperatures 15°C lower. It is implausible that rye merely has a better transcription factor for cold-inducible expression. Differences in the cold-induced proteins themselves provide a more credible explanation for its superiority in hardiness. Again, this argues that an understanding at the level of individual lesions and tolerance mechanisms will be necessary for inter-species transfer of freezing tolerance.

In the next few years, genetics and molecular biology will contribute mutants, proteins, antibodies and transgenic lines to this field. It is conceivable that reporter systems will be developed to study particular facets of freezing pathology (for example, levels of oxidative stress). When employed in rigorous biochemical and physiological experiments, such tools must yield a detailed molecular understanding of how freezing injury is prevented or repaired in hardy plants.

REFERENCES

Abe, H., Yamaguchi-Shinozaki, K., Urao, T., Iwasaki, T., Hosokawa, D. and Shinozaki, K. 1997. Role of *Arabidopsis* MYC and MYB homologs in drought- and abscisic acid-regulated gene expression. *Plant Cell* 9, 1859-1868.

Artus, N.N., Uemura, M., Steponkus, P.L., Gilmour, S.J., Lin, C. and Thomashow, M.F. 1996. Constitutive expression of the cold-regulated *Arabidopsis thaliana COR15a* gene affects both chloroplast and protoplast freezing tolerance. *Proc. Natl. Acad. Sci. USA* 93, 13404-13409.

Baker, S.S., Wilhelm, K.S. and Thomashow, M.F. 1994. The 5'-region of *Arabidopsis thaliana cor15a* has cis-acting elements that confer cold-, drought- and ABA-regulated gene expression. *Plant Mol. Biol.* 24, 701-713.

Berberich, T. and Kusano, T. 1997. Cycloheximide induces a subset of low temperature-inducible genes in maize. *Mol. Gen. Genet.* 254, 275-283.

Bilodeau, P., Udvardi, M.K., Peacock, W.J. and Dennis, E.S. 1999. A prolonged cold treatment-induced cytochrome P450 gene from *Arabidopsis thaliana*. *Plant Cell Envir.* 22, 791-800.

Boothe, J.G., De Beus, M.D. and Johnson-Flanagan, A.M. 1995. Expression of a low-temperature-induced protein in *Brassica napus*. *Plant Physiol.* 108, 795-803.

Boothe, J.G., Sonnichsen, F.D., de Beus, M.D. and Johnson-Flanagan, A.M. 1997. Purification, characterization, and structural analysis of a plant low-temperature-induced protein. *Plant Physiol.* 113, 367-376.

Bridger, G.M., Yang, W., Falk, D.E. and McKersie, B.D. 1994. Cold acclimation increases tolerance of activated oxygen in winter cereals. *J. Plant Physiol.* 144, 235-240.

Capel, J., Jarillo, J., Salinas, J. and Martinez-Zapater, J. 1997. Two homologous low-temperature-inducible genes from *Arabidopsis* encode highly hydrophobic proteins. *Plant Physiol.* 115, 569-576.

Carpenter, C.D., Kreps, J.A. and Simon, A.E. 1994. Genes encoding glycine-rich *Arabidopsis thaliana* proteins with RNA-binding motifs are influenced by cold treatment and an endogenous circadian rhythm. *Plant Physiol.* 104, 1015-1025.

Carpenter, J.F. and Crowe, J.H. 1988. The mechanism of cryoprotection of proteins by solutes. *Cryobiology* 25, 244-255.

Christie, P.J., Alfenito, M.R. and Walbot, V. 1994. Impact of low-temperature stress on general phenylpropanoid and anthocyanin pathways: enhancement of transcript abundance and anthocyanin pigmentation in maize seedlings. *Planta* 194, 541-549.

Christie, P.J., Hahn, M. and Walbot, V. 1991. Low-temperature accumulation of alcohol dehydrogenase-1 mRNA and protein activity in maize and rice seedlings. *Plant Physiol.* 95, 699-706.

Chun, J.U., Yu, X.M. and Griffith, M. 1998. Genetic studies of antifreeze proteins and their correlation with winter survival in wheat. *Euphytica* 102, 219-226.

Churchill, G.C., Reaney, M.J.T., Abrams, S.R. and Gusta, L.V. 1998. Effects of abscisic acid and abscisic acid analogs on the induction of freezing tolerance of winter rye (*Secale cereale* L.) seedlings. *Plant Growth Reg.* 25, 35-45.

Close, T.J. 1996. Dehydrins: emergence of a biochemical role of a family of plant dehydration proteins. *Physiol. Plant.* 97, 795-803.

Conley, T.R., Peng, H.P. and Shih, M.C. 1999. Mutations affecting induction of glycolytic and fermentative genes during germination and environmental stresses in *Arabidopsis*. *Plant Physiol.* 119, 599-607.

Cossins, A.R. 1994. "Homeoviscous adaptation of biological membranes and its functional significance." In: *Temperature Adaptation of Biological Membranes*, ed. A.R. Cossins, pp. 63-76. Portland Press, London.

Crowe, J.H., Hoekstra, F.A. and Crowe, L.M. 1992. Anhydrobiosis. *Ann. Rev. Physiol.* 54, 570-599.

Cudd, A. and Steponkus, P.L. 1988. Lamellar-to-hexagonal$_{II}$ phase transitions in liposomes of rye plasma membrane lipids after osmotic dehydration. *Biochim. Biophys. Acta* 941, 278-286.

de Bruxelles, G.L., Peacock, W.J., Dennis, E.S. and Dolferus, R. 1996. Abscisic acid induces the alcohol dehydrogenase gene in *Arabidopsis*. *Plant Physiol.* 111, 381-391.

De Nisi, P. and Zocchi, G. 1996. The role of calcium in the cold shock responses. *Plant Sci.* 121, 161-166.

Delauney, A.J. and Verma, D.P.S. 1993. Proline biosynthesis and osmoregulation in plants. *Plant J.* 4, 215-223.

Dolferus, R., Jacobs, M., Peacock, W.J. and Dennis, E.S. 1994. Differential interactions of promoter elements in stress responses of the *Arabidopsis ADH* gene. *Plant Physiol.* 105, 1075-1087.

Doucet, C., Byass, L., Elias, L., Worrall, D., Smallwood, M. and Bowles, D.J. 2000. Distribution and characterization of recrystallization inhibitor activity in plant and lichen species from the UK and maritime Antarctic. *Cryobiology* 40, 218-227.

Dowgert, M.F., Wolfe, J. and Steponkus, P.L. 1987. The mechanics of injury to isolated protoplasts following osmotic contraction and expansion. *Plant Physiol.* 83, 1001-1007.

Duman, J.G. and Olsen, T.M. 1993. Thermal hysteresis protein activity in bacteria, fungi, and phylogenetically diverse plants. *Cryobiology* 30, 322-328.

Dunn, M.A., Brown, K., Lightowlers, R. and Hughes, M.A. 1996. A low temperature-responsive gene from barley encodes a protein with single-stranded nucleic acid binding activity which is phosphorylated in vitro. *Plant Mol. Biol.* 30, 947-959.

Dunn, M.A., Goddard, N.J., Zhang, L., Pearce, R.S. and Hughes, M.A. 1994. Low-temperature-responsive barley genes have different control mechanisms. *Plant Mol. Biol.* 24, 879-888.

Dunn, M.A., White, A.J., Vural, S. and Hughes, M.A. 1998. Identification of promoter elements in a low-temperature-responsive gene (*blt*4.9) from barley (*Hordeum vulgare* L.). *Plant Mol. Biol.* 38, 551-564.

Fukuchi-Mizutani, M., Tasaka, Y., Tanaka, Y., Ashikari, T., Kusumi, T. and Murata, N. 1998. Characterization of delta-9 acyl-lipid desaturase homologues from *Arabidopsis thaliana*. *Plant Cell Physiol.* 39, 247-253.

Galiba, G., Quarrie, S.A., Sutka, J., Morounov, A. and Snape, J.W. 1995. RFLP mapping of the vernalization (Vrn1) and frost resistance (Fr1) genes on chromosome 5A of wheat. *Theor. Appl. Genet.* 90, 1174-1179.

Gibson, S., Arondel, V., Iba, K. and Somerville, C. 1994. Cloning of a temperature-regulated gene encoding a chloroplast omega-3 desaturase from *Arabidopsis thaliana*. *Plant Physiol.* 106, 1615-1621.

Gilmour, S.J., Artus, N.N. and Thomashow, M.F. 1992. cDNA Sequence analysis and expression of two cold-regulated genes of *Arabidopsis thaliana*. *Plant Mol. Biol.* 18, 13-22.

Gilmour, S.J. and Thomashow, M.F. 1991. Cold acclimation and cold-regulated gene expression in ABA mutants of *Arabidopsis thaliana*. *Plant Mol. Biol.* 17, 1233-1240.

Gilmour, S.J., Zarka, D.G., Stockinger, E.J., Salazar, M.P., Houghton, J.M. and Thomashow, M.F. 1998. Low temperature regulation of the *Arabidopsis CBF* family of AP2 transcriptional activators as an early step in cold-induced *COR* gene expression. *Plant J.* 16, 433-442.

Goddard, N.J., Dunn, M.A., Zhang, L., White, A.J., Jack, P.L. and Hughes, M.A. 1993. Molecular analysis and spatial expression of a low-temperature-specific barley gene, *blt101*. *Plant Mol. Biol.* 23, 871-879.

Gordon-Kamm, W.J. and Steponkus, P.L. 1984. Lamellar-to-hexagonal(II) phase transitions in the plasma membrane of isolated protoplasts after freeze-induced dehydration. *Proc. Natl. Acad. Sci. USA* 81, 6373-6377.

Gorham, J. 1995. "Betaines in higher plants - biosynthesis and role in stress metabolism." In: *Amino Acids and their Derivatives in Higher Plants*, ed. R.M. Wallsgrove, pp. 171-203. Cambridge University Press, Cambridge.

Graham, D. and Patterson, B.D. 1982. Responses of plants to low, nonfreezing temperatures: proteins, metabolism, and acclimation. *Ann. Rev. Plant Physiol.* 33, 347-372.

Griffith, M., Ala, P., Yang, D.S.C., Hon, W.C. and Moffatt, B.A. 1992. Antifreeze protein produced endogenously in winter rye leaves. *Plant Physiol.* 100, 593-596.

Griffith, M., Antikainen, M., Hon, W.C., Pihakaski Maunsbach, K., Yu, X.M., Chun, J.U. and Yang, D.S.C. 1997. Antifreeze proteins in winter rye. *Physiol. Plant.* 100, 327-332.

Gulick, P.J., Shen, W. and An, H. 1994. *ESI3*, a stress-induced gene from *Lophopyrum elongatum*. *Plant Physiol.* 104, 799-800.

Guy, C., Haskell, D. and Li, Q.B. 1998. Association of proteins with the stress 70 molecular chaperones at low temperature: evidence for the existence of cold labile proteins in spinach. *Cryobiology* 36, 301-314.

Guy, C. and Li, Q.B. 1998. The organization and evolution of the spinach stress 70 molecular chaperone gene family. *Plant Cell* 10, 539-556.

Guy, C.L. 1990. Cold acclimation and freezing stress tolerance: role of protein metabolism. *Ann. Rev. Plant Physiol. Plant Mol. Biol.* 41, 187-223.

Guy, C.L., Niemi, K.J. and Brambl, R. 1985. Altered gene expression during cold acclimation of spinach (*Spinacia oleracea* cultivar Bloomsdale). *Proc. Natl. Acad. Sci. USA* 82, 3673-3677.

Hayden, D.B. and Baker, N.R. 1990. Damage to photosynthetic membranes in chilling-sensitive plants: maize, a case study. *Crit. Rev. Biotechnol.* 9, 321-341.

Hayes, P.M., Blake, T., Chen, T.H.H., Tragoonrung, S. and Chen, F. 1993. Quantitative trait loci on barley (*Hordeum vulgare* L.) chromosome 7 associated with components of winterhardiness. *Genome* 36, 66-71.

Heino, P., Sandman, G., Lang, V., Nordin, K. and Palva, E.T. 1990. Abscisic acid deficiency prevents development of freezing tolerance in *Arabidopsis thaliana* L. Heynh. *Theor. Appl. Genet.* 79, 801-806.

Hightower, R., Baden, C., Penzes, E., Lund, P. and Dunsmuir, P. 1991. Expression of antifreeze proteins in transgenic plants. *Plant Mol. Biol.* 17, 1013-1021.

Hirayama, T., Ohto, C., Mizoguchi, T. and Shinozaki, K. 1995. A gene encoding a phosphatidylinositol-specific phospholipase C is induced by dehydration and salt stress in *Arabidopsis thaliana*. *Proc. Natl. Acad. Sci. USA* 92, 3903-3907.

Hölmstrom, K.O., Mäntylä, E., Welin, B., Mandal, A., Tunnela, O.E., Londesborough, J. and Palva, E.T. 1996. Drought tolerance in tobacco. *Nature* 379, 683-684.

Hölmstrom, K.O., Somersalo, S., Mandal, A., Palva, T.E. and Welin, B. 2000. Improved tolerance to salinity and low temperature in transgenic tobacco producing glycine betaine. *J. Exp. Bot.* 51, 177-185.

Holstege, F.C.P., Jennings, E.G., Wyrick, J.J., Lee, T.I., Hengartner, C.J., Green, M.R., Golub, T.R., Lander, E.S. and Young, R.A. 1998. Dissecting the regulatory circuitry of a eukaryotic genome. *Cell* 95, 717-728.

Hong, S., Jon, J., Kwak, J. and Nam, H. 1997. Identification of a receptor-like protein kinase gene rapidly induced by abscisic acid, dehydration, high salt, and cold treatments in *Arabidopsis thaliana*. *Plant Physiol.* 113, 1203-1212.

Horvath, D.P., McLarney, B.K. and Thomashow, M.F. 1993. Regulation of *Arabidopsis thaliana* L. (Heyn) *cor78* in response to low temperature. *Plant Physiol.* 103, 1047-1053.

Hughes, M.A. and Dunn, M.A. 1996. The molecular biology of plant acclimation to low temperature. *J. Exp. Bot.* 47, 291-305.

Huner, N.P.A., Oquist, G. and Sarhan, F. 1998. Energy balance and acclimation to light and cold. *Trends Plant Sci.* 3, 224-230.

Ichimura, K., Mizoguchi, T., Irie, K., Morris, P., Giraudat, J., Matsumoto, K. and Shinozaki, K. 1998. Isolation of ATMEKK1 (a MAP kinase kinase kinase)-interacting proteins and analysis of a MAP kinase cascade in *Arabidopsis*. *Biochem. Biophys. Res. Comm.* 253, 532-543.

Ingram, J. and Bartels, D. 1996. The molecular basis of dehydration tolerance in plants. *Ann. Rev. Plant Physiol. Plant Mol. Biol.* 47, 377-403.

Ishitani, M., Xiong, L., Lee, B., Stevenson, B. and Zhu, J.K. 1998. *HOS1*, a genetic locus involved in cold-responsive gene expression in *Arabidopsis*. *Plant Cell* 10, 1151-1161.

Ishitani, M., Xiong, L., Stevenson, B. and Zhu, J.K. 1997. Genetic analysis of osmotic and cold stress signal transduction in *Arabidopsis*: interactions and convergence of abscisic acid-dependent and abscisic acid-independent pathways. *Plant Cell* 9, 1935-1949.

Jaglo-Ottosen, K.R., Gilmour, S.J., Zarka, D.G., Schabenberger, O. and Thomashow, M.F. 1998. *Arabidopsis CBF1* overexpression induces *COR* genes and enhances freezing tolerance. *Science* 280, 104-106.

Jarillo, J.A., Capel, J., Leyva, A., Martinez-Zapater, J.M. and Salinas, J. 1994. Two related low-temperature-inducible genes of *Arabidopsis* encode proteins showing high homology to 14-3-3 proteins, a family of putative kinase regulators. *Plant Mol. Biol.* 25, 693-704.

Jarillo, J.A., Leyva, A., Salinas, J. and Martinez Zapater, J.M. 1993. Low temperature induces the accumulation of alcohol dehydrogenase mRNA in *Arabidopsis thaliana*, a chilling-tolerant plant. *Plant Physiol.* 101, 833-837.

Jiang, C., Iu, B. and Singh, J. 1996. Requirement of a CCGAC cis-acting element for cold induction of the *BN115* gene from winter *Brassica napus*. *Plant Mol. Biol.* 30, 679-684.

Jonak, C., Kiegerl, S., Ligterink, W., Barker, P., Huskisson, N.S. and Hirt, H. 1996. Stress signaling in plants: a mitogen-activated protein kinase pathway is activated by cold and drought. *Proc. Natl. Acad. Sci. USA* 93, 11274-11279.

Kanaya, E., Nakajima, N., Morikawa, K., Okada, K. and Shimura, Y. 1999. Characterization of the transcriptional activator CBF1 from *Arabidopsis thaliana*: evidence for cold denaturation in regions outside of the DNA binding domain. *J. Biol. Chem.* 274, 16068-16076.

Kasuga, M., Liu, Q., Yamaguchi-Shinozaki, K. and Shinozaki, K. 1999. Improving plant drought, salt, and freezing tolerance by gene transfer of a single stress-inducible transcription factor. *Nature Biotechnol.* 17, 287-291.

Kawczynski, W. and Dhindsa, R.S. 1996. Alfalfa nuclei contain cold-responsive phosphoproteins and accumulate heat-stable proteins during cold treatment of seedlings. *Plant Cell Physiol.* 37, 1204-1210.

Kaye, C., Neven, L., Hofig, A., Li, Q.B., Haskell, D. and Guy, C. 1998. Characterization of a gene for spinach CAP160 and expression of two spinach cold-acclimation proteins in tobacco. *Plant Physiol.* 116, 1367-1377.

Kenward, K.D., Brandle, J., McPherson, J. and Davies, P.L. 1999. Type II fish antifreeze protein accumulation in transgenic tobacco does not confer frost resistance. *Transgenic Res.* 8, 105-117.

Kishitani, S., Watanabe, K., Yasuda, S., Arakawa, K. and Takabe, T. 1994. Accumulation of glycinebetaine during cold acclimation and freezing tolerance in leaves of winter and spring barley plants. *Plant Cell Envir.* 17, 89-95.

Kishor, P.B., Hong, Z., Miao, G.H., Hu, C.A.A. and Verma, D.P.S. 1995. Overexpression of pyrroline-5-carboxylate synthetase increases proline production and confers osmotolerance in transgenic plants. *Plant Physiol.* 108, 1387-1394.

Kitigawa, Y. and Yoshizaki, K. 1998. Water stress-induced chilling tolerance in rice: putative relationship between chilling tolerance and Ca^{2+} flux. *Plant Sci.* 137, 73-85.

Kiyosue, T., Abe, H., Yamaguchi-Shinozaki, K. and Shinozaki, K. 1998. *ERD6*, a cDNA clone for an early dehydration-induced gene of *Arabidopsis*, encodes a putative sugar transporter. *Biochim. Biophys. Acta* 1370, 187-191.

Kiyosue, T., Yamaguchi-Shinozaki, K. and Shinozaki, K. 1993. Characterization of two cDNAs (*ERD11* and *ERD13*) for dehydration-inducible genes that encode putative glutathione S-transferases in *Arabidopsis thaliana* L. *FEBS Lett.* 335, 189-192.

Kiyosue, T., Yamaguchi-Shinozaki, K. and Shinozaki, K. 1994. Characterization of 2 cDNAs (*ERD10* And *ERD14*) corresponding to genes that respond rapidly to dehydration stress in *Arabidopsis thaliana*. *Plant Cell Physiol.* 35, 225-231.

Kiyosue, T., Yoshiba, Y., Yamaguchi-Shinozaki, K. and Shinozaki, K. A 1996. Nuclear gene encoding mitochondrial proline dehydrogenase, an enzyme involved in proline metabolism, is upregulated by proline but downregulated by dehydration in *Arabidopsis*. *Plant Cell* 8, 1323-1335.

Knight, C.A., Hallett, J. and DeVries, A.L. 1988. Solute effects on ice recrystallization: an assessment technique. *Cryobiology* 25, 55-60.

Knight, H., Trewavas, A.J. and Knight, M.R. 1996. Cold calcium signaling in *Arabidopsis* involves two cellular pools and a change in calcium signature after acclimation. *Plant Cell* 8, 489-503.

Knight, H., Veale, E.L., Warren, G.J. and Knight, M.R. 1999. The *sfr6* mutation in *Arabidopsis* suppresses low-temperature induction of genes dependent on the CRT/DRE sequence motif. *Plant Cell* 11, 875-886.

Knight, M.R., Campbell, A.K., Smith, S.M. and Trewavas, A.J. 1991. Transgenic plant aequorin reports the effects of touch and cold-shock and elicitors on cytoplasmic calcium. *Nature* 352, 524-526.

Kodama, H., Horiguchi, G., Nishiuchi, T., Nishimura, M. and Iba, K. 1995. Fatty acid desaturation chilling acclimation is one of the factors involved in conferring low-temperature tolerance to young tobacco leaves. *Plant Physiol.* 107, 1177-1185.

Kudla, J., Xu, Q., Harter, K., Gruissem, W. and Luan, S. 1999. Genes for calcineurin B-like proteins in *Arabidopsis* are differentially regulated by stress signals. *Proc. Natl. Acad. Sci. USA* 96, 4718-4723.

Kurkela, S. and Borg-Franck, M. 1992. Structure and expression of *kin2*, one of two cold- and ABA-induced genes of *Arabidopsis thaliana. Plant Mol. Biol.* 19, 689-692.

Kurkela, S. and Franck, M. 1990. Cloning and characterization of a cold and ABA-inducible *Arabidopsis* gene. *Plant Mol. Biol.* 15, 137-144.

Kwon, S.W. and Markhart, A.H. 1997. III Fatty acid unsaturation of membrane and linoleate desaturase gene expression during root acclimation to chilling temperature in canola (*Brassica napus* L.). *J. Korean Soc. Hort. Sci.* 38, 43-46.

Lang, V., Mantyla, E., Welin, B., Sundberg, B. and Palva, E.T. 1994. Alterations in water status, endogenous abscisic acid content, and expression of *rab18* gene during the development of freezing tolerance in *Arabidopsis thaliana. Plant Physiol.* 104, 1341-1349.

Lang, V. and Palva, E.T. 1992. The expression of a rab-related gene, *RAB18*, is induced by abscisic acid during the cold acclimation process of *Arabidopsis thaliana* (L.) Heynh. *Plant Mol. Biol.* 20, 951-962.

Lee, H., Xiong, L., Ishitani, M., Stevenson, B. and Zhu, J.K. 1999. Cold-regulated gene expression and freezing tolerance in an *Arabidopsis* thaliana mutant. *Plant J.* 17, 301-308.

Levitt, J. ed. 1980. *Responses of Plants to Environmental Stresses*, Vol. 1. *Chilling, Freezing and High Temperature Stresses*. Academic Press, UK.

Leyva, A., Jarillo, J.A., Salinas, J. and Martinez Zapater, J.M. 1995. Low temperature induces the accumulation of *phenylalanine ammonia-lyase* and *chalcone synthase* mRNAs of *Arabidopsis thaliana* in a light-dependent manner. *Plant Physiol.* 108, 39-46.

Limin, A.E., Danyluk, J., Chauvin, L.P., Fowler, D.B. and Sarhan, F. 1997. Chromosome mapping in low temperature-induced Wcs120 family genes and regulation of cold-tolerance expression in wheat. *Mol. Gen. Genet.* 253, 720-727.

Lin, C., Guo, W.W., Everson, E. and Thomashow, M.F. 1990. Cold acclimation in *Arabidopsis* and wheat. A response associated with expression of related genes encoding 'boiling-stable' polypeptides. *Plant Physiol.* 94, 1078-1083.

Lin, C. and Thomashow, M.F. 1992. DNA sequence analysis of a complementary DNA for cold-regulated *Arabidopsis* gene *COR15* and characterization of the COR15 polypeptide. *Plant Physiol.* 99, 519-525.

Lindow, S.E., Arny, D.C. and Upper, C.D. 1982. Bacterial ice nucleation: a factor in frost injury to plants. *Plant Physiol.* 70, 1084-1089.

Liu, Q., Kasuga, M., Sakuma, Y., Abe, H., Miura, S., Yamaguchi-Shinozaki, Y. and Shinozaki, K. 1998. Two transcription factors, *DREB1* and *DREB2*, with an EREBP/AP2 DNA binding domain separate two cellular signal transduction pathways in drought- and low-temperature-responsive gene expression, respectively, in *Arabidopsis. Plant Cell* 10, 1391-1406.

Llorente, F., Martinez-Zapater, J.M. and Salinas, J. 1998. Molecular characterization of a new gene from *Arabidopsis* encoding for a peroxidase: *Arabidopsis thaliana* peroxidase precursor (RCI3A) mRNA. Genbank locus ATU97684, accession no. U97684.

Lynch, D.V. and Steponkus, P.L. 1987. Plasma membrane lipid alterations associated with cold acclimation of winter rye seedlings *Secale cereale* L. Cultivar Puma. *Plant Physiol.* 83, 761-767.

Lyons, J.M. 1973. Chilling injury in plants. *Ann. Rev. Plant Physiol.* 24, 445-466.

Maia, I.G., Benedetti, C.E., Leite, A., Turcinelli, S.R., Vercesi, A.E. and Arruda, P. 1998. *AtPUMP*: an *Arabidopsis* gene encoding a plant uncoupling mitochondrial protein. *FEBS Lett.* 429, 403-406.

Mäntylä, E., Lång, V. and Palva, E.T. 1995. Role of abscisic acid in drought-induced freezing tolerance, cold acclimation, and accumulation of LTI78 and RAB18 proteins in *Arabidopsis thaliana. Plant Physiol* 107, 141-148.

McKersie, B.D., Chen, Y., DeBeus, M., Bowley, S.R., Bowler, C., Inze, D., D'Halluin, K. and Botterman, J. 1993. Superoxide dismutase enhances tolerance of freezing stress in transgenic alfalfa (*Medicago sativa* L.). *Plant Physiol.* 103, 1155-1163.

McKown, R., Kuroki, G. and Warren, G. 1996. Cold responses of *Arabidopsis* mutants impaired in freezing tolerance. *J. Exp. Bot.* 47, 1919-1925.

Medina, J., Bargues, M., Terol, J., Perez-Alonso, M. and Salinas, J. 1999. The *Arabidopsis CBF* gene family is composed of three genes encoding AP2 domain-containing proteins whose expression is regulated by low temperature but not by abscisic acid or dehydration. *Plant Physiol.* 119, 463-469.

Meskiene, I., Ligterink, W., Bogre, L., Jonak, C. and Kiegerl, S. 1998. The SAM kinase pathway: an integrated circuit for stress signaling in plants. *J. Plant Res.* 111, 339-344.

Mizoguchi, T., Hayashida, N., Yamaguchi Shinozaki, K., Kamada, H. and Shinozaki, K. 1995. Two genes that encode ribosomal-protein S6 kinase homologs are induced by cold or salinity stress in *Arabidopsis thaliana. FEBS Lett.* 358, 199-204.

Mizoguchi, T., Ichimura, K., Irie, K., Morris, P., Giraudat, J., Matsumoto, K. and Shinozaki, K. 1998. Identification of a possible MAP kinase cascade in *Arabidopsis thaliana* based on pairwise yeast two-hybrid analysis and functional complementation tests of yeast mutants. *FEBS Lett.* 437, 56-60.

Mizoguchi, T., Irie, K., Hirayama, T., Hayashida, N., Yamaguchi-Shinozaki, K., Matsumoto, K. and Shinozaki, K. 1996. A gene encoding a mitogen-activated protein kinase kinase kinase is induced simultaneously with genes for a mitogen-activated protein kinase and an S6 ribosomal protein kinase by touch, cold, and water stress in *Arabidopsis thaliana. Proc. Natl. Acad. Sci. USA* 93, 765-769.

Monroy, A.F. and Dhindsa, R.S. 1995. Low-temperature signal transduction: induction of cold acclimation-specific genes of alfalfa by calcium at 25 degrees C. *Plant Cell* 7, 321-331.

Monroy, A.F., Castonguay, Y., Laberge, S., Sarhan, F., Vezina, L.P. and Dhindsa, R.S. 1993a. A new cold-induced alfalfa gene is associated with enhanced hardening at subzero temperature. *Plant Physiol.* 102, 873-879.

Monroy, A.F., Sarhan, F. and Dhindsa, R.S. 1993b. Cold-induced changes in freezing tolerance, protein phosphorylation and gene expression: evidence for a role of calcium. *Plant Physiol.* 102, 1227-1235.

Monroy, A.F., Labbe, E. and Dhindsa, R.S. 1997 Low temperature perception in plants: Effects of cold on protein phosphorylation in cell-free extracts. *FEBS Lett.* 410, 206-209.

Monroy, A.F., Sangwan, V. and Dhindsa, R.S. 1998. Low temperature signal transduction during cold acclimation: protein phosphatase 2A as an early target for cold-inactivation. *Plant J.* 13, 653-660.

Murata, N., Ishizaki-Nishizawa, O., Higashi, S., Hayashi, H., Tasaka, Y. and Nishida, I. 1992. Genetically engineered alteration in the chilling sensitivity of plants. *Nature* 356, 710-713.

Murata, N. and Los, D. 1997. Membrane fluidity and temperature perception. *Plant Physiol.* 115, 875-879.

Nanjo, T., Kobayashi, M., Yoshiba, Y., Kakubari, Y, Yamaguchi-Shinozaki, K. and Shinozaki, K. 1999. Antisense suppression of proline degradation improves tolerance to freezing and salinity in *Arabidopsis thaliana. FEBS Lett.* 461, 205-210.

Nishida, I. and Murata, N. 1996. Chilling sensitivity in plants and cyanobacteria: the crucial contribution of membrane lipids. *Ann. Rev. Plant Physiol. Plant Mol. Biol.* 47, 541-568.

Nordin, K., Vahala, T. and Palva, E.T. 1993. Differential expression of two related low temperature induced genes in *Arabidopsis thaliana* L. Heynh. *Plant Mol. Biol.* 21, 641-653.

Olien, C.R. and Smith, M.N. 1977. Ice adhesions in relation to freeze stress. *Plant Physiol.* 60, 499-503.

Palta, J.P., Whitaker, B.D. and Weiss, L.S. 1993. Plasma membrane lipids associated with genetic variability in freezing tolerance and cold acclimation of *Solanum* species. *Plant Physiol.* 103, 793-803.

Pan, A., Hayes, P.M., Chen, F., Chen, T.H.H., Blake, T., Wright, S., Karsai, I. and Bedo, Z. 1994. Genetic analysis of the components of winterhardiness in barley (*Hordeum vulgare* L.). *Theoret. Appl. Genet.* 89, 900-910.

Pearce, R.S. 1985. The membranes of slowly drought-stressed wheat seedlings: a freeze-fracture study. *Planta* 166, 1-14.

Pearce, R.S. 1988. Extracellular ice and cell shape in frost-stressed cereal leaves: a low temperature scanning electron microscope study. *Planta* 175, 313-324.

Pearce, R.S. 1999. Molecular analysis of acclimation to cold. *Plant. Growth. Reg.* 29, 47-76.

Pearce, R.S. and Ashworth, E.N. 1992. Cell shape and localisation of ice in leaves of overwintering wheat during frost stress in the field. *Planta* 188, 324-331.

Pearce, R.S. and Willison, J.H.M. 1985. A freeze-etch study of the effect of extracellular freezing on the cellular membranes of wheat. *Planta* 163, 304-316.

Phillips, J.R., Dunn, M.A., Hughes, M.A. 1997. mRNA stability and localisation of the low-temperature-responsive barley gene family *blt14*. *Plant Mol. Biol.* 33, 1013-1023.

Pilon-Smits, E.A.H., Ebskamp, M.J.M., Paul, M.J., Jeuken, M.J.W., Weisbeek, P.J. and Smeekens, S.C.M. 1995. Improved performance of transgenic fructan-accumulating tobacco under drought stress. *Plant Physiol.* 107, 125-130.

Plieth, C., Knight, H. and Knight, M.R. 1999. Temperature sensing in plants: measuring and modelling the primary mechanisms of signal perception. *Plant J.* 18,491-497.

Polisensky, D.H. and Braam, J. 1996. Cold-shock regulation of the *Arabidopsis TCH* genes and the effects of modulating intracellular calcium levels. *Plant Physiol.* 111, 1271-1279.

Pollock, C.J. and Jones, T. 1979. Seasonal patterns of fructan metabolism in forage grasses. *New Phytol.* 83, 8-15.

Prasad, T.K. 1997. Role of catalase in inducing chilling tolerance in pre-emergent maize seedlings. *Plant Physiol.* 114, 1369-1376.

Raynal, M., Guilleminot, J., Gueguen, C., Cooke, R., Delseny, M. and Gruber, V. 1999. Structure, organization and expression of two closely related novel LEA (late-embryogenesis-abundant) genes in *Arabidopsis thaliana. Plant Mol. Biol.* 40, 153-165.

Ristic, Z. and Ashworth, E.N. 1993. Changes in leaf ultrastructure and carbohydrates in *Arabidopsis thaliana* L. (Heyn) cv. Columbia during rapid cold acclimation. *Protoplasma* 172, 111-123.

Robertson, A.J., Reaney, M.J.T., Wilen, R.W., Lamb, N., Abrams, S.R. and Gusta, L.V. 1994. Effects of abscisic acid metabolites and analogs on freezing tolerance and gene expression in bromegrass (*Bromus inermis* Leyss) cell cultures. *Plant Physiol.* 105, 823-830.

Rouse, D.T., Marotta, R. and Parish, R.W. 1996. Promoter and expression studies on an *Arabidopsis thaliana* dehydrin gene. *FEBS Lett.* 381, 252-256.

Sakai, A. and Larcher, W., eds. 1987. "Frost survival of plants". In: *Responses and Adaptation to Freezing Stress.* Springer-Verlag, Berlin.

Sakamoto, A., Valverde, R., Alia, N.N., Chen, T.H.H. and Murata, N. 2000. Transformation of *Arabidopsis* with the *codA* gene for choline oxidase enhances freezing tolerance of plants. *Plant J.* 22, 449-453.

Sheen, J. 1996. Ca^{2+}-dependent protein kinases and stress signal transduction in plants. *Science* 274, 1900-1902.

Shinozaki, K. and Yamaguchi-Shinozaki, K. 1997. Gene expression and signal transduction in water-stress response. *Plant Physiol.* 115, 327-334.

Shinozaki, K. and Yamaguchi-Shinozaki, K. 2000. Molecular responses to dehydration and low temperature: differences and cross-talk between two stress signaling pathways. *Curr. Opin. Plant Biol.* 3, 217-223.

Smallwood, M., Worrall, D., Byass, L., Elias, L., Ashford, D., Doucet, C.J., Holt, C., Telford, J., Lillford, P. and Bowles, D.J. 1999. Isolation and characterization of a novel antifreeze protein from carrot (*Daucus carota*). *Biochem J.* 340, 385-391.

Somerville, C.R. and Browse, J. 1991. Plant lipids: metabolism, mutants, and membranes. *Science* 252, 80-87.

Steponkus, P.L. 1984. Role of the plasma membrane in freezing injury and cold acclimation. *Ann. Rev. Plant Physiol.* 35, 543-584.

Steponkus, P.L. and Gordon-Kamm, W.J. 1985. Cryoinjury of isolated protoplasts: a consequence of dehydration or the fraction of the suspending medium that is frozen? *Cryo. Lett.* 6, 217-226.

Steponkus, P.L., Uemura, M., Balsamo, R.A., Arvinte, T. and Lynch, D.V. 1988. Transformation of the cryobehavior of rye protoplasts by modification of the plasma membrane lipid composition. *Proc. Natl. Acad. Sci. USA* 85, 9026-9030.

Steponkus, P.L., Uemura, M., Joseph, R.A., Gilmour, S.J. and Thomashow, M.F. 1998. Mode of action of the *COR15a* gene on the freezing tolerance of *Arabidopsis thaliana. Proc. Natl. Acad. Sci. USA* 95, 14570-14575.

Steponkus, P.L., Uemura, M. and Webb, M.S. 1993. Membrane destabilization during freeze-induced dehydration. *Curr. Top. Plant Physiol.* 10, 37-47.

Stockinger, E.J., Gilmour, S.J. and Thomashow, M.F. 1997. *Arabidopsis thaliana CBF1* encodes an AP2 domain-containing transcriptional activator that binds to the C-repeat/DRE, a cis-acting DNA regulatory element that stimulates transcription in response to low temperature and water deficit. *Proc. Natl. Acad. Sci. USA* 94, 1035-1040.

Storlie, E.W., Allan, R.E. and Walker-Simmons, M.K. 1998. Effect of the Vrn1-Fr1 interval on cold hardiness levels in near-isogenic wheat lines. *Crop Sci.* 38, 483-488.

Strauss, G. and Hauser, H. 1986. Stabilization of lipid bilayer vesicles by sucrose during freezing. *Proc. Natl. Acad. Sci. USA* 83, 2422-2426.

Strizhov, N., Abraham, E., Okresz, L., Blickling, S., Zilberstein, A., Schell, J., Koncz, C. and Szabados, L. 1997. Differential accumulation of two *P5CS* genes controlling proline accumulation during salt stress requires ABA and is regulated by *ABA1*, *ABI1* and *AXR2* in *Arabidopsis*. *Plant J.* 12, 557-569.

Sutton, F., Ding, X. and Kenefick, D.G. 1992. Group 3 LEA gene HVA1 regulation by cold acclimation and deacclimation in two barley cultivars with varying freeze resistance. *Plant Physiol.* 99, 338-340.

Tahtiharju, S., Sangwan, V., Monroy, A.F., Dhindsa, R.S. and Borg, M. 1997. The induction of *KIN* genes in cold-acclimating *Arabidopsis thaliana*. Evidence of a role for calcium. *Planta* 203, 442-447.

Tarczynski, M.C., Jensen, R.G. and Bohnert, H.J. 1993. Stress protection in transgenic tobacco producing a putative osmoprotectant, mannitol. *Science* 259, 508-510.

Teutonico, R.A. and Osborn, T.C. 1994. Mapping of RFLP and qualitative trait loci in *Brassica rapa* and comparison to the linkage maps of *B. napus*, *B. oleracea*, and *Arabidopsis thaliana*. *Theor. Appl.Genet.* 89, 885-894.

Teutonico, R.A., Yandell, B., Satagopan, J.M., Ferreira, M.E., Palta, J.P. and Osborn, T.C. 1995. Genetic analysis and mapping of genes controlling freezing tolerance in oilseed *Brassica*. *Mol. Breeding* 1, 329-339.

Thomashow, M. 1999. Plant cold acclimation: freezing tolerance genes and regulatory mechanisms. *Ann. Rev. Plant Physiol. Plant Mol. Biol.* 50, 571-599.

Thomashow, M.F. 1998. Role of cold-responsive genes in plant freezing tolerance. *Plant Physiol.* 118, 1-7.

Thorlby, G.J., Veale, E., Butcher, K. and Warren, G. 1999. Map positions of *SFR* genes in relation to other freezing-related genes of *Arabidopsis thaliana*. *Plant J.* 17, 445-452.

Tokuhisa, J.G., Feldmann, K.A., La Brie, S.T. and Browse, J. 1997. Mutational analysis of chilling tolerance in plants. *Plant Cell Envir.* 20, 1391-1400.

Tokuhisa, J.G., Vijayan, P., Feldmann, K.A. and Browse, J.A. 1998. Chloroplast development at low temperatures requires a homolog of DIM1, a yeast gene encoding the 18S rRNA dimethylase. *Plant Cell* 10, 699-711.

Uemura, M., Joseph, R.A. and Steponkus, P.L. 1995. Cold acclimation of *Arabidopsis thaliana*. Effect on plasma membrane lipid composition and freeze-induced lesions. *Plant Physiol.* 109, 15-30.

Uemura, M. and Steponkus, P.L. 1989. Effect of cold acclimation on the incidence of two forms of freezing injury in protoplasts isolated from rye leaves. *Plant Physiol.* 91, 1131-1137.

Uemura, M. and Steponkus, P.L. 1994. A contrast of the plasma membrane lipid composition of oat and rye leaves in relation to freezing tolerance. *Plant Physiol.* 104, 479-496.

Urao, T., Yabukov, B., Yamaguchi-Shinozaki, K. and Shinozaki, K. 1998. Stress-responsive expresson of genes for two-component response regulator-like proteins in *Arabidopsis thaliana*. *FEBS Lett.* 427, 175-178.

Van Nocker, S. and Vierstra, R. 1993. Two cDNAs from *Arabidopsis thaliana* encode putative RNA binding proteins containing glycine-rich domains. *Plant Mol. Biol.* 21, 695-699.

Van Zee, K., Chen, F.Q., Hayes, P.M., Close, T.J. and Chen, T.H.H. 1995. Cold-specific induction of a dehydrin gene family member in barley. *Plant Physiol.* 108, 1233-1239.

Verbruggen, N., Hua, X.J., May, M. and Van Montagu, M. 1996. Environmental and developmental signals modulate proline homeostasis: evidence for a negative transcriptional regulator. *Proc. Natl. Acad. Sci. USA* 93, 8787-8791.

Walker, M.A. and McKersie, B.D. 1993. Role of the ascorbate-glutathione antioxidant system in chilling resistance of tomato. *J. Plant Physiol.* 141, 234-239.

Wallis, J.G., Wang, H. and Guerra, D.J. 1997. Expression of a synthetic antifreeze protein in potato reduces electrolyte release at freezing temperatures. *Plant Mol. Biol.* 35, 323-330.

Wanner, L.A. and Junttila, O. 1999. Cold-induced freezing tolerance in *Arabidopsis*. *Plant Physiol.* 120, 391-399.

Warren, G., McKown, R., Marin, A. and Teutonico, R. 1996. Isolation of mutations affecting the development of freezing tolerance in *Arabidopsis thaliana* (L.) Heynh. *Plant Physiol.* 111, 1011-1019.

Warren, G., McKown, R., Teutonico, R., Kuroki, G., Veale, E. and Sagen, K. 1997. "*Arabidopsis* mutants impaired in freezing tolerance after cold acclimation." In: *Plant Cold Hardiness*, eds. P. Li and T. Chen, pp. 45-56. Plenum, New York.

Warren, G.J. 1998. Cold stress: manipulating freezing tolerance in plants. *Current Biol.* 8, R514-R516.

Welin, B.V., Olson, A., Nylander, M. and Palva, E.T. 1994. Characterization and differential expression of dhn/lea/rab-like genes during cold acclimation and drought stress in *Arabidopsis thaliana*. *Plant Mol. Biol.* 26, 131-144.

Welin, B.V., Olson, A. and Palva, E.T. 1995. Structure and organization of two closely-related low-temperature-induced dhn/lea/rab-like genes in *Arabidopsis thaliana*. *Plant Mol. Biol.* 29, 391-395.

White, A.J., Dunn, M.A., Brown, K. and Hughes, M.A. 1994. Comparative analysis of genomic sequence and expression of a lipid transfer protein gene family in winter barley. *J. Exp. Bot.* 45, 1885-1892.

Wilen, R.W., Gusta, L.V., Lei, B., Abrams, S.R. and Ewan, B.E. 1994. Effects of abscisic acid (ABA) and ABA analogs on freezing tolerance, low-temperature growth, and flowering in rapeseed. *J. Plant Growth Reg.* 13, 235-241.

Wilhelm, K.S. and Thomashow, M.F. 1993. *Arabidopsis thaliana COR15b*, an apparent homologue of *COR15a*, is strongly responsive to cold and ABA, but not drought. *Plant Mol. Biol.* 23, 1073-1077.

Williams, J., Bulman, M., Huttly, A., Phillips, A. and Neill, S. 1994. Characterization of a cDNA from *Arabidopsis thaliana* encoding a potential thiol protease whose expression is induced independently by wilting and abscisic acid. *Plant Mol. Biol.* 25, 259-270.

Williams, W.P. 1990. Cold-induced lipid phase transitions. *Phil. Trans. Royal Soc. Lond. B.: Biol. Sci.* 326, 555-570.

Xin, Z. and Browse, J. 1998. *eskimo1* mutants of *Arabidopsis* are constitutively freezing-tolerant. *Proc. Natl. Acad. Sci. USA* 95, 7799-7804.

Xiong, L., Ishitani, M., Lee, H. and Zhu, J. 1999a. *Hos5* - a negative regulator of osmotic stress-induced gene expression in *Arabidopsis thaliana*. *Plant J.* 19, 569-578.

Xiong, L., Ishitani, M. and Zhu, J.-K. 1999b. Interaction of osmotic stress, temperature, and abscisic acid in the regulation of gene expression in *Arabidopsis*. *Plant Physiol.* 119, 205-211.

Yamaguchi-Shinozaki, K. and Shinozaki, K. 1993. Characterization of the expression of a desiccation-responsive *RD29* gene of *Arabidopsis thaliana* and analysis of its promoter in transgenic plants. *Mol. Gen. Genet.* 236, 331-340.

Yamaguchi-Shinozaki, K. and Shinozaki, K. 1994. A novel cis-acting element in an *Arabidopsis* gene is involved in responsiveness to drought, low-temperature, or high-salt stress. *Plant Cell* 6, 251-264.

Yang, K.Y., Nam, S.H., Kim, Y.H., Eun, M.Y., Kim, K.C., Ki, W.K., Song, D.U. and Cho, B.H. 1995. An *Arabidopsis* transcript homologous to the carrot DC 1.2 cDNA is induced by several environmental stresses. *Molecules Cells* 5, 539-543.

Yokoi, S., Higashi, S., Kishitani, S., Murata, N. and Toriyama, K. 1998. Introduction of the cDNA for *Arabidopsis* glycerol-3-phosphate acyltransferase (GPAT) confers unsaturation of fatty acids and chilling tolerance of photosynthesis on rice. *Mol. Breeding* 4, 269-275.

Yoshiba, Y., Kiyosue, T., Katagiri, T., Ueda, H., Mizoguchi, T., Yamaguchi-Shinozaki, K., Wada, K., Harada, Y. and Shinozaki, K. 1995. Correlation between the induction of a gene for delta-1-pyrroline-5-carboxylate synthetase and the accumulation of proline in *Arabidopsis thaliana* under osmotic stress. *Plant J.* 7, 751-760.

Chapter 11

PLANT RESPONSES TO NUTRITIONAL STRESSES

Frank W. Smith

CSIRO Tropical Agriculture,Long Pocket Laboratories, Indooroopilly, Qld 4068, Australia
frank.smith@pi.csiro.au

INTRODUCTION

Early successes in breeding crop plants for increased yields resulted in selection for high nutrient requirements and high soil fertility. This often necessitated high inputs of fertilizers and soil ameliorants in order to achieve optimum yields and product quality. The costs involved and the environmental damage that often resulted from this approach has led to attempts to reduce mineral nutrient inputs and better fit crop plants to soils. This has been particularly important in agricultural production from infertile tropical and sub-tropical soils. The approach requires plant genotypes with high nutrient efficiency and tolerance to nutrient stresses. To help in production of such genotypes, the physiological responses of plants to nutrient stresses and the genetic controls involved in adaptation to those stresses need to be understood by those involved in plant improvement programs.

The essential mineral nutrients play key roles in many aspects of plant metabolism, growth and development. There is therefore a very wide spectrum of responses to nutrient stresses. Typically, the nature of responses to a particular nutrient stress depends upon which plant processes are most sensitive to that stress and the severity of the stress. These may vary between species. They may also vary between genotypes within the same species and so reflect genotypic differences in nutrient efficiency related to differences in nutrient acquisition and/or utilisation. Such differences provide plants

M.J. Hawkesford and P. Buchner (eds.),
Molecular Analysis of Plant Adaptation to the Environment, 249–269.

with avenues that enable them to adapt to moderate nutrient stresses. Plant breeders are successfully exploiting these adaptive traits to produce more nutrient efficient genotypes. As might be expected from complex traits, they are commonly multigenic.

Although most nutrient stresses decrease plant growth rate, the effects on metabolic processes differ between nutrients. For some nutrients, changes in the rates of specific assimilatory and biochemical reactions are the predominant responses to a deficiency of that nutrient (Marschner, 1995). But, plants have adapted to some nutrient stresses by also increasing their capacity to acquire more nutrient from the soil. These adaptations have arisen during evolution so apply to nutrients that have been poorly available to plants for a long time. They may give rise to major morphological and biochemical adaptations. For nutrients such as phosphate, zinc and nitrogen, many plant species have also acquired or retained the capacity to form symbiotic associations with other organisms in order to gain that essential nutrient in stress situations.

In this review, phosphate deficiency is used as an example of a nutrient stress in which plants embrace significant morphological, biochemical and symbiotic responses in order to acquire additional nutrient and improve the internal utilization of that nutrient which has been taken up. This example illustrates coordination between morphological and metabolic responses in stressed plants and poses questions regarding the complex regulatory mechanisms underlying these co-ordinated responses. Finally, the review reflects upon how recent technological advances may provide opportunities for unravelling these complex regulatory processes and, perhaps, lead to strategies for manipulating them in ways that improve the adaptation of agricultural plants to less fertile soils.

ADAPTATION OF PLANTS TO PHOSPHORUS STRESS

The low availability of phosphate in many soils is a common constraint on crop production. Plants derive their immediate phosphorus requirements from inorganic P (P_i) present in the soil solution that surrounds their roots. P_i in soil solution is in equilibrium with P_i sorbed onto iron and aluminium compounds and soil organic matter in the soil solid phase. This equilibrium, described by phosphate sorption isotherms (Fox and Kamprath, 1970), defines both the concentration of P_i in solution and the capacity of the soil to maintain or buffer P_i in solution. The high absorptive capacity and reactivity for phosphate of the components of the solid phase result in only a very small proportion of total soil P_i being in solution. Further, a large fraction of soil total P may be in organic forms in many soils. These forms are not directly available to plants. They must first be mineralised to yield P_i, which immediately becomes part of the overall P_i pool, the availability of which is defined by the phosphate sorption isotherm for that soil. As a result, plant roots are exposed to quite low P_i concentration in soil solution. A survey of representative soils of the USA revealed that, even in the most fertile soils, the P_i concentration in soil solution did not exceed 8 µM (Barber *et al.*, 1962). In sandy soils, alkaline soils and the highly weathered soils of the tropics and subtropics concentrations of P_i in soil solution are more likely to be less than 1 µM (Reisenauer, 1966).

Stress arising from phosphate deficiency is therefore very common in native vegetation and in agricultural and forestry production systems. Farming systems in many developed countries have sought to overcome these deficiencies by applying fertilisers and manures. However, the nature of the sorption and fixation reactions in the soil inevitably result in inefficient use of applied phosphatic fertilisers and a large bank of P can accumulate in heavily fertilised soils. In some instances this P has found its way into waterways and lakes through leaching and erosion and caused eutrophication, algal blooms and environmental damage. Concern for the environment, the cost to farmers of applying phosphatic fertilisers and the inaccessibility of phosphatic fertilisers to many farmers in developing countries have led to renewed interest in the adaptive mechanisms that plants use to improve phosphorus acquisition and utilisation when subjected to P-stress.

PARTITIONING OF CARBON AND PHOSPHORUS IN P-STRESSED PLANTS

The overall agronomic and physiological response of plants to P-deficiency is reduced growth rate. However, many developmental, physiological and biochemical changes usually underlie this reduced growth rate. Carbon fixation is often sub-optimal due to reduced leaf expansion (Fredeen *et al.*, 1989), fewer leaves (Lynch *et al.*, 1991) and lower photosynthetic efficiency (Lauer *et al.*, 1989). Flowering can be delayed (Rossiter, 1978) and seed formation restricted (Barry and Miller, 1989). There is also a change in the partitioning of carbon that manifests itself as an increase in root/shoot ratio. This change, which can even lead to sucrose accumulation in the roots of P-deficient plants (Khamis *et al.*, 1990), provides a relatively larger root system for more efficient P acquisition by P-stressed plants. As indicated below, structural, physiological and biochemical modifications and symbiotic associations may further enhance P uptake by this relatively larger root system.

In addition to changes in carbon partitioning there are alterations in the partitioning of P between roots and shoots during the onset of P-deficiency. The dynamics of these flows have been illustrated in experiments with the tropical forage legume *Stylosanthes hamata* (Smith *et al.,* 1990). In those experiments, plants were grown in dilute flowing culture systems in which the external P_i was maintained at concentrations just optimal for unrestricted growth (1 μM), sub-optimal for growth (0.2 μM), or supra-optimal for growth (5 μM), before being deprived of an external P_i supply. Plants already stressed for P or having a just adequate supply responded almost immediately by further restricting shoot growth and transferring P from the shoots to roots where it was used to prolong further root growth. Plants previously growing at the supra-optimal P supply had stored more P in their roots than was required to support root growth. Those plants initially responded to P deprivation by mobilising the stored P in the root and transferring it to the shoot where it was used to maintain further shoot growth. However, after a few days shoot growth P concentrations were reduced and these plants sensed that P-stress was now developing. They then transferred P from shoot to root to try and maintain root growth. These experiments illustrate how rapidly P can be remobilised and repartitioned in order to support key meristems that promote either shoot growth and increased productive capacity when P is adequate, or root growth and potentially increased P acquisition capacity when P is limiting. These changes undoubtedly involve regulatory circuits that respond quite rapidly to changes in the P status of the plant.

MORPHOLOGICAL MODIFICATIONS TO ROOTS IN P-STRESSED PLANTS

Along with developing a larger root mass relative to the total plant mass when the P availability is limiting, some species alter the architecture of their root systems. Studies with *Phaseolus* indicate that genotypes that have highly branched root systems and more root apices are efficient in acquiring P_i. P-efficient genotypes also change the geotropic response of their roots such that lateral roots grow out from the basal roots at angles that enable them to better explore the upper layers of the soil (Lynch, 1995; Lynch and Beebe, 1995). Interception of available soil P_i may also be optimised by enhanced lateral root development in localised zones of the soil that are rich in P_i (Drew and Saker, 1978; Jackson *et al.,* 1990). The diameter of roots may also increase under P-stress thereby increasing the surface area of the root in contact with the soil (Ma *et al.,* 2001).

Root hair development may be dramatically influenced by P-stress in many species (Foehse and Jungk, 1983; Bates and Lynch, 1996). This response has been studied in detail in *Arabidopsis* (Bates and Lynch, 2000; Ma *et al.,* 2001). In those studies, the density of root hairs was not influenced by P availability during the early stage of seedling growth. However, 9 days after germination, root hair density declined under high-P supply and increased under low-P supply. Root hair density increased logarithmically from almost no root hairs at 2000 μM P in the external medium to 60 root hairs mm^{-1} root at 1 μM P. These changes in root hair density were associated with changes in the diameter and root anatomy of P stressed plants. Larger root diameter arose from increases in the number of cortical cells in P-stressed plants, a developmental change that also indicates that P-stress influences differentiation in the root meristem. Epidermal cells were also smaller in P-stressed plants and this, together with the larger number of underlying cortical cells and larger root diameter, increased the number of files of trichoblasts (those epidermal cells that normally bear root hairs) from 8 to 12 in *Arabidopsis*. Further, in P-stressed plants only 10% of trichoblasts did not produce root hairs, whereas 76% of trichoblasts did not produce root hairs in high-P plants.

Low P availability also stimulates the elongation of root hairs (Bates and Lynch, 1996). The average length of fully expanded root hairs in 15-day old *Arabidopsis* plants grown on 1000 μM P was 0.3 mm, whereas plants grown on 1 μM P had fully expanded root hairs averaging 0.9 mm in length. This effect on root hair elongation was found to be a localised response. The length of the root hairs was related to the external P_i concentration in the medium immediately surrounding that section of the root rather than to the overall phosphate status of the plant.

The net effect of these modifications is that, under P-stress, plants respond with structural changes to their root systems that enable them to maximise their capacity to acquire P_i from the soil. They develop relatively larger root systems. Efficient genotypes develop an architecture that places active roots in regions of the soil more likely to contain available P. They enhance lateral root development in localised zones of soil rich in P_i, and they increase the volume of soil explored by extending into the soil a larger number of root hairs and longer root hairs.

In some species P-stress elicits the development of specialised root structures known as cluster or proteoid roots (Dinkelaker *et al.,* 1995). These are features of many species of the Proteaceae and some legumes such as the tree *Casuarina* and the annual herbaceous legume *Lupinus albus*. These structures are composed of dense clusters to determinate lateral roots through which organic acids are released to increase the availability of soil P (Gardner *et al.,* 1983). The P_i released in the localised regions of the cluster roots is then readily taken up through the dense mat of short lateral roots that form the cluster. The formation of cluster roots responds to the P status of the plant. Examination of *Lupinus albus* root systems revealed an average of 53 cluster roots on P-stressed plants, but only 15 on plants supplied with adequate P (Marschner *et al.*, 1987).

BIOCHEMICAL RESPONSES TO P-STRESS

The central role that P_i and P compounds play in plant metabolism results in a wide range of biochemical responses to P-stress. Some of these are quite specific, others the result of secondary effects on plant growth and development. Underlying most of these responses are requirements to maintain carbon and energy flows in basic physiological processes and provide essential phosphate compounds for growth, development and reproduction. Remobilisation of internal P and acquisition of additional external P_i contribute to this in P-stressed plants.

Responses of plant processes to P-stress

All major plant processes are affected by P-stress. A detailed discussion of these wide-ranging effects is beyond the scope of this Chapter. However such a discussion can be found in the reviews of Plaxton (1996) and Plaxton and Carswell (1999). Photosynthesis is reduced by inhibition of several enzymes of the Calvin cycle and both direct and indirect effects on energy transduction in the thylakoids. Reductions in P_i, ATP and ADP associated with P-stress can impair carbon flow through key enzymes of the glycolytic pathway. This can lead to activation of alternative enzymes with lower P_i or nucleotide phosphate requirements that bypass some of those normally involved in carbon and energy flows. Starch may accumulate in P-stressed plants as a result of the removal of P_i inhibition of key enzymes in the starch biosynthesis pathway (Stark *et al.*, 1992). Secondary metabolism is also influenced by P-stress leading to accumulation of flavonoids, anthocyanins and certain aromatic compounds.

Regulation of cytoplasmic P_i concentration

P_i concentrations in the cytoplasm of cells needs to be maintained for the optimal biochemical and physiological functioning of those cells. P_i accumulated in the vacuoles of cells provides a buffer against fluctuations in the cytoplasmic P_i concentration. This provides plant cells with a mechanism for maintaining P_i homeostasis (Mimura, 1999). Vacuolar P_i concentrations as high as 120 mM have been measured in plants well supplied with external P_i (Mimura *et al.*, 1990). P_i can be withdrawn from this pool under conditions of P-stress until it is virtually exhausted. Measurements of cytoplasmic P_i suggest P_i influx and efflux is regulated so as to try and maintain cytoplasmic P_i concentrations of the order of 5 to 17 mM (Mimura *et al.*, 1996). Transfer of P_i between the vacuole and the cytoplasm is regulated by the tonoplast, but the molecular mechanisms involved in P_i transfer across this membrane have not yet been identified. However, it is known that P_i influx through the tonoplast is stimulated by ATP (Mimura *et al.*, 1990) so may involve an active transport mechanism. Very little is known about vacuolar P_i efflux mechanisms. There are flux measurements indicating that, under P-sufficient conditions the tonoplast has low permeability to P_i and this permeability increases markedly under P-stress (Mimura, 1995).

Kinases, phosphatases and ribonucleases

Phosphate plays a major regulatory role in many metabolic processes through protein phosphorylation. The activation or deactivation of proteins through the actions of specific kinases or phosphatases controls the rate of many biochemical reactions and transport processes. Some insight into the extent of these means of regulating plant metabolism can be gleaned from an examination of the *Arabidopsis* genome. This genome contains more than 250 genes that encode phosphatases, the majority of which appear to be involved in the dephosphorylation of specific proteins. It should be expected therefore that P-stress would influence phosphorylation of proteins with subsequent effects on plant metabolism. Evidence for this is emerging from experiments in which plant cell suspension cultures have been deprived of P_i (Carswell *et al.*, 1997).

Another range of plant phosphatases are important in both the internal recycling of P in plants and the release of P_i from soil organic compounds. Acid phosphatases, generally with low substrate specificity, may be particularly important in the release of P_i from phosphate esters. Intracellular acid phosphatases have been found in most plant organs (Duff *et al.*, 1994) where they are commonly found in cell vacuoles (Nishimura and Beevers, 1978). Compartmentation in the vacuole protects essential organic P compounds in the cytoplasm from catalysis. The intracellular phosphatases are involved in remobilising P during developmental changes, senescence or P-stress. Other phosphatases are excreted to the cell wall and into the rhizosphere. These extracellular phosphatases play a role in hydrolysing P_i from organic P compounds in the soil (Goldstein *et al.*, 1989; Lefebvre *et al.*, 1990). There are numerous reports of large increases in acid phosphatase activities when plants are subjected to P-stress (see review of Duff *et al.*, 1994). These increases provide important adaptive mechanisms for remobilizing internal P and assisting in the acquisition of additional external P_i. The regulatory mechanisms underlying the increases in phosphatase activities in P-stressed plants are likely to be varied and complex because of the range of phosphatases involved, their compartmentation and their diverse substrate specificity. Trancriptional regulation by plant P-status of genes that encode specific phosphatases has been demonstrated (Plaxton and Carswell 1999), but post-transcriptional and post-translational regulation are also likely to be involved. Interestingly, most phosphatases are glycosylated, leading to the suggestion that differential glycosylation may provide a regulatory mechanism for some of these enzymes (Duff *et al.*,1994).

RNA provides another source of P_i that can be remobilised during P-stress. Specific RNases that result in the release of P_i from RNA have been found to be induced during P-starvation (Bariola *et al.*, 1994; Dodds *et al.*, 1996). The genes encoding at least some of these RNases are transcriptionally regulated in response to the P-status of the cells (Kock, 1998).

Phosphate transport

Both the increased capacity for P_i acquisition and the internal remobilisation of P in P-stressed plants are facilitated by changes in P_i transport through plant membranes. Two gene families that encode P_i transporter proteins have been identified in plants (Smith *et al.*, 2000). Members of the *Pht1* family, cloned first from *Arabidopsis* (Muchhal *et al.*, 1996, Smith *et al.*, 1997) have been isolated from potato (Leggewie *et al.*, 1997), tomato (Daram *et al.*, 1998; Liu *et al.*, 1998a), *Catharanthus* (Kai *et al.*, 1997), *Medicago* (Liu *et al.*, 1998b) and barley (Smith *et al.*, 1999). There are 9 members of the *Pht1* family in the *Arabidopsis* genome and at least 8 in barley (Smith, 2001). The *Pht1* transporters are composed of 520 to 550 amino acids and are approximately 58 kDa in size. They contain 12 hydrophobic membrane spanning domains (MDSs) arranged in the 6+6 topology with a long central loop that is characteristic of the major facilitator superfamily (Pao *et al.*, 1998). *Pht1* transporters act as $H^+/H_2PO_4^-$ symporters, relying on protons "pumped" to the outer surface of the membrane to maintain the necessary electrochemical gradient across the membrane to energise the cotransport function (Smith, 2001).

A transporter protein with similar topology to the mammalian Na^+-dependant phosphate transporters has been isolated from *Arabidopsis* (Daram *et al.*, 1999). Proteins with this topology will make up the *Pht2* family of plant phosphate transporters. The *Ph2* family appears to be represented by a single member in the *Arabidopsis* genome. Although similar to Na^+/ $H_2PO_4^-$ cotranporters in other organisms, the *Arabidopsis* *Pht2;1* transporter appears to function as a $H^+/H_2PO_4^-$ co-transporter.

Those molecular studies that have been done on the *Pht1* family of P transporters to date have concentrated on members expressed in roots (Raghothama, 1999). The initial work with *Arabidopsis* showed that at least 3 of the 9 *Arabidopsis* genes are expressed in roots. Functional expression of one of these genes in cultured tobacco cells established that it encodes a high-affinity phosphate transporter with a K_m for phosphate of 3.1 μM (Mitzukawa *et al.*, 1997). Regrettably, heterologous expression of plant *Pht1* transporters in yeast has been somewhat disappointing so kinetic data obtained by this technique for other plant phosphate transporters is not particularly reliable (Smith, 2001). *In situ* hybridisation studies using the *LePt1* sequence from tomato as a probe has established that this member of the *Pht1* family is primarily expressed in the outer cell layers of roots – epidermal cells, outer cortical cells and the root cap (Daram *et al.*, 1998). Immunolabelling of MtPT1, a very similar P transporter isolated from *Medicago truncatula*, also indicates that this protein is located in the plasma membrane of cells in outer layers of the root and primarily in epidermal cells, including the root hairs (Chiou *et al.*, 2001). An interesting observation from this immunolabelling work is that the MtPT1 protein may be more concentrated towards the tips of the root hairs. These transporters occur in the plant membranes in closest contact with the soil solution so are most likely to be the high-affinity P_i transporters responsible for uptake of soil P_i into the root symplast.

Expression of these genes in plant roots is transcriptionally regulated by the P status of the plant (Muchhal and Ragothama, 1999; Smith *et al.*, 1997). For instance, steady state levels of transcripts corresponding to both the *APT1* and *APT2* genes of *Arabidopsis* are low in plants adequately supplied with external P_i (Smith *et al.*, 1997). When these plants were deprived of external P_i and became P-stressed, there was a 4 to 5-fold increase in these transcripts. Alleviation of P-stress by re-supplying external P_i led to rapid down-regulation of these genes. Such feedback regulation of transcription results in greater numbers of P transporter proteins in the plasma membranes of epidermal cells of P-stressed plants. It accounts for the observed increases in the capacity of plants to take up P_i when they are deprived of P (Clarkson and Scattergood, 1982; Clarkson and Lüttge, 1991; Cogliatti and Clarkson, 1983; Lefebvre and Glass, 1982). Coupled with more numerous and longer root hairs, relatively more root, more favourable root architecture and increased availability of soil P_i through excreted organic acids and phosphatases, the large increase in the numbers of P_i transporters in P-stressed plants maximises the capacity of such plants to acquire external P_i.

Should P_i become readily available to such plants they need to restrict P uptake quickly in order to maintain P_i homeostasis and prevent P toxicity. This is achieved through rapid turnover of the transporter proteins and corresponding mRNA transcripts and by down-regulating the expression of genes encoding root P transporters.

Physiological studies on plants grown in a manner that enables different roots to be supplied with different levels of external P_i, indicate, that the capacity of all roots to take up P responds to the overall P status of the plant (Drew and Saker, 1978), not the localised supply to the root. The P-responsiveness of transcription of the *Pht1* genes is therefore a systemic control rather than a localised control and probably involves signals derived through plant shoots. The nature of these signals and the transduction pathways involved has not yet been delineated. The *pho2* mutant of *Arabidopsis,* which accumulates very high concentrations of P_i in its shoots appears to have lost the capacity to down-regulate the transfer of P from roots to shoots (Delhaize and Randall, 1995). It may therefore be a mutation in the P-sensing mechanism in the shoot or a component of the signal transduction pathway.

Other nutrient deficiencies can influence the transcription of some of the *Pht1* genes expressed in plant roots. Experiments with barley have indicated that the expression of high-affinity phosphate transporters in roots is up-regulated in zinc deficient plants even in plants grown on high external P_i (Huang *et al.,* 2000). This leads to accumulation of very high levels of P in zinc-deficient plants, a phenomenon that has been observed for many years (Welch *et al.,* 1982). This effect of zinc was specific and could not be replaced by manganese, a similar divalent cation. Thus, zinc appears to play a specific role in the signal transduction pathway that regulates the expression of these *Pht1* genes in roots. Other nutrient deficiencies may result in the expression of *Pht1* genes not being up-regulated during P-stress. This phenomenon has been noted in plants deprived of either nitrogen and phosphate or sulfate and phosphate (Smith *et al.,* 1999) and appears to be a non-specific effect of the second nutrient deficiency. Under these circumstances it appears that, when growth is also limited by a second nutrient deficiency, plants may fail to up-regulate transcription of *Pht1* genes so as to avoid uptake of P_i that cannot be utilised at that time.

The transfer of P throughout the plant involves transport across membranes other than those involved in uptake of P_i from the soil into the root symplast. Members of the *Pht1* family are also likely to play roles in some of the transport processes that occur in the vascular tissues of the root and shoot. Remobilisation of P from senescing tissues, particularly during P-stress, will require transporters expressed in shoot tissues. The single member of the *Pht2* family of transporters in the *Arabidopsis* genome appears to at least partially fulfil this function (Daram *et al.,* 1999). However, *Pht2* appears to be constitutively expressed in leaves so does not respond strongly to P-stress. Loading of P into leaf mesophyll cells and into flowers and fruiting structures during plant development are also likely to require transport of P_i across membranes from apoplastic spaces. The identities of the genes that encode these transporter proteins have not been reported to date. The functioning of critical organelles such as chloroplasts and mitochondria also necessitates transfer of P across membranes. A different family of genes encodes proteins that transport P into some of these plastids (Kammerer *et al.,* 1998).

Transfer of P_i around the plant also requires unloading or efflux mechanisms. Very little is known about the molecular mechanisms involved in P_i efflux across the plasma membrane. The electrochemical gradients involved suggest that ion channels could mediate P_i efflux from cells. Radial transport of P_i in the root requires unloading from cells within the stele in order for that P_i to enter xylem for long-distance transport to shoots (Clarkson, 1993). The *pho1* mutant of *Arabidopsis* is deficient in this process (Poirier *et al.,* 1991) so presumably carries a mutation in either the gene encoding the protein through which efflux occurs or a gene that regulates the unloading process. The location, deep within the stele, of the cells that are involved in unloading has made physiological studies difficult, but availablity of the *pho1* mutant provides opportunities to use molecular and genetic approaches to unravel the mechanisms of unloading from the root symplast and loading into the xylem.

Efflux of P_i from plant roots is a well known phenomenon (Lefebvre and Clarkson, 1984; Mimura, 1999). This efflux provides another mechanism for plant cells to maintain P_i homeostasis. Measurements with barley roots indicated that P_i efflux rates could be 25% of P_i influx rates in plants adequately supplied with external P_i (Lefebvre and Clarkson, 1984). P_i efflux decreases in P-stressed plants, as illustrated by studies with *Lemna* and *Spirodela.* The efflux rates in plants adequately supplied with P_i were 10% of the influx rate, but this declined to 1% in P-stressed plants (McPharlin and Bieleski, 1989). In severely P-stressed plants, efflux could be similar to influx, resulting in no net P_i uptake (Bieleski and Lauchli, 1992).

Organic acid excretion

Many plant species excrete organic acids from their roots into the rhizosphere. These compounds, present in root exudates, aid in the release of P_i from Fe and Al compounds in the soil. This release may occur as a result of both desorption of P_i from sesquioxide surfaces and chelation of the iron and aluminium from Fe-phosphates and Al-phosphates (Gerke, 1992). Citric acid is claimed to be the most effective organic acid in mobilising P_i (Staunton and Leprince, 1996), although several other carboxylic acids, including malic, succinic, fumaric, malonic and oxalic, are also found in root exudates (Marschner, 1995). The range of organic acids released is dependant on the species (Ohwaki and Hirata, 1992) with certain species excreting specialised organic acids such as piscidic acid by pigeon pea (Ae *et al.,* 1990). When grown under P-deficient conditions, there is a dramatic increase in excretion of organic acids by the roots of those species that exhibit this trait (Dinkelaker *et al.,* 1989, 1995; Hoffland *et al.,* 1989). Common legumes such as peanut and chickpea excrete large amounts of organic acids when P-stressed (Ohwaki and Hirata, 1992) and even *Medicago sativa* doubles the amount of citric acid excreted by its roots when P-deficient (Lipton *et al.,* 1987). This P-stress induced excretion is of particular importance in increasing the availability of P to plants with cluster roots. High levels of citric acid are excreted from the cluster roots of *Lupinus albus* and this results in increased availability of P_i in the soil surrounding these structures (Dinkelaker *et al.,* 1995; Marschner *et al.,* 1987). Organic acid excretion has also been shown to protect the roots of some plants from Al toxicity (Ma *et al.,* 2000; Ryan *et al.,* 1995), thereby enabling root proliferation and increased foraging capacity for P_i in acid soils.

Increased organic acid excretion by P-stressed plants requires additional carboxylate synthesis. Thus increased phosphoenolpyruvate carboxylase activity has been noted in P-starved tomato (Pilbeam *et al.,* 1993), pea (Rolland *et al.,* 1996), lupin (Johnson *et al.,* 1996) and rape (Hoffland *et al.,* 1992). The importance of organic acid excretion by roots to the alleviation of P-stress and improved Al tolerance has led to attempts to enhance this trait in plants by molecular manipulation. It has been reported that constitutive expression in tobacco plants of a gene from *Pseudomonas aeruginosa* that encodes the citrate synthase enzyme has resulted in one to three-fold higher citrate synthase activity in transgenic plants. Further, the resulting increase in citrate biosynthesis has led to increased excretion of citrate by the roots of those transgenic plants (De la Fuente-Martinez *et al.,* 1997). Transgenic plants over-expressing citrate synthase are reported to have grown better on both alkaline soils with low P-availability (Lopez-Bucio *et al.,* 2000a) and on acid soils in which Al-phosphates were the primary source of P (Lopez-Bucio *et al.,* 2000b). However, the inability of another research group to repeat this work (Delhaize *et al.,* 2001) suggests that, at present, this strategy may not be sufficiently robust for improving the availability of soil P to plants. This group found that, in spite of generating transgenic plants with very high levels of citrate synthase, they could not detect either increased citrate concentrations in the roots or increased citrate efflux from these transgenic lines. Further, they were also unable to show increased citrate efflux from transgenic tobacco lines that had previously been reported to exhibit this character. Clearly, further verification of the reliability of increasing citrate excretion as a means of alleviating P-stress and increasing Al-tolerance through molecular manipulation of citrate synthase in plants is required. Strategies that seek to manipulate the regulation of anion channels in roots through which the organic anions are excreted may prove more reliable.

SYMBIOTIC ASSOCIATIONS

An important adaptation that up to 80% of flowering plant species employ to acquire soil P is a symbiotic association with mycorrhizal fungi (Harley and Smith, 1983). Whilst many plants form ectomycorrhizal associations, plants of agricultural importance generally form arbuscular endomycorrhizal (AM) symbiotic associations. In these associations the fungi derive their carbon from plants by colonising the roots. They also develop a network of external hyphae through which P_i and other nutrients are taken up from the soil and transferred back to the host plant roots. AM fungi form highly branched structures known as arbuscules within root cortical cells. The plant plasma membrane invaginates around this structure so that a close association between the fungal and plant membranes is formed. It is thought that these membranes are involved in the exchange of phosphate and carbon by the symbionts. The processes involved have been well reviewed by Harrison (1997; 1999a; 1999b) and Smith and Read (1997).

The P-status of the soil can effect sporulation of AM fungi and the species of AM fungi present (Douds and Schenck, 1990). High availability of soil P_i can reduce colonisation of plant roots by AM fungi and hence the effectiveness of the symbiotic association. The role of the P-status of the host plant in regulating the symbiosis is poorly understood at present (Smith and Read, 1997). Nevertheless, the presence of this additional mechanism for P uptake by mycorrhizal plants poses important questions about the regulation of the alternative pathways for P uptake by mycorrhizal plants. A gene encoding a high-affinity P transporter active in the external hyphae of the mycorrhizal fungus *Glomus versiforme* has been cloned (Harrison and van Buuren, 1995). It is thought that this transporter is responsible for uptake of P_i into the fungus from the soil solution. However, it is not known whether the expression of this gene or the activity of the transporter in the external hyhae is regulated by the P-status of the host plant. Genes involved in the transfer of P across the arbuscule/cortical cell interface have not been identified unequivocally to date. No molecular information at all is available on the transport system of fungal origin that is responsible for efflux of phosphate from the fungus at this interface. There is an indication that a member of the *Pht1* family of P transporters may be responsible for P uptake by the plant at the peri-arbuscular membrane. *In situ* hybridisation studies (Rosewarne *et al.*, 1999) have shown transcripts that hybridise to a probe from the *LePT*1 gene, a member of the *Pht1* family expressed in root hairs and epidermal cells, to be present in cortical cells of mycorrhizal tomato plants (Daram *et al.*, 1998). However, the presence of 9 very similar *Pht1* genes in the *Arabidopsis* genome, suggests more specific identification techniques are required. This caution is enhanced by careful studies with two very similar genes from the *Pht1* family isolated from *Medicago truncatula*. Neither of these genes were expressed at the cortical cell/arbuscule interface (Liu *et al.*, 1998b). Importantly these studies demonstrated that these genes, one of which has recently been shown to be a primary transporter involved in P uptake through root hairs and epidermal cells (Chiou *et al.*, 2001), were down-regulated in mycorrhizal plants. Down-regulation of genes encoding primary P transporters in the plant root is consistent with measurements indicating that, in some instances, direct phosphate uptake by plant roots may become almost inactive in mycorrhizal plants and the plant may rely almost entirely upon P delivered through the fungal hyphae (Pearson and Jocobsen 1993). A particular point of interest arising from these results is to what extent this down-regulation of the plant root phosphate transport system is due to the enhanced P-status of mycorrhizal plants or to some signal associated with mycorrhizal colonisation of the roots. Studies with another gene, *Mt4*, that is induced in plant roots under P-stress have shown that it is also down-regulated in a mutant of *Medicago truncatula* that does not form a mycorrhizal symbiosis effective in P transfer to the host. These studies have demonstrated the existence of another regulatory pathway in mycorrhizal plants that is associated with the presence of the fungi rather than the transfer of P. The presence of such a pathway may have important implications for expression of other genes that are responsive to plant P-status in mycorrhizal plants.

CONCLUSION

The co-ordinated morphological, biochemical and symbiotic responses of plants to P-stress point to complex regulatory networks. Fine control over specific biochemical reactions or transport processes is likely to involve post-transcriptional regulation. The direct involvement of P_i in protein phosphorylation and dephosphorylation has already been highlighted. Modifications through glycosylation of proteins provide other avenues for regulating their activity. Many of the biochemical reactions involved in key plant processes such as photosynthesis, glycolysis, sugar metabolism and nitrogen assimilation are subject to feedback or allosteric regulation by P_i or organic P compounds. These regulatory mechanisms can be perturbed during P-stress. Some of the consequences of these perturbations are alluded to in this review and in the material referred to, but many are unknown at present.

Coarser, but longer term control of adaptive responses of plants to nutrient stresses is brought about through transcriptional regulation of genes. Examples of how the expression of genes is regulated during P-stress have been cited and attention drawn to the present lack of knowledge of the signaling and transduction pathways involved. The complexity of such pathways and the interactions between the various genes involved can be gleaned from the PHO-regulon of the unicellular yeast *Saccharomyces cerevisiae*. A cascade of both positive and negative regulatory proteins control the expression of genes encoding phosphate transporters and phosphatases in this relatively simple eukaryotic organism (Ogawa *et al.*, 2000; Oshima *et al.*, 1996; see http://cmgm.stanford.edu/pbrown/phosphate/ for a brief summary of the PHO-regulon). Regulatory mechanisms in multicellular plants in which growth, developmental, temporal and adaptive processes must be co-ordinated are likely to be far more complex.

Technological advances are now providing the tools needed to address this complexity in plants. Large libraries of Expressed Sequence Tags (ESTs) can be generated from nutrient stressed plants. Using differential display or subtractive library techniques, genes with altered expression under conditions of nutrient stress can be identified in such libraries. Micro-array "chips" are beginning to provide a more powerful tool. Potentially, these "chips" enable most of the genes within a genome whose expression is significantly altered by a particular nutrient stress to be identified. They thus provide opportunities for identifying and studying the network of genes that interact as plants adapt to particular nutrient stresses. Information from micro-arrays can be supplemented with that from analysis of the suite of proteins synthesised during adaptation to a nutrient stress. Large scale profiling of key metabolites in nutrient stressed plants is also feasible and provides valuable pointers to the effects of a particular stress on metabolic processes and potential regulatory mechanisms.

The utility of these genomic approaches to the study of nutritional physiology is illustrated in a recent project on nitrate nutrition (Wang *et al.*, 2000). The diversity of the responses to nitrate nutrition at the mRNA level were highlighted by simultaneously examining the expression of 5524 clones from *Arabidopsis*. Information in the Stanford Micro-array Database (http://genome-www4.stanford.edu/Micro-array/SMD/) indicates that similar "chips" have been prepared to study networks of genes involved in potassium nutrition and sulfate nutrition. The availability of the *Arabidopsis* genome data and large EST libraries will undoubtedly result in the production of "chips" containing larger arrays and targeted to other nutrient stresses. These large scale screening techniques for identifying genes and gene clusters involved in adaptation to nutrient stresses generate a daunting amount of data. Lack of the necessary bioinformatics capabilities to handle these data restrict the realisation of their full potential at present, but it can be expected that demand will be filled by further advances in this area.

Verification of the concepts derived from genomic approaches to studying adaptation is essential. Close examination of the phenotype and studyng the responses of mutant and tagged transgenic lines of model plants such as *Arabidopsis* can provide this verification. Libraries of these mutants are now available and techniques exist for silencing specific genes identified through screening procedures. The integration of genomic, proteomic, metabolite profiling and mutant analysis approaches is being successfully adopted in a large program dealing with tolerance to salt stress (Bohnert *et al.,* 2001). Rapid advances in understanding the complex responses of plants to nutrient stresses can be expected from similar integrated approaches.

Acknowledgements

Research at CSIRO Tropical Agriculture is supported in part by the Australian Grains Research and Development Corporation.

REFERENCES

Ae, N., Arihara, J., Okada, K., Yoshihara, T. and Johansen. C. 1990. Phosphorus uptake by pigeon pea and its role in cropping systems of the Indian subcontinent. *Science* 248, 477-480.

Barber, S.A., Walker, J.M., and Vasey, E.H. 1962. Principles of ion movement through the soil to the plant root. Trans. Joint Meeting Commission IV & V Internat. Soil Sci., New Zealand 1962. pp. 121-124.

Bariola, P.A., Howard, C.J., Taylor, C.B., Verburg, M.T., Jaglan, V.D. and Green, P.J. 1994. The *Arabidopsis* ribonuclease gene RNS1 is tightly controlled in response to phosphorus starvation. *Plant J.* 6, 673-685.

Barry, D.A.J. and Miller, M.H. 1989. Phosphorus nutritional requirements of maize seedlings for maximum yield. *Agron J.* 81, 95-99.

Bates, T.R. and Lynch, J.P. 1996 Stimulation of root hair elongation in *Arabidopsis thaliana* by low phosphorus availability. *Plant Cell Environ.* 19, 529-538.

Bates, T.R. and Lynch, J.P. 2000. The efficiency of *Arabidopsis* root hairs in phosphorus acquisition. *Amer. J. Bot.* 87, 964-970.

Bieleski, R.L. and Lauchli, A. 1992. Phosphate uptake, efflux and deficiency in the water fern, *Azolla. Plant Cell Environ.* 15, 665-673.

Bohnert, H.J., Ayoubi, P., Borchert, P., Bressan, R.A., Burnap, R.L., Cushman, J.C., Cushman, M.A., Deyholos, M., Fischer, R., Galbraith, D.W., Hasegawa, P.M., Jenks, M., Kawasaki, S., Koiwa, H., Kore-eda, S., Lee, B-H., Michalowski, C.B., Misawa, E., Nomura, M., Ozturk, N., Postier, B., Prade, R., Song, C-P., Tanaka, Y., Wang, H. and Zhu, J-K. 2001. A genomics approach towards salt tolerance. *Plant Physiol. Biochem.* 39, 295-311.

Carswell, M.C., Grant, B.R. and Plaxton, W.C. 1997. Disruption of the phosphate-starvation response of oilseed rape suspension cells by the fungicide phosphonate. *Planta* 203, 67-74.

Chiou, T-J., Liu, H. and Harrison, M.J. 2001. The spatial expression patterns of a phosphate transporter (*MtPt1*) from *Medicago truncatula* indicate a role in phosphate transport at the root/soil interface. *Plant J.* 25, 281-293.

Clarkson, D.T. and Scattergood, C.B. 1982. Growth and phosphate transport in barley and tomato plants during development of and recovery from phosphate stress. *J. Exp. Bot.* 33, 865-875.

Clarkson, D.T. and Lüttge, U. 1991. Mineral nutrition: inducible and repressible nutrient transport systems. *Progress in Bot.* 52, 61-83.

Clarkson, D.T. 1993. Roots and the delivery of solutes to the xylem. *Phil. Trans. R. Soc. Lond.* B 341, 5-17.

Cogliatti, D.H. and Clarkson, D.T. 1983. Physiological changes in phosphate uptake by potato plants during development of and recovery from phosphate deficiency. *Physiol. Plant.* 58, 287-294.

Daram, P., Brunner, S., Persson, B.L., Amrhein, N. and Bucher, M. 1998. Functional analysis and cell-specific expression of a phosphate transporter from tomato. *Planta* 206, 225-233.

Daram, P., Brunner, S., Rausch, C., Steiner, C., Amrhein, N. and Bucher, M. 1999. *Pht2;1* encodes a low-affinity phosphate transporter from *Arabidopsis. Plant Cell* 11, 2153-2166.

De la Fuente-Martinez, J.M., Ramirez-Rodriguez, V., Cabrera-Ponce, J.L. and Herrera-Estrella, L. 1997. Aluminium tolerance in transgenic plants by alteration of citrate synthesis. *Science* 276, 1566-1568.

Delhaize, E. and Randall, P. 1995. Characterisation of a phosphate-accumulator mutant of *Arabidopsis thaliana. Plant Physiol.* 107, 207-213.

Delhaize, M., Hebb, D.M., and Ryan, P.R. 2001. Expression of a *Pseudomonas aeruginosa* citrate synthase gene in tobacco is not associated with either enhanced citrate accumulation or efflux. *Plant Physiol.* 125, 2059-2067.

Dinkelaker, B., Romheld, V. and Marschner, H. 1989. Citric acid excretion and precipitation of calcium citrate in the rhizosphere of white lupin (*Lupinus albus* L.). *Plant Cell Environ.* 12, 285-292.

Dinketaker, B., Hengeler, C. and Marschner, H. 1995. Distribution and function of proteoid roots and other root clusters. *Bot. Acta* 108, 183-200.

Dodds, P.N., Clarke, A.E. and Newbigin, E. 1996. Molecular characterization of an S-like RNase of *Nicotiana alata* that is induced by phosphorus starvation. *Plant Mol. Biol.* 31, 227-238.

Douds, D.D. and Schenck, N.C. 1990. Relationship of colonisation and sporulation by VA mycorrhizal fungi to plant nutrient and carbohydrate contents. *New Phytol.* 116, 621-627.

Drew, M.C. and Saker, L.R. 1978. Nutrient supply and growth of the seminal root system in barley. III. Compensatory changes in growth of lateral roots and in rates of phosphate uptake in response to a localised supply of phosphate. *J. Exp. Bot.* 29, 435-451.

Duff, S.M.G., Sarath, G. and Plaxton, W.C. 1994. The role of acid phosphatases in plant phosphorus metabolism. *Physiol Plant.* 90, 791-800

Foehse, D. and Jungk, A. 1983. Influence of phosphate and nitrate supply on root hair formation of rape, spinach and tomato plants. *Plant Soil* 74, 359-368.

Fox, R.L. and Kamprath, E.J. 1970. Phosphate sorption isotherms for evaluating the phosphorus requirement of soils. *Soil Sci. Soc. Amer. Proc.* 34, 902-907.

Fredeen, A.L., Rao, I.M. and Terry, N. 1989. Influence of phosphorus nutrition on growth and carbon partitioning in *Glycine max*. *Plant Physiol.* 89, 225-230.

Gardner, W.K., Barber, D.A., Parbery, D.G. 1983. The acquisition of phosphorus by *Lupinus albus* L.: III. The probable mechanism by which phosphorus movement in the soil/root interface is enhanced. *Plant Soil* 70, 107–124.

Gerke, J. 1992. Phosphate, aluminium and iron in the soil solution of three different soils in relation to varying concentrations of citric acid. *Z. Pflanzenernähr. Bodenk.* 155, 339-343.

Goldstein, A.H., Mayfield, S.P., Danon, A. and Tibbot, N.K. 1989. Phosphate starvation inducible metabolism in *Lycopersicon esculentum*. III. Changes in protein secretion under nutrient stress. *Plant Physiol.* 91, 175-182.

Harrison, M.J. 1997. The arbuscular mycorrhizal symbiosis: an underground association. *Trends Plant Sci.*2, 54-60.

Harrison, M.J. 1999a. Molecular and cellular aspects of the mycorrhizal symbiosis. *Annu. Rev. Plant Physiol. Plant Mol. Biol.* 50, 361-389.

Harrison, M.J. 1999b. Biotrophic interfaces and nutrient transport in plant/fungal symbioses. *J. Exp. Bot.* 50, 1013-1022.

Harrison, M.J. and van Buuren, M.L. 1995. A phosphate transporter from the mycorrhizal fungus *Glomus versiforme*. *Nature* 378, 626-629.

Harley, J.L. and Smith, S.E. 1983. *Mycorrhizal Symbiosis*, Academic Press, London.

Hoffland, E., Boogaard, R.V.D., Nelemans, J. and Findenegg, G. 1992. Biosynthesis and root exudation of citric and malic acids in phosphate-starved rape plants. *New Phytol.* 122, 675-680.

Hoffland, E., Findenegg, G.R. and Nelemans, J.A. 1989. Solubilization of rockphosphate by rape. 2. Local root exudation of organic acids as a response to P starvation. *Plant Soil* 113, 161-165.

Huang, C., Barke, S.J., Langridge, P. Smith, F.W. and Graham, R.D. 2000. Zinc deficiency up-regulates expression of high-affinity transporter genes in both phosphate-sufficient and deficient barley roots. *Plant Physiol.* 124, 415-422.

Jackson, R.B., Manwaring, J.H. and Caldwell, M.M. 1990. Rapid physiological adjustment of roots to localised soil enrichment. *Nature* 344, 58-60.

Johnson, J.F., Allan, D.L. and Vance, C.P. 1996. Phosphorus deficiency in *Lupinus albus*. Altered lateral root development and enhanced phosphoenolpyruvate carboxylase. *Plant Physiol.* 112, 31-41.

Kai, M., Masuda, Y., Kikuchi, Y., Osaka, M. and Tadano, T. 1997. Isolation and characterization of a cDNA from *Catharanthus roseus* which is highly homologous with phosphate transporter. *Soil Sci. Plant Nutr.* 83, 227-235.

Kammerer, B., Fischer, K., Hilpert, B., Schubert, S., Gutensohn, M., Weber, A. and Flügge, U.I. 1998. Molecular characterization of a carbon transporter in plastids from heterotrophic tissues: the glucose 6-phosphate/phosphate antiporter. *Plant Cell* 10, 105-107

Khamis, S., Chaillou, S. and Lamaze, T. 1990. CO_2 assimilation and partitioning of carbon in maize plants deprived of orthophosphate. *J. Exp. Bot.* 41, 1619-1625.

Köck, M., Theierl, K., Stenzel, I. and Glund, K. 1998. Extracellular administration of phosphate-sequestering metabolites induces ribonucleases in cultured tomato cells. *Planta* 204, 404-407.

Lauer, M.J., Blevins, D.G., and Sierzputowska-Grazc, H. 1989. ^{31}P-nuclear magnetic resonance determination of phophate compartmentation in leaves of reproductive soybeans (*Glycine max* L.) as affected by phosphate nutrition. *Plant Physiol.* 89, 1331-1336.

Lefebvre, D.D. and Clarkson, D.T. 1984. Compartmental analysis of of phosphate in roots of intact barley seedlings. *Can. J. Bot.* 65, 1504-1508.

Lefebvre, D.D., Duff, S.M.G., Fife, C. Julien-Inalsingh, C. and Plaxton, W.C. 1990. Response to phosphate deprivation in *Brassica nigra* suspension cells. Enhancement of intracellular, cell surface and secreted phosphatase activities compared to increases in P_i-adsorption rate. *Plant Physiol.* 93, 504-511.

Lefebvre, D.D. and Glass, A.D.M. 1982. Regulation of phosphate influx in barley roots: effects of phosphate deprivation and reduction in influx with provision of orthophosphate. *Physiol. Plant.* 54, 199-206.

Leggewie, G., Willmitzer, L. and Reismeier, J.W. 1997. Two cDNAs from potato are able to complement a phosphate uptake-deficient yeast mutant: identification of phosphate transporters from higher plants. *Plant Cell* 9, 381-392.

Lipton, D.S., Blancher, R.W. and Blevins, D.G. 1987. Citrate, malate and succinate concentrations in exudates from P-starved *Medicago sativa* L. seedlings. *Plant Physiol.* 85, 315-317.

Liu, C., Muchhal, U.S., Uthappa, M., Kononowicz, A.K. and Raghothama, K.G. 1998a. Tomato phosphate transporter genes are differentially regulated in plant tissues by phosphorus. *Plant Physiol.* 116, 91-99.

Liu, H., Trieu, A.T., Blaycock, L.A. and Harrison, M.J. 1998b. Cloning and characterization of two phosphate transporters from *Medicago truncatula* roots: regulation in response to phosphate and to colonization by arbuscular mycorhizal (AM) fungi. *Mol. Plant Microbe. Interact.* 11, 14-22.

Lopez-Bucio, J., Martinez de la Vega, O., Guevara-Garcia, A. and Herrera-Estrella, L. 2000a. Enhanced phosphorus uptake in transgenic tobacco that overproduce citrate. *Nature Biotech.* 18, 450-453.

Lopez-Bucio, J., Ramirez-Rodriguez, V. and Herrera-Estrella, L. 2000b. Improving phosphate acquisition efficiency in transgenic plants by citrate overproduction. *AgBiotechNet* 2, 1-3.

Lynch, J., Lauchli, A. and Epstein, E. 1991. Vegetative growth of the common bean in response to phosphorus nutrition. *Crop Sci.* 31. 380-387.

Lynch, J.P. 1995. Root architecture and plant productivity. *Plant Physiol.* 109, 7-13.

Lynch. J.P. and Beebe, S.E. 1995. Adaptation of beans (*Phaseolus vulgaris* L.) to low phosphorus availablility. *Hortscience* 30, 1165-1171.

Ma, J.F. 2000. Role of organic acids in detoxification of aluminium in higher plants. *Plant Cell Physiol.* 41, 383-390.

Ma, Z., Bielenberg, D.G., Brown, K.M. and Lynch, J.P. 2001. Regulation of root hair density by phosphorus availability in *Arabidopsis thaliana. Plant Cell Environ.* 24, 459-467.

Marschner, H. 1995 *Mineral Nutrition of Higher Plants.* 2nd edition. Academic Press, London.

Marschner, H., Romheld, V. and Cakmak, I. 1987. Root-induced changes of nutrient availability in the rhizosphere. *J. Plant Nutr.* 10, 1175-1184.

McPharlin, I.R. and Bieleski, R. 1989. P_i efflux and influx by P-adequate and P-deficient *Spirodela* and *Lemna. Aust. J. Plant Physiol.* 16, 391-399.

Mimura, T. 1995. Homeostasis and transport of inorganic phosphate transport in plants. *Plant Cell Physiol.* 36, 1-7.

Mimura, T. 1999. Regulation of phosphate transport and homeostasis in plant cells. *Int. Rev. Cytology* 191, 149-200.

Mimura, T., Dietz, K-J., Kaiser, W., Schramm, M.J., Kaiser, G. and Heber, U. 1990. Phosphate transport across biomembranes and cytosolic phosphate homeostasis in barley leaves. *Planta* 180, 139-146.

Mimura, T., Sakano, K. and Shimmen, T. 1996. Studies on distribution, re-translocation and homeostasis of inorganic phosphate in barley leaves. *Plant Cell Environ.* 19, 311-320.

Mitsukawa, N., Okumura, S., Shirano, Y., Sato, S., Kato, T., Harashima, S. and Shibata, D. 1997. Overexpression of an *Arabidopsis thaliana* high-affinity phosphate transporter gene in tobacco cultured cells enhances cell growth under phosphate-limited conditions. *Proc. Natl. Acad Sci. USA* 94, 7098-7102.

Muchhal, U.S., Pardo, J.M. and Raghothama, K.G. 1996. Phosphate transporters from the higher plant *Arabidopsis thaliana. Proc. Nat. Acad. Sci. USA* 93, 10519-10523.

Muchhal, U.S. and Raghothama, K.G. 1999. Transcriptional regulation of plant phosphate transporters. *Proc. Natl. Acad. Sci USA* 96, 5868-5872.

Nishimura, M. and Beevers, H. 1978. Hydrolases in vacuoles from castor bean endosperm. *Plant Physiol.* 62, 44-48.

Ogawa, N., DeRisi, J. and Brown, P.O. 2000. New components of a system for phosphate accumulation and polyphosphate metabolism in *Saccharomyces cerevisiae* revealed by genomic expression analysis. *Molec. Biol. Cell* 11, 4309-432

Ohwaki, Y. and Hirata, H. 1992. Differences in carboxylic acid exudation among P-starved leguminous crops in relation to carboxylic acid contents in plant tissues and phospholipid levels in roots. *Soil Sci. Plant Nutr.* 38, 235-243.

Oshima, Y., Ogawa, N. and Harashima, S. 1996. Regulation of phosphatase synthesis in *Saccharomyces cerevisiae* – a review. *Gene* 179, 171-177.

Pao, S.S., Paulsen, I.T. and Saier, M.H. 1998. Major facilitator superfamily. *Microbiol. Molec. Biol. Rev.* 62, 1-34.

Pearson, J.N. and Jakobsen, I. 1993. Symbiotic exchange of carbon and phosphorus between cucumber and three arbuscular mycorrhizal fungi. *New Phytol.* 124, 481-488.

Pilbeam, D.J., Cakmak, I., Marschner, H. and Kirkby, E.A. 1993. Effect of withdrawal of phosphorus on nitrate assimilation and PEP carboxylase activity in tomato. *Plant Soil* 154, 111-117.

Plaxton, W.C. and Carswell, M.C. 1999. "Metabolic Aspects of the phosphate starvation response in plants". In: *Plant Responses to Environmental Stresses; from Phytohormones to Genome Reorganisation,* ed. H. R. Lerner. pp. 349-372. Marcel Dekker, New York.

Plaxton, W.C. 1996. The organisation and regulation of plant glycolysis. *Annu. Rev. Plant Physiol. Plant Mol. Biol.* 47, 185-214.

Poirier, Y., Thomas, S., Somerville, C. and Scheifelbein, J. 1991. A mutant of *Arabidopsis* deficient in xylem loading of phosphate. *Plant Physiol.* 97, 1087-11093

Raghothama, K.G. 1999. Phosphate acquisition. *Annu. Rev. Plant Physiol. Plant Mol. Biol.* 50, 665-693.

Reisenauer, H.M. 1966. "Mineral nutrients in soil solution". In: *Environmental Biology,* eds P.L. Altman and D.S. Dittmer. pp 507-508. Fed. Amer. Soc. Exp. Biol., Bethesda.

Rolland, R.H., Contard, P. and Betsche, T. 1996. Adaptation of pea to elevated atmospheric CO_2: rubisco, phosphoenolpyruvate carboxylase and chloroplast phosphate translocator at different levels of nitrogen and phosphorus nutrition. *Plant Cell Environ.* 19, 109-117.

Rosewarne, G.M., Barker, S.J., Smith, S.E., Smith, F.A. and Schachtman, D.P. 1999. A *Lycopersicon esculentum* phosphate transporter (LePT1) involved in phosphorus uptake from a vesicular-arbuscular mycorrhizal fungus. *New Phytol.* 144, 507-516.

Rossiter, R.C. 1978. Phosphorus deficiency and flowering in subterranean clover *(Tr. subterraneum* L.). *Ann. Bot.* 42, 325-329.

Ryan, P.R., Delahaize, E. and Randall, P.J. 1995. Characterisation of Al-stimulated malate efflux from the root apices of Al-tolerant genotypes of wheat. *Planta* 196, 103-110.
Smith, F.W., Jackson, W.A. and Van den Berg, P.J. 1990. Internal phosphorus flows during development of phosphorus stress in *Stylosanthes hamata. Aust. J. Plant Physiol.* 17, 451-464.
Smith, F.W., Ealing, P.M., Dong, B. and Delhaize, E. 1997. The cloning of two *Arabidopsis* genes belonging to a phosphate transporter family. *Plant J.* 11, 83-92.
Smith, F.W., Cybinski, D. and Rae, A.L. 1999. "Regulation of expression of genes encoding phosphate transporters in barley roots". In: *Plant Nutrition – Molecular Biology and Genetics*, eds. G. Gissel-Nielsen and A. Jensen. pp. 145-150. Kluwer Academic Publishers, Dordrecht.
Smith, F.W., Rae, A.L. and Hawkesford, M.J. 2000. Molecular mechanisms of phosphate and sulphate transport in plants. *Biochim. Biophys. Acta* 1465, 236-245.
Smith, F.W. 2001. Sulphurus and phosphorus transport systems in plants *Plant Soil* 232, 109-118.
Smith, S.E. and Read, D.J. 1997. *Mycorrhizal Symbiosis*. Academic Press, San Diego.
Stark, D.M., Timmerman, K.P., Barry, G.F., Priess, J. and Kilshore, G.M. 1992. Regulation of the amount of starch in tissues by ADP glucose pyrophosphorylase. *Science* 258, 341-351.
Staunton, S. and Leprince, F. 1996. Effect of pH and some organic anions on the solubility of soil phosphate: implications for P bio-availability. *Eur. J. Soil Sci.* 47, 231-239.
Wang, R., Guegler, K., LaBrie, S.T. and Crawford, N.M. 2000. Genomic analysis of a nutrient response in *Arabidopsis* reveals diverse expression patterns and novel metabolic and potential regulatory genes induced by nitrate. *Plant Cell* 12, 1491-1509.
Welch, R.W., Webb, M.J. and Loneragan, J.F. 1982. "Zinc in membrane function and its role in phosphorus toxicity". In: *Plant Nutrition 1982*, Proc 9th Intern. Plant Nutrition Colloq., Warwick, UK, ed. A Scaife. pp. 710-715. CAB International, Farnham Royal, UK.

Index

MIX
Papier aus verantwortungsvollen Quellen
Paper from responsible sources
FSC® C105338

If you have any concerns about our products,
you can contact us on
ProductSafety@springernature.com

In case Publisher is established outside the EU,
the EU authorized representative is:
Springer Nature Customer Service Center GmbH
Europaplatz 3, 69115 Heidelberg, Germany

Printed by Libri Plureos GmbH
in Hamburg, Germany